S0-BIF-470

An Introduction to Nuclear Astrophysics

An Introduction to Nuclear Astrophysics

Richard N. Boyd

THE UNIVERSITY OF CHICAGO PRESS • CHICAGO AND LONDON

RICHARD N. BOYD is science director at the National Ignition Facility in Livermore, California.

The University of Chicago Press, Chicago 60637
The University of Chicago Press, Ltd., London

Printed in the United States of America
16 15 14 13 12 11 10 09 08 07 1 2 3 4 5
ISBN-13: 978-0-226-06971-5 (cloth)
ISBN-10: 0-226-06971-0 (cloth)

Library of Congress Cataloging-in-Publication Data

Boyd, Richard N.
An introduction to nuclear astrophysics / Richard N. Boyd.
p. cm.
Includes bibliographical references and index.
ISBN-13: 978-0-226-06971-5 (cloth : alk. paper)
ISBN-10: 0-226-06971-0 (cloth : alk. paper) 1. Nuclear astrophysics—Textbooks. I. Title.
QB464.B69 2007
523.01′97–dc22
2007029462

∞ The paper used in this publication meets the minimum requirements of the American National Standard for Information Sciences—Permanence of Paper for Printed Library Materials, ANSI Z 39.48-1992.

Contents

Preface

In writing a textbook, it is difficult to overstate the importance of the textbooks that have preceded the current effort. This certainly applies to the excellent books of D. D. Clayton (1983) and of C. Rolfs and W. S. Rodney (1988); both books have provided immense help in generating the current textbook. At the same time, the current effort has attempted to provide textbook discussions of some topics that were simply not a part of the body of knowledge of nuclear astrophysics at the time the previous textbooks were written. Of course, each textbook also has its personal emphasis. Clayton's book contains much material on stellar structure, and that of Rolfs and Rodney includes an unusually extensive chapter on experimental techniques; the interested reader is urged to go to those books for the information on those subjects that is either not included in the current textbook or is dealt with in considerably less detail. There are certainly other books that contain vast amounts of information that is relevant to nuclear astrophysicists; those by B. E. J. Pagel (1997) on galactic chemical evolution and by D. Arnett (1996) on supernovae certainly come to mind in this context, but there others that are referenced where relevant in the current textbook.

Of course, any textbook will be dated to some extent, but nowhere is this more the case than in astrophysics. The rapid advances that are being made in all fields of astrophysics instrumentation will render chapter 2 obsolete within a few years, and it should be to the delight of all of us that this is the case. However, the fields of science that are studied with those instruments will also undergo "revolutions" at comparable rates, so the other chapters will also require revision on the same timescale. Indeed, it has been difficult to impose a deadline on the material that was covered in this book just because of the rapid advances that are occurring on an almost daily basis. This, of course, is testimony to the strength and level of excitement in the field of astrophysics.

With a little luck, however, there will be subsequent editions of this book. I certainly intend to keep up to speed with developments as to both the instruments used in astrophysics and the experimental/observational and theoretical results to which they are related. Indeed, I invite the practitioners in the field to keep me up to date with their developments; it will be my pleasure to read their papers and include the most relevant results in the book's next edition. In this context, I also invite comments on the book's discussions, especially as to any errors that may exist in the current edition. For those, I apologize and promise to correct them at the earliest opportunity.

This book is addressed to advanced seniors in college or first-year graduate students, depending somewhat on their level of preparation. It has been assumed that the students who use this book have had at least a year of quantum mechanics, although I have tried to write the book in such a way that an advanced course is not required. Courses in thermodynamics and electricity and magnetism are also required as background, although an advanced undergraduate treatment of those subjects should be adequate. Beyond these basic courses, I have tried to be sufficiently complete in the discussions presented that the interested student can gain a useful understanding of the subjects from this textbook but can follow up on the references given so as to gain more details of the subject if that is desired. This applies to scattering theory and to neutrino oscillations in the basic physics chapters, and to supernovae and novae, the s- and r-processes, and to cosmology in subsequent chapters. Problems are suggested where relevant, although modern computers have greatly altered the structure of problems that can be assigned. This certainly applies to the network codes that have been written to accommodate the nucleosynthesis that occurs in some of the processes described in this book. One of the superb services provided by some of the authors of some these codes is that versions of them have been made available on the Internet so that students can actually run at least simplified simulations of some of the processes of interest. I have attempted to include references to the Web sites for these codes where relevant.

The structure of the book is as follows. It is directed primarily toward physics students who are approaching the subject of nuclear astrophysics from the perspective of physicists. Thus the book begins with some very basic facts and nomenclature of astronomy for the benefit of students who may have no background in astronomy. Chapter 2 presents discussions of many of the instruments, both terrestrial and spaceborne, that are currently (or in some cases, were until recently) being used to observe the properties of different features of the universe, or of the stars, or of the nuclei that are the focus of this book. Chapters 3 and 4 discuss the background topics—some nuclear and neutrino physics, scattering formalism and thermonuclear reaction rates, and galactic chemical evolution—that are necessary to understand nuclear astrophysics at more than a cursory level. Some compromise has been made in these chapters between completeness of the presentation and the desire not to have the entire book deal just with the introductory topics. The student will find this to be especially true of scattering theory, where an entire course could have been spent dealing with the issues associated with that topic. My intention was to provide enough information to motivate the equations that are needed to produce the thermonuclear reaction rates used extensively in astrophysical calculations that involve nucleosynthesis.

Having thus dispensed with the introductory material necessary for an understanding of nuclear astrophysics, Chapter 5 discusses the Sun and the nucleosynthesis and energy generation that occur therein. It also discusses the very important subject of solar neutrinos, both as to their impact on our understanding of the Sun and as to their basic properties. Chapter 6 discusses the stages of stellar burning—through nuclear reactions—that represent what occurs in massive stars that have consumed their core hydrogen and have advanced beyond their hydrogen-burning phase. It also describes some of the issues of neutrino astrophysics that are associated with supernovae. Also discussed in that chapter is the subject of gamma-ray bursts, an exciting and still somewhat enigmatic topic of great current interest in astrophysics. Chapter 7 discusses the processes of nucleosynthesis that produce most of the nuclides heavier than iron, the s-process and the r-process, as well as some additional processes that produce refinements on the abundances of the heavier nuclides. Chapter 8 introduces some modern topics in nuclear astrophysics, most notably, novae and the different aspects of the rp-process. Finally, chapter 9 introduces modern cosmology. This topic used to involve, at least in the context of nuclear astrophysics, a thorough discussion of big bang nucleosynthesis. However, in the past decade, the measurements of cosmology have rather suddenly gone far beyond that topic; thus I have tried to present some of the excitement of the current status of observational cosmology.

Copious references have been provided, not just to satisfy my colleagues in nuclear astrophysics, but to provide the students who use this book with a strong starting point with which to get themselves up to speed on the subjects that interest them most. I have included many of the original references on many subjects; I am convinced that the best way to convey the excitement of the field is to let the original authors tell their own stories. I am certain that I have missed some important references; it is my hope that my colleagues will note this so that they can be included in subsequent editions of the book.

There will certainly be some material that is more, or less, important in a course directed at a specific group of students. In this context, it has been suggested that I indicate material that I think could be skipped over in the interest of the more important material, should that be appropriate. Certainly, the detail given on resonance reactions in chapter 3 is more than has generally been included, so that might be considered a "luxury" for the first pass through for students of nuclear astrophysics. This might also apply to the detailed discussions of galactic chemical evolution in chapters 4, 7, and 9. Chapter 2 has been written so as to include some science associated with each of the projects included, but that could be left to the student's own effort. The subject of neutrino oscillations, at least at the detail given in chapter 3, has not generally been included as part of the subject matter of nuclear astrophysics. However, my belief is that the natural evolution of that subject now requires that level of discussion and understanding so I would not suggest weakening the discussion of this subject. Similarly, I have included more details of astronomical observations and of grains, especially in chapter 7, than is usual, but I regard those subjects as providing the "truth" to which nuclear astrophysics must answer so I would also not suggest weakening those discussions.

Much gratitude is due the many people in the field who have provided commentary on this edition of this book. This certainly includes John Cowan, Lars Bildsten, Pat Osmer, Hendrik Schatz, Bill Donnelly, Brad Meyer, Rob Hoffman, Hans Geissel, Roberto Gallino, Rene Ong, Cary Davids, and Brian Fields, each of whom reviewed at least one chapter for me. However, it also includes the many graduate students at Ohio State University who took my course there on nuclear astrophysics and who made many constructive comments on the evolving notes that were used in that course. A special thanks goes to Nigel Sharp, who helped greatly in preparing some of the figures.

A special thanks is also due the three anonymous reviewers of the manuscript for the University of Chicago Press. Although they were obviously from different scientific backgrounds, each did a wonderfully detailed job in their review, and each made many constructive comments. Nearly all of them have been adopted. It has also been a great pleasure working with my publisher, Jennifer Howard, at the University of Chicago Press. She has done her best to make my job as tractable as possible and always with good cheer. Any author should be delighted to have so helpful a publisher as Jennifer.

I wish to dedicate this book to three men who have guided my career and my life. The first is my Father, Virgil Boyd, and the second is my uncle, Raymond Shannon. Both served as extraordinary role models, in many different ways, for perseverance and integrity. The third is William Fowler, who dominated the field of nuclear astrophysics for several decades, as well as my entry into that area of science, and is in a very real sense the "father" of every nuclear astrophysicist.

Finally, I note that as I was in the finishing stages of writing this book, I was saddened to learn of the deaths of three giants in the field of nuclear astrophysics. The first is John Bahcall. As one can immediately discern from reading chapter 5, John's contributions to solar and neutrino physics grace practically every page in that chapter. The second is Ray Davis. His perseverance to detecting solar neutrinos went far beyond the level of dedication that should be expected of any scientist. The third is Al Cameron, who came up with a description of the processes of nucleosynthesis that was essentially the same as that of Burbidge, Burbidge, Fowler, and Hoyle, and did so completely independently. The contributions to the field of nuclear astrophysics of all three of these giants will be long remembered, and their inputs will be sorely missed.

Richard N. Boyd
Falls Church, Virginia

1
NUCLEAR ASTROPHYSICS BACKGROUND

1.1 Introduction and General Perspectives

Nuclear Astrophysics integrates the information that nuclear physics provides—isotopic abundances, masses, half-lives, and nuclear reaction cross sections—with models of stellar and other cosmic environments to test those models and use them to understand the processes of nucleosynthesis, that is, those processes by which the nuclides are produced. Thus it necessarily involves the interactions of a variety of subfields of science. Nuclear physics is clearly central to its endeavors, but the reality against which its successes are measured involves several fields of astronomy, covering an extraordinarily broad range of wavelengths from radio waves to extremely high energy gamma rays. It also involves many aspects of astrophysics, including stellar evolution and galactic chemical evolution, high-energy physics, cosmology, neutrino physics, and others. In this book, we will deal with all of these subjects, at least at the level at which they interface with our considerations of nuclear astrophysics.

The central guiding fact of nuclear astrophysics, however, is that nuclear reactions drive the processes of nucleosynthesis, the results of which we observe in stars and on Earth. Indeed, these reactions

- provide the energy for the stars to shine for, in many instances, billions of years,
- synthesize most of the elements of the chart of the nuclides,
- determine the evolution of stars, and
- describe big bang nucleosynthesis.

As noted above, there are several ways in which nuclear astrophysicists obtain the truths against which they must measure their understanding. The processes that occur in the cosmos do so both inside stars and in the interstellar medium. For the most part, the nuclei that we observe were synthesized inside stars via the nuclear reactions that are involved in stellar evolution, but some of them are made in collisions of high-energy nuclei that occur on the surfaces of stars or even in the regions between stars. Thus it is important to understand a wide variety of nuclear processes, ranging from the very low energy reactions that occur inside stars to those involving the highest energy particles found in the universe. Although it is not possible to observe these reactions as they occur in stars, their signatures are to be measured in the products of nucleosynthesis, the photons and neutrinos that are emitted by stars, and somewhat more directly by observing the high-energy cosmic rays that exist in the cosmos.

The photons that are produced range in energy over many orders of magnitude, from the radio waves (which have wavelengths as large as about 10 m or photon energies as small as 1×10^{-7} eV) that tell us, for example, about the distribution of hydrogen in the universe; to the infrared photons (with energies less than 1 eV) that tell us about some of the "cooler" objects in our Galaxy and in other galaxies; to observable photons (with energies of several eV) that are emitted by stars like our own Sun; to ultraviolet photons from hotter stars; to X-rays (with energies of order 10 keV) that can be produced in explosive processes occurring on the surfaces of stars; to gamma rays (of energies of several MeV) that can be signatures of some of the nuclei synthesized in stars but that are unstable (which indicates that nucleosynthesis truly is occurring in stars!); to very high energy gamma rays that are produced in some of the most violent processes that can occur in the universe (with energies as high as 10 TeV). However, the photons for the most part cannot traverse much matter, although those at some energies can penetrate more than those at others. Thus they specifically cannot tell us about the processes that occur in the cores of stars. Samples of each type of observatory of these different types of photons are discussed in chapter 2, and their implications are, of course, discussed throughout the book. The lowest energy photons, those detected by radio telescopes, are produced in the gas clouds that pervade the galaxies. The lower energy stellar processes tend to produce photons with energies in the infrared or optical; these may characterize the processes that are occurring in quiescent stellar "burning." (In this book, the term "burning" will refer to the nuclear reactions that are occurring that generate energy and synthesize the nuclei with which we are concerned.) The higher energy photons will produce the signatures of the explosive processes that occur in the final phases of stellar evolution, as described in chapters 6 and 7, or in the processes that can occur on the surfaces of very dense stars, as described in chapter 8. In some cases, the sources of the photons are not understood; these are obviously the foci of active areas of research. The instruments that measure photons and that are described in chapter 2, ranging from radio astronomy telescopes to observatories of the highest energy gamma rays, span about 20 orders of magnitude in energy.

By contrast, the neutrinos that are produced in stars can penetrate large quantities of matter (but this also makes them very difficult to detect!), so they can produce direct information about the reactions and other processes that are occurring in the cores of stars. Thus they have become very important in recent years for providing the information that only they can. Detectors of neutrinos from quiescent stellar burning, in which the star remains in that stage of its evolution for a long period of time, are described in chapter 2 (*Super-Kamiokande* and the *Sudbury Neutrino Observatory*, [*SNO*]) and in chapter 5 (the Homestake solar neutrino detector and the detectors of the lower energy neutrinos from the Sun). Neutrinos of higher energies can also be emitted in the final phase of stellar evolution, as described in chapter 6, or in some of the ultra-high-energy processes, which are not understood at present, that might emit other extremely high energy particles. Their detectors are also described in chapter 2; they include *AMANDA* (the *Antarctic Muon and Neutrino Detector Array*), *IceCube* (the *Cubic Kilometer Antarctic Ice Detector*), *ANTARES* (*Astronomy with a Neutrino Telescope and Abyss Environmental RESearch in the Mediterranean*, off France), and *NESTOR* (the *Neutrino Extended Submarine Telescope with Oceanographic Research* in the Bay of Navarino, off Greece). Because neutrinos are so difficult to detect, their observation necessitates huge detectors. Unfortunately, these detectors would, because of their size, also detect huge numbers of the cosmic rays that rain down continuously on Earth, so neutrino detectors have to be sited far underground, or deep in the water, so that the overburden can shield out most of the cosmic rays. Only then can the neutrinos be observed above the background. One such laboratory, the *SNOLab* in Sudbury, Ontario, Canada, is described in chapter 2, although several similar laboratories exist around the world, and more are planned.

Another class of photons has been instrumental in providing incredible definition of the processes that occurred in our universe's birth event, the big bang. These are the cosmic microwave background photons, detection of which requires very special detectors. These extremely low-energy photons are the whispers that are left over from the big bang; they and the fluctuations they exhibit have provided us with an unprecedented look at the distribution of matter even before galaxies and stars formed, at several hundred thousand years after the big bang, which is now known to have occurred 13.7 billion years ago. These measurements have also provided us with information about the distribution of other entities, for example, the "dark matter" that is known to pervade the universe and the mysterious "dark energy" that appears to have a huge impact on the expansion and evolution of our universe. One of the detectors of these cosmic microwave background photons (the *Wilkinson Microwave Anisotropy Probe* [*WMAP*]) will be discussed in chapter 2, and the implications for cosmology of the results obtained from *WMAP* will be discussed in detail in chapter 9. Because *WMAP* detects differences in temperatures of the order of a part in 10^6 of an already small number, it extends our photon detection range down to 10^{-9} eV, increasing the overall range in energy of photon detection to 22 orders of magnitude.

However, more direct measurements of the nuclei that exist in our universe (theorists tell us that there may be others!) can also be revealing. This has led to measurements of the high-energy nuclei that pervade the interstellar medium as well as the intergalactic medium. These nuclei are sometimes detected by satellites that observe directly their collisions with spaceborne detectors, sometimes by observing secondary emission, for example, gamma rays, that they might emit and sometimes by observing the interactions that occur when extremely high energy cosmic rays interact with the Earth's atmosphere. It is not only important to understand the abundances of these nuclei, but their observation can also elucidate the processes by which they are produced. The detectors for these cosmic rays are discussed in chapter 2 (the *High Resolution Fly's Eye detector* [*HiRes*] and the *Pierre Auger Observatory*). Since these observatories extend the energy range of observed entities up to 10^{20} eV, or 10^7 above that of the highest energy gamma rays we can detect, the observable energy range now extends over 29 orders of magnitude. Neutrinos might extend that even further, as they do not have the same energy cutoff that is believed to exist for cosmic rays (see chap. 2), but that is not yet determined.

Finally, some of the most extraordinary processes that can occur in the universe might be observed from the gravitational radiation they would emit. One of the detectors for these observations (the *Laser Interferometer Gravitational Wave Observatory* [*LIGO*]) is described in chapter 2. Gravity wave astronomy is in its infancy, but some of the signals detectors of gravitational radiation might observe would be produced from collisions of neutron stars or from a neutron star and a black hole; these apparently fall under the general rubric of gamma-ray bursts and are described in chapter 6.

As noted above, other endeavors are also essential to describing the nucleosynthesis that occurs in our Galaxy and our universe. Stellar evolution, of course, is essential to describing the different nuclides that are synthesized in the different stages of a star's life and characterizing how the stars expel their nuclei into the interstellar medium. Note that only the nuclei that exist outside stars are "counted" in the cosmic abundances that we consider; indeed, many of those that exist in stars never do escape into the interstellar medium. The theoretical description of how stars seed the interstellar medium with the material that they synthesize and expel, and how that material gets mixed with the preexisting material, goes under the name "galactic chemical evolution"; this is also essential in our understanding of nucleosynthesis. It is discussed in chapters 4, 7, and 9.

One point to emphasize is that the facts with which nature presents us can only be those resulting from a snapshot in time. For example,

- It will be seen that stars are grouped into various types in a plot of the luminosity versus the surface temperature (see sec. 1.5). This does not necessarily tell us that stars do not ever occupy other parts of that plot but only that they spend most of their time in phases that occupy the parts of the plot that are observed to be populated.
- Furthermore, the nuclei that we find occupying the periodic table are only the stable nuclei that are produced in stars, but the processes that produce stable nuclei may involve many more nuclei, which in some cases are very short-lived and very far from stability. This can be especially true of some of the processes of nucleosynthesis that occur at high temperatures, which are discussed in the latter chapters of this book, and that can occur on timescales of the order of a second. As a specific example, the r-process, which is thought to synthesize half the nuclei heavier than iron, is thought to occur on such a timescale.
- As a final example, we see through the photons that we detect what is happening on the periphery of a star, but this may tell us little about what is actually going on in its interior. The periphery of a massive star may remain essentially unchanged as it passes through its several final stages of stellar evolution, right up to the time it explodes as a supernova and the shock wave from that event finally does explode the periphery of the star.

Thus astrophysicists in general must not limit their perceptions to the realities of the present but must also recognize that, even though some of the conditions and nuclei that exist in stars may be difficult for us to produce in the laboratory, nature often has no difficulty whatsoever in producing them. Simulating the conditions and producing these nuclei will constitute a great challenge to nuclear astrophysicists.

Thus another very important aspect of nuclear astrophysics is the accelerators by which the nuclear reactions and some of the extremely unstable nuclei that we will encounter are studied. Several of these accelerator facilities are also described in chapter 2. They include small accelerators to study the very low energy reactions that occur in stars, typically at energies of tens of keV, but they also include the much higher energy facilities that are required to produce and study the properties of the radioactive nuclides that are encountered in some of the explosive processes of nucleosynthesis; these include the *National Superconducting Cyclotron Laboratory* in the United States, the *Gesellschaft fur Schwerionenforschung* [*GSI*], in Germany, and the *Center for Nuclear Study Radioactive Ion Beam Facility* [*CRIB*] at the Institute for Physical and Chemical Research in Japan. These facilities are described in some detail in chapter 2, but results obtained from these and other accelerators that exist around the world are sprinkled throughout the subsequent chapters. These accelerators and their ancillary facilities are at the core of nuclear astrophysical research.

Historically modern nuclear astrophysics can be thought of as beginning with the observation of Sir Arthur Eddington, who observed in the early twentieth century that "What is possible in the *Cavendish Laboratory* [i.e., nuclear reactions] may not be too difficult in the Sun." This realization was based on the observation that the immense power produced by the Sun simply could not be the result of chemical reactions or gravitational collapse but had to be the result of the enormous energy that can result from nuclear reactions. Subsequently, in 1939, Hans Bethe identified at least some of the nuclear reactions that actually do power the Sun, which require that four hydrogen nuclei combine to form ^{4}He; the complete set of reactions is discussed in detail in chapter 5. In subsequent years, John Bahcall and Roger Ulrich and their collaborators developed the "standard solar model," by which the processes that occur in the Sun and which synthesize the Sun's nuclei and produce its neutrinos occur. This work has resulted in a number of publications in *Reviews of Modern Physics*, but the benchmark paper was that of Bahcall and Ulrich in 1988.

A critical advance in understanding nucleosynthesis was also made by Hoyle in 1954 in describing the triple-α reaction, by which three ^{4}He nuclei, or α-particles, fuse to form ^{12}C. Salpeter (1952, 1953) and Opik (1951) had hypothesized earlier that this process was responsible for the creation of ^{12}C, since there are no stable mass 5 or mass 8 nuclei, thereby necessitating some way of circumventing those masses to synthesize heavier nuclei from the lighter ones that are made in the big bang or that are synthesized in hydrogen burning. Salpeter had noted that, even though ^{8}Be is not stable, it does live long enough ($\sim 10^{-16}$ s) for some abundance of it to be built up, occasionally allowing it, at sufficiently high densities, even in the short time it lives, to capture an additional ^{4}He nucleus and form ^{12}C. That nucleus then serves as the launching point for the synthesis of all heavier nuclei through a variety of processes. The triple-α reaction perhaps illustrates the beautiful relationship between nuclear physics and astrophysics better than any other. In order to understand how this reaction synthesized the amount of ^{12}C that is observed, Hoyle had to guess that a critical nuclear level existed at exactly the right excitation energy to create a resonance large enough to synthesize the amount of ^{12}C that is made in stars. Nuclear physicists required several decades of hard work to confirm the properties of that nuclear state; some of this, as well as that involved in understanding the very important reaction $^{12}\text{C} + {}^4\text{He} \rightarrow {}^{16}\text{O} + \gamma$, is described in detail in chapter 6.

These separate pieces were assembled in 1957 in a benchmark paper by Burbidge, Burbidge, Fowler, and Hoyle, which is generally denoted in the frequent references to it as B^{2}FH. However, that paper also included descriptions of additional processes of nucleosynthesis, most notably, the neutron capture processes by which most of the nuclei heavier than iron are thought to be synthesized. This paper correctly identified many of the details of stellar nucleosynthesis and related them to the stages of stellar evolution in which different nuclides are produced, although the authors obviously did not have nearly as many facts about the physical universe as we now have, and they certainly did not have the advantages (and disadvantages!) of modern computers to aid in their endeavors. Remarkably, a similar description of the processes of nucleosynthesis was derived by Cameron, also in 1957. These papers were updated in 1997 by Wallerstein et al., who provided an elaboration and modernization of most of the details of nucleosynthesis and added much information about the stages of stellar evolution that are inextricably linked to the nucleosynthesis produced therein. Indeed, the study of these processes and the nucleosynthesis that results from them will form most of the subject matter for this book. However, before we begin discussing nuclear reactions and their effects in stars, we need a framework, both nuclear and astronomical, within which to discuss the stars themselves.

The reactions that produced the nuclei that resulted from big bang nucleosynthesis were characterized by George Gamow and his collaborators around 1950 (Alpher, Bethe, and Gamow 1948 [the famous "$\alpha\beta\gamma$ paper." Bethe did not contribute to the original paper, but his name was added by Gamow as a play on the first three letters of the Greek alphabet]; Alpher and Herman 1949; Gamow 1952). Gamow and his associates tried to describe the synthesis of all the nuclides, all the way to uranium, as products of the big bang. However, they found that the same mass 5 and 8 gaps that hindered synthesis of nuclides heavier than helium in stars also made it impossible for the big bang to synthesize all the nuclides. Alpher and Herman (1949) calculated the temperature of the cosmic microwave background to be 5 K, a harbinger of the *WMAP* project to measure the cosmic microwave background radiation and a correction of an earlier estimate by Gamow. The processes of the big bang, and the modern updates of which Gamow could not possibly have been aware, are discussed in chapter 9. A number of important papers have been produced describing these processes, but that of Wagoner, Fowler, and Hoyle in 1967 certainly set the stage for many of the subsequent studies of big bang nucleosynthesis.

The range in time spanned by nuclear astrophysics covers nearly as many orders of magnitude as the energies of the photons that are detected. Following the big bang, as the universe cooled, protons and neutrons are thought to have formed after about 10^{-6} s. Further expansion of the universe occurred to produce further cooling, until nuclei began to be assembled after roughly 10 s, and big bang nucleosynthesis was completed after several minutes. As the universe cooled further, electrons were captured on the nuclei that had formed after about 380,000 years. The first stars formed after about 10^8 years, and galaxies formed after about 10^9 years. The stars in those galaxies were born and died, seeding their galaxy with the products of their nucleosynthesis and enriching the interstellar medium. The results of *WMAP* have shown that the universe is 13.7 billion years old, so these processes of stellar evolution and galactic chemical evolution have been going on for roughly the last 12.7 billion years. Thus the range of time that is relevant to nuclear astrophysics spans 23 orders of magnitude.

The range in spatial extent that our studies will encompass is even more dramatic than those of energy and time. They begin with nuclei, which are of order 10^{-15} m, to atoms and molecules, at 10^{-10} m, to meteorites, at 10^{-6} m, to stars, at 10^8 m. These are included in galaxies, which are typically 10^{22} m in size, countless numbers of which are included in the visible universe, the extent of which is given by the speed of light times the age of the universe, 10^{26} m. Thus our considerations span 41 orders of magnitude in size!

1.2 Nuclear Abundances

The basic data against which nuclear astrophysicists measure their success are the abundances of the nuclides. The quest to understand these abundances necessitates an understanding of the processes by which nuclei are synthesized and, hence, of the conditions within the stellar or other environments in

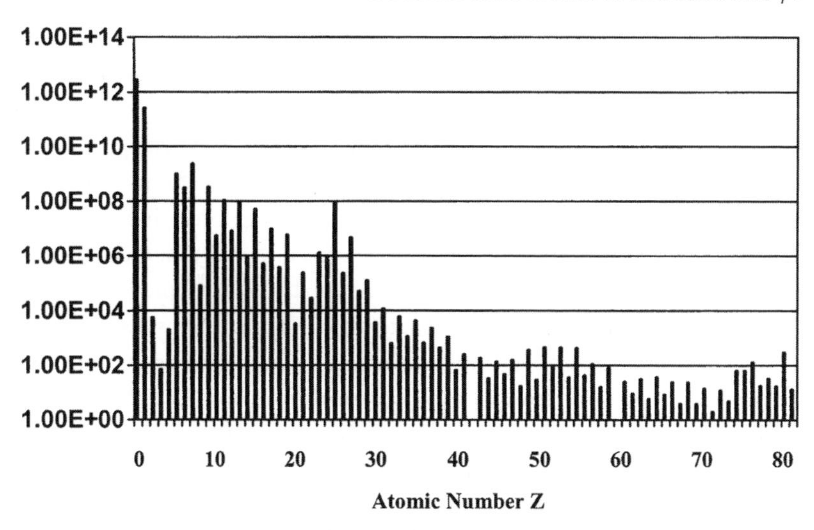

Fig. 1.1. Relative elemental abundances of the nuclides (Anders and Grevesse 1989). The abundances are relative to Si, which is arbitrarily set to 1×10^8.

which the nucleosynthesis takes place. The mechanisms that synthesize both the elements and their isotopes are required, since there are many instances in which the neutron-rich isotopes of an element are synthesized under very different conditions from those in which the neutron-poor isotopes are synthesized. Perhaps the most obvious source of information on these abundances is found in measurements of the abundances in stars; this has been the challenge for astronomers for decades. The data produced in their spectrographs, the instruments that separate the light into its different wavelength components, are fitted with codes that include all known atomic transitions of both atoms and ions in various ionization states. The codes themselves include a vast amount of atomic physics. However, the level of detail that can be obtained by astronomical observations on stars can provide isotopic abundances for only a few elements, although this is changing, as is discussed in chapter 7. Thus, the abundance information must be derived from a variety of other sources, including observations of solar elemental abundances, analysis of meteorites, and, for many of the isotopic abundances, analysis of terrestrial rock samples.

In their compilation of the abundances, Anders and Grevesse (1989) found that the elemental abundances of stable nuclides from H to Pb span 12 orders of magnitude. The elemental abundances from Anders and Grevesse (1989) are plotted in figure 1.1. Hydrogen and helium, the most abundant elements, are thought to have been made in the big bang and modified only slightly since. The nuclides from carbon to the most massive observed in nature are made in stars. The spikes for every other element from about $Z = 6$ to $Z = 20$ are a result of the relatively large binding energies (see chap. 3) of the "α-nuclei" (nuclei having an even proton number Z and an even neutron number N with $N = Z$): ^{12}C, ^{16}O, ^{20}Ne, ^{24}Mg, ^{28}Si, and so on, up to ^{40}Ca. Alternating abundances in other regions reflect the nuclear pairing force, that is, the fact that protons and neutrons in nuclei prefer to be paired. The peak at iron reflects the fact that the nuclei around iron have the highest binding energies per nucleon (see chap. 3) in the periodic table. There are no stable isotopes of Tc ($Z = 43$) or Pm ($Z = 61$). Interesting peaks also exist beyond iron; they are associated with the processes of nucleosynthesis by which the nuclides heavier than iron are made: primarily the s-process and the r-process. Both of these involve successive neutron captures, either "slowly" (the s-process) or "rapidly" (the r-process). These are discussed in detail in chapter 7. Thus, the elemental abundances are only a small part of the problem to be understood. Indeed, it is necessary to understand the nuclear physics of each nuclide in the periodic table as well as its relationship to other nuclides in order to fully understand the astrophysical processes by which they are all synthesized.

The following table gives a few of the relative abundances (with the abundance of Si set to 1×10^8):

Nucleus	Relative Abundance	Nucleus	Abundance
H	2.7×10^{12}	Zr	1.1×10^3
He	1.8×10^{11}	Sn	3.8×10^2
C	1.1×10^9	Sm	2.6×10
O	1.8×10^9	Pb	2.5×10^2
Si	1.0×10^8	U	2.6
Fe	1.0×10^8		

These huge variations, spanning 12 orders of magnitude, must tell us something about how these elements are made; that is half of the subject of this book. The other half deals with how we use that abundance information to infer details about the processes that go on in stars.

The additional detail provided by the isotopic abundances is extremely important to nuclear astrophysics because different isotopes of a single element may be produced by completely different processes of nucleosynthesis. Sometimes these involve relatively long-term processes in which the star exists for millions or billions of years in quasi-equilibrium. In other cases, the entire process can occur on a timescale of seconds and may involve nuclei with half-lives as short as tens of milliseconds. In still others, the processes may not even occur in the interiors of stars. Furthermore, the isotopic abundances even within a single element may span orders of magnitude, depending on the processes responsible for their synthesis.

The isotopic abundances (Anders and Grevesse 1989) are shown in figure 1.2. There it can be seen that ^{1}H and ^{4}He are clearly the most abundant nuclides; they were formed in the big bang (although some He is also made in stars). The Li, Be, and B nuclides are very rare; this is the result of there not being a stable mass 5 nucleus through which the mass 5–11 atomic mass units (abbreviated "u") nuclides can be synthesized. Indeed, those nuclides are formed primarily by cosmic-ray spallation (interactions between high-energy projectiles in which one or both are split into fragments) in the interstellar medium, but to some extent also by somewhat circuitous stellar burning processes that will be described in subsequent chapters. Note, however, that it is fortunate for us that there is no stable mass 5 u nucleus; our Sun would have consumed all its nuclear fuel long ago had that been the case. As noted above, the even Z and even N nuclides, and especially the α-nuclides, are generally more abundant than their neighbors; this is a result of the nuclear pairing force making the even-even nuclides more stable than either the even-odd or odd-odd nuclides (see sec. 3.1). Indeed, there are only four stable odd-odd nuclei in the entire chart of the nuclides: ^{2}H, ^{6}Li, ^{10}B, and ^{14}N.

While the isotopic abundances of ^{1}H and ^{4}He are at one extreme, at the other extreme are the abundances of ^{138}La ($Z = 57$) and ^{180}Ta ($Z = 73$); the latter has an abundance that is nearly 15 orders of magnitude below that of ^{1}H. In between those extremes are many interesting features, which will provide the clues we need to understand how the various isotopes of the periodic table are synthesized in stars and, in turn, how those abundances allow us to determine the conditions in the stars that sustained the various processes of nucleosynthesis.

Figure 1.2 confirms that the α-isotopes of the α-nuclei are indeed strongly populated. However, there are some other interesting features to note. For example, some elements appear to have much larger numbers of stable isotopes than their neighbors; this applies to Ca, Ni, Zn, Zr, Sn, and Ba. However, this also applies to some other nuclides. This is related to nuclear properties; the former elements turn out to be particularly tightly bound, as are the α-nuclei, so that a large variation in their number of neutrons can occur. However, many of the others can have many isotopes because their structure is such that adding or removing neutrons does not have a large effect on their binding energy. In any event, these are all nuclei with even proton numbers.

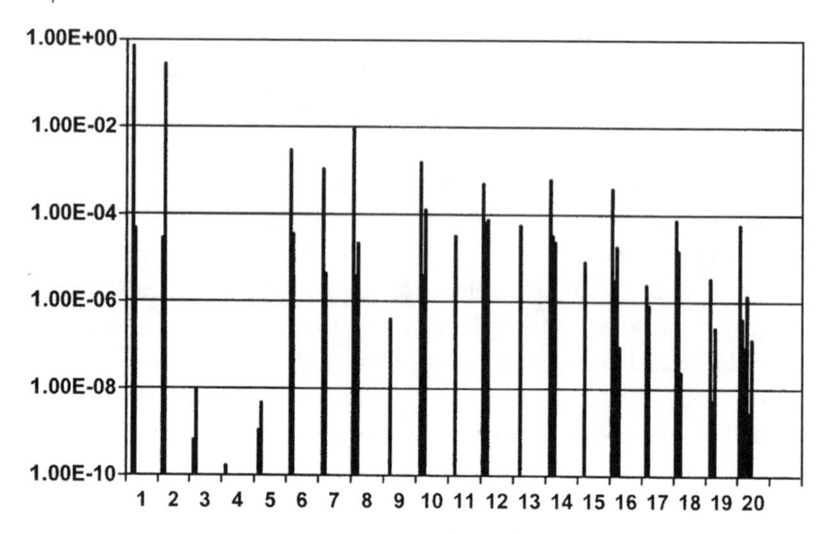

Fig. 1.2a. Abundances of the stable isotopes from ^{1}H to ^{48}Ca as a function of *Z*, their atomic number. The isotopic abundances are given in order of increasing mass for each element.

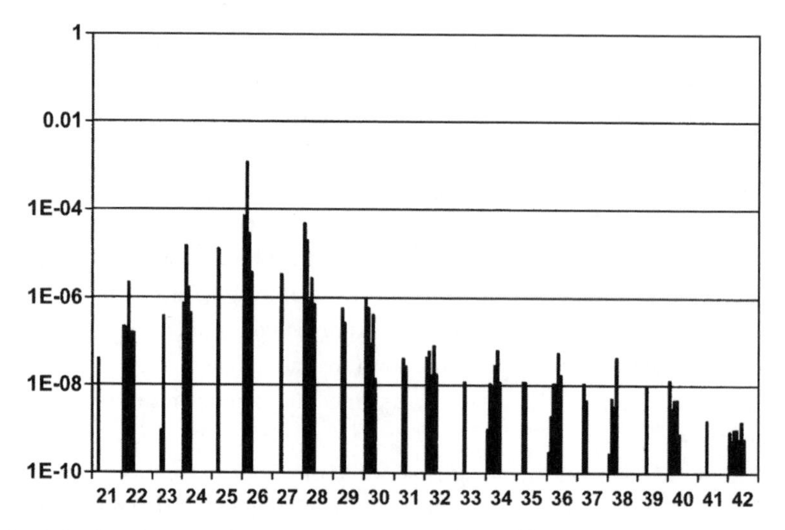

Fig. 1.2b. Abundances of the stable isotopes from ^{41}Sc to ^{100}Mo as a function of *Z*, their atomic number. Note the scale change from fig. 1.2a.

It should also be noted that the table of the abundances involves several hundred nuclides but that there are thought to be several thousand nuclides that occur between the proton and neutron "drip lines." These define the points beyond which the resulting nuclei will decay by baryon emission, rather than by β-decay: electron emission, positron emission, or electron capture. Nuclear

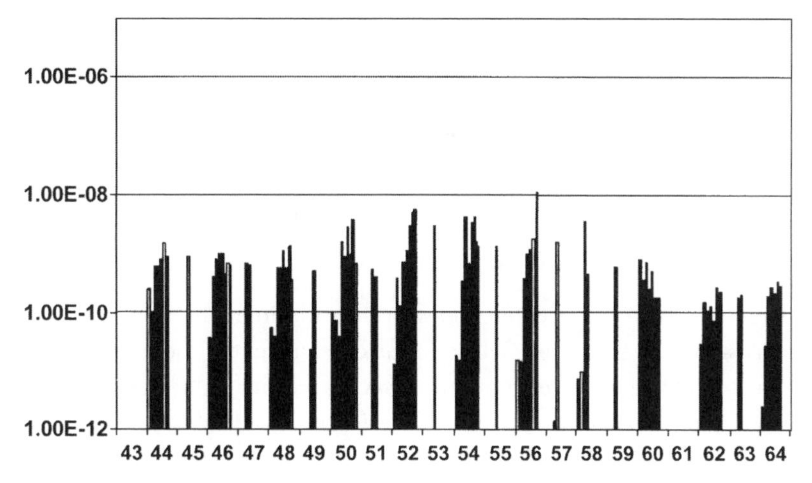

Fig. 1.2c. Abundances of the stable isotopes from Tc ($Z = 43$, no stable isotopes) to ^{160}Gd as a function of Z, their atomic number. Pm ($Z = 61$) also has no stable isotopes. Note the scale change from fig. 1.2b.

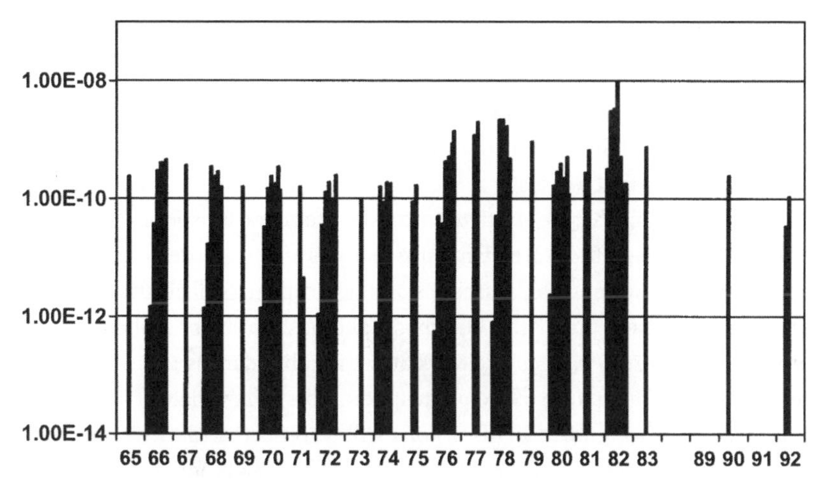

Fig. 1.2d. Abundances of the stable isotopes from ^{165}Ho ($Z = 67$) to ^{238}U as a function of Z, their atomic number. Note that, although they are radioactive, Th and U are sufficiently long-lived that they have a nonzero natural abundance. Note the scale change from fig. 1.2c.

physicists have been confined to studies of stable or nearly stable nuclei for most of the years the field has existed. Of course, nature has never been so confined. However, in the past decade or so facilities have been developed that have allowed nuclear physicists to begin to study the nuclei that approach the drip lines and some of the reactions involving them that are important to astrophysics; some of these facilities are discussed in chapter 2. While these have initiated a revolution in nuclear astrophysics, the results of those studies are just beginning to be obtained, and the most extreme nuclear examples that are relevant to astrophysics are yet to be studied because present facilities are limited in their capabilities to produce nuclei far from stability.

1.3 Some Basics of Astronomy

Astronomy has been a scientific endeavor since the time of the ancient Greeks. It was advanced greatly by Galileo in his development of the telescope, but it has seen huge advances in recent years on account of the advent of telescopes of order 10 m in diameter, of the *Hubble Space Telescope* (*HST*), and of adaptive optics, which are used on ground-based telescopes. Concomitant developments in spectral analysis have pushed the information produced by astronomers to its current state. All of these topics are discussed in some detail below and in chapter 2.

Measuring elemental abundances is the task of astronomers. When light from a star's photosphere—its visible surface—is viewed with a high-resolution spectrograph, the light is resolved into its different spectral components. The intensities of the light of different wavelengths so observed reflect both the temperature of the star and the abundances of the various atomic and ionic species, as the surface of a star is a conglomeration of atoms, ions, and electrons. The photons that are emitted from a star are characteristic of a blackbody spectrum, although there are distinctive features superimposed on that basic spectrum. The ions and atoms in the star's photosphere will absorb radiation at specific wavelengths of the blackbody photons. In a star's photosphere, atoms and ions may exist in excited states with abundances that are characteristic of the temperature of the photosphere, so a wide variety of absorption lines resulting from those atoms or ions will be observed. Clearly, stars with hotter photospheres will exhibit features characteristic of more highly excited atoms and ions and more highly ionized ions than will those with cooler photospheres; indeed the relative abundances of these characteristic states allow astronomers to determine the temperature of the photosphere.

A portion of a series of spectra, shown for the region slightly larger than the visible sprectrum for thirteen stars, is shown in figure 1.3. The stars selected cover a range of surface temperatures from that of O stars (at the top), which range from 28000 to 50000 K, to M stars (at the bottom), which range from 2000 to 3500 K. The spectral identification of a star having the temperature represented by each spectrum is indicated at the left-hand side. The evolution

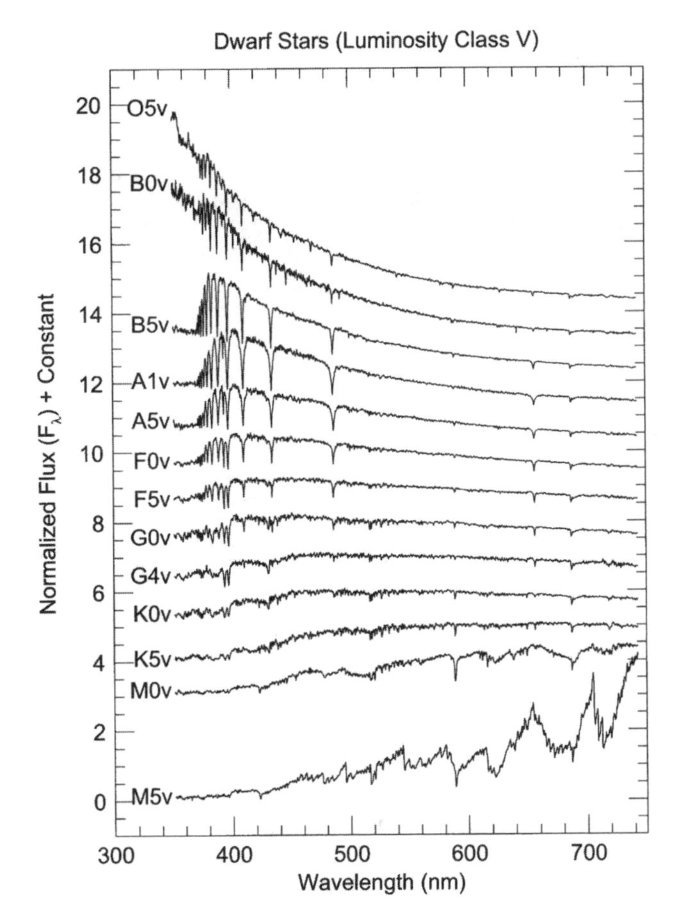

Fig. 1.3. A portion of the spectra from 13 stars. From these data the intensities of the various absorption lines are determined, which subsequently leads to a determination of the relative abundances of the elements. Courtesy of R. Pogge, http://www.astronomy.ohio-state.edu/~pogge/Ast162/Unit1/sptypes.html. This figure originated with Jacoby, Hunter, and Christian (1984).

of the spectral features with temperature can be seen in these spectra from O and B bright stars through the mid-temperature stars to the relatively cool M stars. The spectra can be seen to evolve dramatically with temperature, with the number of short wavelength lines increasing with temperature, and of long wavelength structures increasing with lower temperature.

As noted above, analysis of a star's spectrum provides the information from which one determines the elemental and, in some cases, isotopic abundances of the star's photosphere. The sample spectra shown in figure 1.3, are to be

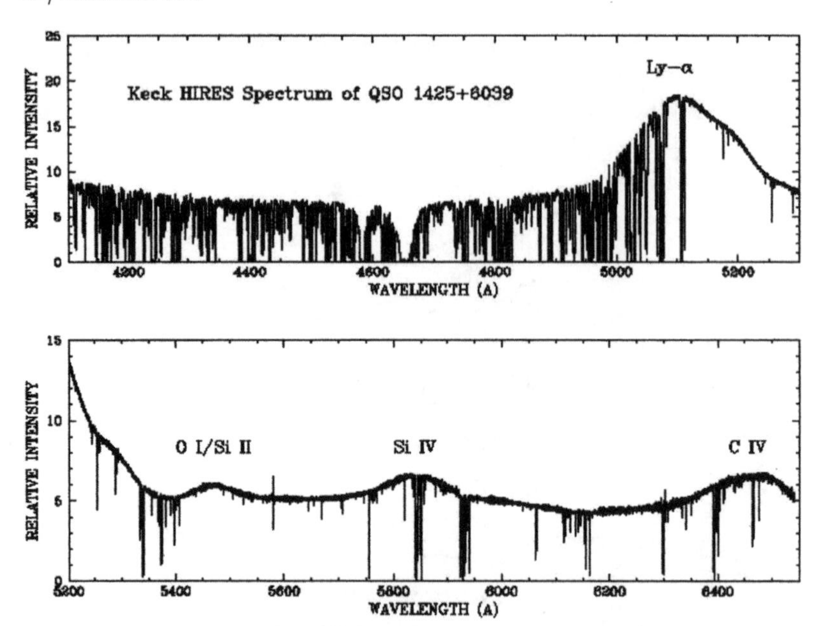

Fig. 1.4. Absorption spectrum of quasar Q1425+6039. As indicated on the figure, many of the absorption features can be identified as originating from particular elements. (The roman numerals following the elemental symbols denote the ionization state: I indicates a neutral atom, II a singly ionized atom, etc.) Quasars have a large Lyman-alpha emission line. This line is at the characteristic wavelength λ_o but is detected as having been shifted because of the relative motion between Earth and the quasar. The line is thus found at a value for $\lambda = (1 + z_o)\lambda_o$, with this z_o (see sec. 1.4) characterizing the position of the quasar, and in this case is $z_o = 3.18$. Absorption features result from intervening gas clouds. From http://www.astro.caltech.edu/~wws/qsoabs.html. Courtesy of W. Sargent.

contrasted with the spectrum from a very different kind of object, a quasar (an object that generally emits large quantities of radiation and that is thought to be powered by accretion of surrounding matter onto a black hole), which is shown in figure 1.4. In both examples, a number of the absorption lines in the spectra are identified, but clearly much more information can be obtained by more careful analysis. This is generally performed with computer codes that have the capability to analyze thousands of spectral lines; the locations of these lines are fixed (if the redshift [see sec. 1.4] of the object is known!), but the line identifications are generally the means by which the redshift is determined. The depths of the lines are also fit; they are directly related to the densities of the respective atoms or ionic species of each element. In principle, this will give the abundance of each element in the star's photosphere, although considerably more effort is actually necessary to obtain this information. Specifically, the widths of the lines must be increased from their intrinsic values in order to reproduce the spectra, owing at least to the amount of Doppler broadening resulting from thermal motion of the atoms and ions in the stellar photosphere, to collisional or "pressure" broadening that results from the scattering of the photons following their emission, and to turbulence. The locations of the lines, indeed their basic character, can be affected by magnetic fields (they can be split by these fields) and by mass loss. Often the causes of these different effects can be difficult to distinguish (but multiple lines provide important clues!), which provides challenges for astronomers.

A vast amount of atomic physics has been an essential input to the codes that relate the observed line strengths to the abundances of the elements from which they arise. Although this could be the subject of a much larger discussion, it is noted here only for completeness. A good discussion of stellar spectra is given by Michael Richmond (http://spiff.rit.edu/classes/phys230/lectures/spec_interp/spec_interp.html).

It should be emphasized that from these spectra astronomers usually determine elemental abundances. As noted above, nuclear astrophysicists also need to know isotopic abundances. Only in special cases has it been possible to determine isotopic abundances from stellar spectra. Thus, to determine the isotopic abundances one utilizes the stellar spectra to determine elemental abundances and, primarily, analysis of terrestrial materials to provide the isotopic abundances. This technique has provided most of the information that has gone into determining the isotopic abundances used in the standard references (Anders and Grevesse 1989). It should be noted, however, that it usually provides values for average abundances, which have been provided by homogenization of the elements that have been expelled into the interstellar medium by stellar explosions and winds, mixed with the preexisting Galactic interstellar medium and then merged into Earthly segments of (chemically selected) elements. Just to note one complication of this type of analysis, however, some chemical selectivity can be isotope dependent, so great care has to be taken to ensure that the results obtained from these analyses have accounted for all such effects. However, recent technological advances now permit analysis of terrestrial samples with greatly enhanced precision compared with that of years past, so much greater sensitivity to some of the subtleties associated with the isotopic abundances of even these samples is now possible. Some of these advances are discussed by Ireland (2004).

Another technique by which isotopic abundances can be determined involves meteoritic analysis, in which one analyzes the meteorites that have fallen to Earth for their isotopic constituents. This utilizes highly developed techniques to analyze tiny inclusions in meteorites to determine the abundances of the various isotopes that reside in the grains. What has been found is that these grains must have formed close to the region of the star where the elements therein were synthesized, thus "freezing in" the local isotopic abundances. Thus the abundances have been found to vary, in some cases dramatically, from those that were homogenized in the interstellar medium. In some cases, huge variations in the isotopic ratios in some elements have been observed in these grains. The data obtained from the grains are providing incredibly detailed information about the conditions of the stars in which nucleosynthesis occurred. In some instances large isotopic variations have been directly attributable to the processes that synthesized the elements, and the isotopes thereof; these are presented in much more detail in chapter 7.

Basic optics teaches us that diffraction limits the angular resolution obtainable with a lens to order λ/d, where λ is the wavelength observed and d is the diameter of the lens. Thus one can obtain the best possible angular resolution with the largest possible telescopes. However, this ignores some of the issues of modern telescopy, for example, that it is challenging to make large mirrors with the perfection necessary to not introduce distortion of the images and that the Earth's atmosphere introduces aberrations on a timescale of hundredths of a second that will also prevent obtaining the best resolution for faint objects, which require long exposures. The first difficulty has been overcome for mirrors up to 10 m in diameter (however, lenses cannot be made nearly that large without introducing distortion), but another solution to the problem has been obtained by making large mirrors out of smaller segments, for example, the *Keck Telescope*, that are then computer adjusted to optimize the image of some test star. A similar approach is also being developed to minimize the perturbations produced by motions of the Earth's atmosphere; this is the "adaptive optics" referred to above (and discussed in greater detail in chap. 2), in which the elements of the telescope are computer adjusted on timescales of thousandths of a second to optimize the image of a test star. A variety of tricks have been developed to produce a bright star on which the telescope can be tuned; some of these are also discussed in chapter 2.

Another effect of the Earth's atmosphere is that it only transmits limited ranges of wavelengths of electromagnetic radiation, as can be seen in figure 1.5. Note that the band of high transmission in the optical wavelengths corresponds to the wavelengths that are visible to the human eye; this may not be a fortuitous coincidence! In any event, this atmospheric transmission certainly imposes constraints on what can be observed by terrestrial telescopes, and on their design. However, this selective transmission of the Earth's atmosphere can be avoided by putting the telescope into space, as was done with the *HST*. This instrument has provided some stunning images, one of which is shown

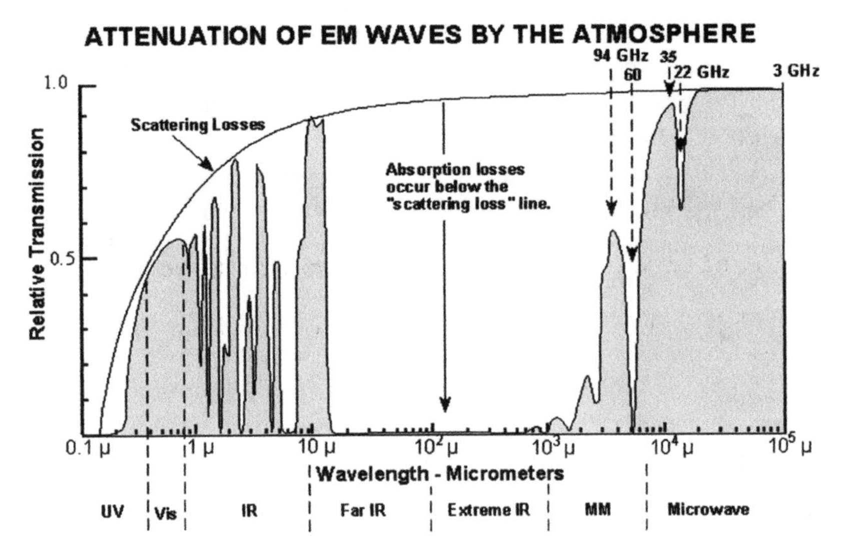

Fig. 1.5. Transmission (and attenuation) percentage through the Earth's atmosphere. Note the visible portion of the spectrum, "Vis," has nearly the maximum transmission possible. From Wikipedia, at https://ewhdbks.mugu.navy.mil/absorb.gif.

in chapter 2, as it can achieve the ultimate resolution for a telescope of its size (2.4 m in diameter). Of course, the difficulties attendant with putting a large technically advanced object into orbit and operating it for many years present a challenge for space-based astronomy, but the results the *HST* has produced have truly revolutionized observational astronomy by opening up parts of the electromagnetic spectrum that were previously not accessible. While there are windows of the electromagnetic spectrum in which one can observe starlight from the Earth's surface over a wide range of wavelength, its atmosphere is simply opaque to much of the electromagnetic spectrum.

A variety of different types of stars will be important to our considerations. Obviously, one important star is our Sun. It has a mass of 2×10^{33} g, and its mean radius is 696,100 km. Its mean density is 1.4 g cm^{-3}. It rotates with a mean period of 25 days (as indicated by sunspots), but its poles rotate less rapidly, that is, with a longer period, than does its equator. The sunspots have a cycle in their activity of about 11 years; this influences the solar activity observed on Earth. Its surface temperature is 5800 K, and it radiates 3.9×10^{33} ergs/s from its entire surface in electromagnetic radiation. The Sun also exhibits standing density waves (see, for example, Bahcall and Pinsonneault 1998), a phenomenon that has produced an extraordinarily detailed analysis of the structure of the Sun, as is discussed more extensively in chapter 5.

Our solar system formed about 4.6 billion years ago. Based on biological evidence (some primitive life-forms existed more than 1 billion years ago on Earth), its luminosity has been fairly constant, at least for that period of time. However, it has not been absolutely constant; this is certainly one of the things we would like to understand, that is, how the properties of stars such as their luminosity, or energy output, their size and surface temperature, and their mode of energy generation change from their time of formation, through their various stages of evolution, until their deaths.

Stars burn (not chemically; as noted above, throughout this book "burn" will refer to nuclear reactions) their nuclear fuels. Our Sun is currently burning its hydrogen; we believe its core has evolved from the 3:1 hydrogen to helium (by mass) mixture with which it began to its present mixture of about 1:1 hydrogen to helium, which is the "ash" that is created by hydrogen burning. The nuclear reactions that perform this hydrogen to helium conversion are discussed in detail in chapter 5.

During our tour of astrophysics, we will encounter some unusual stars, each of which will be important to us in some way. For the moment these will be named and defined in the context of their standard model, which is in most cases well established, but we will return to most of them in subsequent chapters.

- *Eclipsing stars.* About half of the stars we see in the night sky are actually binary systems, which, if we are located in the right plane, can form eclipsing systems. This allows measurement of their masses, which is a difficult parameter to determine in the absence of this sort of system.
- *Eruptive stars—novae.* Novae can produce enormous flashes of energy, of order 10^{45} ergs, usually on timescales of tens of seconds. These result from a close binary system, in which one star, presumably a white dwarf (see below), accretes matter from the other and in which the accreted material undergoes a thermonuclear explosion.
- *Eruptive stars—X-ray bursters.* These are the result of accretion from one member of a binary pair of stars onto a collapsed companion; in the case of X-ray bursters, the companion is a neutron star. The relatively high frequency at which these events occur has allowed them to be studied in great detail; this will be discussed in chapter 8.
- *Eruptive stars—gamma-ray bursters (GRBs).* When the first GRBs were detected, they were initially thought to be the result of nuclear weapons tests. However, they were subsequently found to be cosmic objects that produced gamma rays. The amount of energy they emit is prodigious, especially since they have been found to be at cosmological distances.

Unusually energetic, and somewhat bizarre, core collapse supernovae (see chap. 6) seem to be plausible candidates for the origins of many bursts, although there are other possibilities, notably, collisions of neutron stars; GRBs are currently an active area of research (see chap. 6).

- *Red giants.* These are stars that have evolved beyond their first phase of nuclear reactions in their cores into successive phases that emit sufficient radiation to expand the photosphere of the star. Thus, even though the total energy output of the star is increased, its energy output per unit surface area has been reduced, and the star appears reddened, that is, its surface temperature is lowered, as a result.
- *White dwarfs.* These are dense stars that resemble huge atoms in that they are supported by electron degeneracy pressure (see sec. 4.7). These are the end product of stars of moderate mass; our Sun will ultimately become a white dwarf. They typically have masses that are the order of that of our Sun and radii of roughly 1/100 that of the Sun, that is, roughly that of the Earth.
- *Neutron stars.* These are stars that have extremely high densities. They resemble huge nuclei in that they are composed almost entirely of neutrons and their sizes are supported by neutron degeneracy pressure. They are the end product of fairly massive stars. Neutron stars typically have masses about 1.5 times that of our Sun and radii of about 10 km, roughly the size of the interstate beltways around most major U.S. cities.
- *Black holes.* These are also possible end products of the evolution of core collapse supernovae. There are no mass limits on black holes; indeed, there is thought to be a black hole that has a mass of 3×10^6 solar masses at the center of our Galaxy, and many galaxies are thought to have similarly sized, or even larger, black holes at their centers.
- *Type Ia supernovae.* Type Ia supernovae result from accretion of sufficient material from a companion star onto a white dwarf that the result is a thermonuclear runaway. The white dwarf then will explode, destroying the entire star in the process. Type Ia supernovae are driven by thermonuclear processes.
- *Core collapse supernovae.* The other type of supernova results from the natural evolution of a massive star through its various burning stages and core compressions until it finally achieves essentially nuclear density and must then either become a neutron star or a black hole. In doing so, it explodes the outer layers of the star into the interstellar medium. These stars are designated type Ib, Ic, or II supernovae, depending on the conditions of their periphery prior to their explosion. Although the

various stages of stellar evolution involve a variety of nuclear reactions, the energy that drives core collapse supernovae comes not from nuclear energy but from gravity.

Historically, all supernovae with spectra that exhibited a large amount of hydrogen were designated type II, and those with a paucity of hydrogen were designated type I. However, some of the type I's have been grouped in this book with the type II's because of the mechanism that drives their explosion. Specifically, type Ib and Ic supernovae have little hydrogen in their spectra because their outer layers—the layers that would normally contain the hydrogen in massive stars—have been expelled from the star by stellar winds. The result is that a deeper lying shell, one that is dominated by helium (type Ib) or carbon (type Ic), becomes the surface of these stars. However, despite their designation as type I supernovae, these stars are sufficiently massive that they will ultimately undergo core collapse and then explode, ending up either as neutron stars or black holes. These are discussed in greater detail in chapter 6.

- *Pulsating stars—pulsars.* These are rotating neutron stars. They appear to Earthly observers to be pulsating, but the pulsation is apparently due to a jet that is rotating, searchlight like, which is observed each time it sweeps by the Earth. Pulsars have periods that approach 1 ms but are typically in the range of a few to hundreds of ms. It has been concluded that the only thing that could be rotating that fast without disintegrating would be an extremely compact object; the only known candidate is a neutron star. Pulsars are the most accurate "clocks" known in nature. Their frequencies have been found to be constant to a level of roughly one part in 10^{15}. Interestingly, frequency changes, when they do occur, seem to be the result of some cataclysmic event at the surface of the neutron star, possibly a "starquake" that changes the moment of inertia of the star and, hence, its rotation frequency.
- *Pulsating stars—Cepheid variables.* These stars pulsate as a result of an instability in their atmospheres. The pulsation frequency is related to their intrinsic luminosity; this makes them important as distance indicators. Cepheid variables are discussed in greater detail below.

1.4 Distance Indicators

Parallax

Clearly, measurements of objects in the sky are two dimensional in nature; one of the abiding problems in observational astronomy has been to obtain

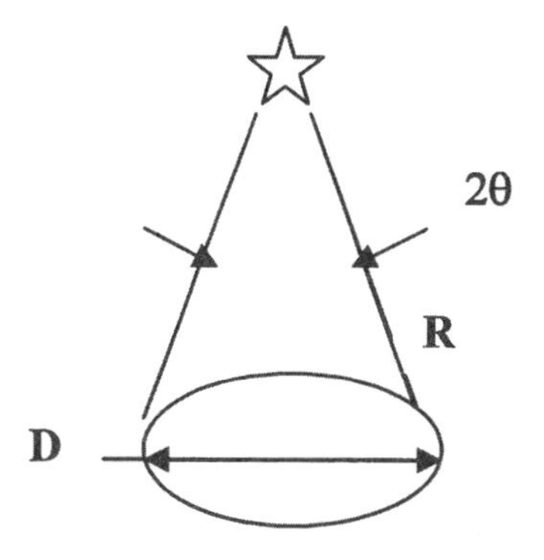

Fig. 1.6. The quantities that define parallax.

the third dimension, the distance to the object. There are several techniques used, the most direct being the parallax technique, in which the apparent motion of a nearby star as the Earth circles the Sun allows a measurement of its distance. While this technique works only for relatively nearby stars, it does give an unequivocal value. For example, Alpha Centauri (the nearest star to us except the Sun) produces 1.52 arcsec, or 0.76 arcsec parallax. As indicated in figure 1.6, the parallax is defined as $\theta = D/2R$.

Definition. A star that produces a parallax of 1 arcsec is said to be 1 parsec (pc) distant from the Earth. 1 pc $= 3 \times 10^{18}$ cm $= 3.3$ light-years.

Cepheid Variables

Another technique of measuring distance involves the use of Cepheid variables, stars that exhibit a regular variability in their luminosity, with the frequency of the variation being directly related to the intrinsic luminosity. The oscillations of the luminosity of a Cepheid variable are the result of alternating expansions and contractions of the star, a result of the way the star's outer layers transmit light. This stellar "opacity" depends on the temperature of the medium, since the fraction of the atoms that are ionized, or at higher temperatures, the mean ionization state of each of the atoms, will depend critically on the temperature. Since this can involve many atoms and ions, the opacity can be extremely complicated to characterize with high accuracy. When a Cepheid contracts, the density in the atmosphere increases and (primarily) its helium ions combine with electrons to form helium atoms. These atoms absorb light very efficiently, so in this state the atmosphere traps the radiant energy inside the star, which reduces its luminosity and allows heat energy to build up. This causes the star to expand, which decreases the density in the atmosphere and allows the helium atoms to reionize because of the emerging radiation. Light now escapes from the star's interior, and the luminosity increases, since there are fewer helium atoms to hold it in, and therefore the pressure that caused the expansion diminishes. Soon the outer layers begin to fall back in, and the star contracts, beginning a new cycle. These cycles continue indefinitely (Snow 1985). Unfortunately, although Cepheid variables are relatively bright stars, selecting them out of the 10^{11} stars in a distant galaxy is simply impossible, so they cannot be used as distance indicators to very large distances. However, they have provided extremely important "standard candles" for distance determinations. One caveat, however, is that there appear to be two distinct classes of Cepheids.

Redshifts

Redshifts have long been used as the prime distance indicators for distant galaxies. However, this statement is based on a tremendous amount of effort by astronomers to establish its credibility. Parallax can be used to establish the viability of the Cepheid variables, which can then be used to calibrate the redshifts, which can then be used to determine the largest distances at which galaxies can be observed. As discussed below, type Ia supernovae have recently been used to provide much brighter standard candles than the Cepheid variables ever could, so they have provided an important check of redshifts as distance indicators. These are discussed further below. The "universal" expansion of the universe was discovered by Hubble, who found that

$$v = \mathrm{H}R, \tag{1.4.1}$$

where v is the velocity of recession between any two distant objects, R is the distance between the two objects, and H is the Hubble constant. What this tells us is that the universe is expanding, with the rate of expansion increasing with the distance, that is, everything in the universe is expanding away from everything else (at least on large distance scales). This is known as Hubble's law, which stood the test of time for many decades but which recently has been shown to be incomplete, as is discussed in chapter 9. This law allowed astronomers to infer the distance to objects from their velocity of recession, which was directly related to their redshifts, z, which can be seen, in equation 1.4.2, to be the fractional increase in wavelength of objects that are moving away from the observer at velocity v. This is defined as

$$z = [\lambda(v) - \lambda(0)]/\lambda(0) = \Delta\lambda/\lambda(0). \tag{1.4.2}$$

It can also be shown that

$$z = [(1 + \beta)/(1 - \beta)]^{1/2} - 1, \tag{1.4.3}$$

where $\beta = v/\mathrm{c}$ and c is the speed of light. For small β, $z \approx \beta$. $\Delta\lambda$, of course, can be determined from the spectrum of the star being observed by observing the wavelength shifts of the characteristic emission or absorption lines.

Type Ia Supernovae

Other techniques have also been applied to determine distances; the most recently developed distance indicator is that of type Ia supernovae. These are readily identifiable and seem to exhibit a remarkable constancy of luminosity (or they are correctable), $10^{9.7}$ times the luminosity of the Sun, virtually independently of a variety of factors that could affect their luminosity. Thus they are good standard candles, and they are very bright. They have been used as the distance indicator of choice for cosmological studies and have produced extraordinary recent results on baryonic matter, dark matter, and dark energy (see chap. 9).

Other Standard Candles

While the above discussion gives a sampler of the techniques used to determine distances, it is not an exhaustive list. Also used, for example, are RR Lyrae stars and the Tully-Fisher approach. RR Lyrae stars are asymptotic giant branch stars (see sec. 1.6), so they have little variation in their luminosities. However, they can be identified because of their relatively rapidly varying luminosity. Thus they can provide a standard candle. Although they are quite numerous, they are relatively faint stars. The Tully-Fisher approach relates the size, and hence the intrinsic luminosity, of a galaxy to the width of the 21-cm line of hydrogen emitted from it. The width of this line is related to the rotation speed of the galaxy. More massive galaxies are likely to be rotating more rapidly, producing a rough galactic standard candle.

1.5 Luminosity and Temperature

In order to understand astronomical observations, one must know some of the basic language. Naturally, astronomers have developed a classification system based on the two most obvious observables: brightness and surface temperature. The total energy output, or luminosity, of a star involves its emitted photons, neutrinos, and mass, or

1.5.1 $$L_{\text{total}} = L_{\gamma} + L_{\nu} + L_{dm/dt}.$$

However, since the observed brightness of a star depends both on its intrinsic brightness and its distance from us, stars are characterized by magnitudes, which are defined so that a difference in apparent brightness of 5 magnitudes corresponds to a ratio of 100 in luminosity. Thus, for two stars characterized by brightnesses b_1 and b_2,

1.5.2 $$b_1/b_2 = 100^{(m_2-m_1)/5} = 2.512^{(m_2-m_1)}.$$

Thus

$$m_2 - m_1 = -2.5 \log(b_1/b_2). \tag{1.5.3}$$

Bolometric magnitude refers to the luminosity that would be observed by perfectly efficient detectors, independent of particle and wavelength, if such could be built. Of course, the observed luminosity also depends on the distance to the star, the $1/r^2$ factor, so it is also important to determine the absolute magnitude of a star—this is its intrinsic brightness. This is defined as the brightness a star would have if it were at a distance of 10 pc from Earth, denoted B, and all the energy emitted were observed:

$$m = -2.5 \log[B(10/r)^2] + \text{Constant} \tag{1.5.4}$$

$$= -2.5 \log B + C + 5 \log r - 5. \tag{1.5.5}$$

Denoting the magnitude at 10 pc by $M = -2.5 \log B$ (now with $C = 0$), this becomes

$$m = M + 5 \log r - 5, \textit{ or} \tag{1.5.6}$$

$$M = m - 5 \log r + 5 = \text{absolute magnitude.} \tag{1.5.7}$$

Note that larger magnitudes denote dimmer objects; for Venus $m = -4$ typically, the unaided visual limit of seeing is about $+6$, and the binocular limit is about $m = +10$.

The surface temperature is determined from the excitation and/or ionization states of the surface atoms/ions, and so it requires atomic spectroscopic measurements. However, the color at the surface has been historically used as a temperature indicator. The hottest stars are O stars, with temperatures ranging as high as 50,000 K and becoming progressively cooler as one progresses through the letter sequence O, B, A, F, G, K, M (the mnemonic is "Oh Be A Fine Girl (or Guy!), Kiss Me"). The letter-temperature correspondence is given below. See also figure 1.7 below for visual depiction of these classes.

Type	Characteristic	Main sequence temperature (K)
O	Hottest blue-white stars with few lines	28,000–50,000
B	Hot blue-white stars	9900–28,000
A	White stars	7400–9900
F	Yellow-white stars	6000–7400
G	Yellow stars	4900–6000
K	Cool orange stars	3500–4900
M	Coolest red stars	2000–3500

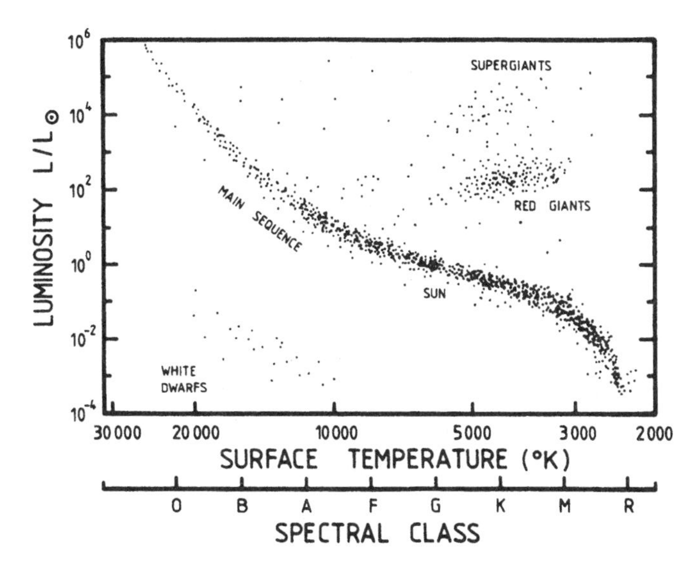

Fig. 1.7. The Hertzsprung-Russell (HR) diagram. The curved diagonal line that moves from the upper left to the lower right is known as the main sequence; most stars, roughly 90 % of them, are found to lie along that line. The group of stars that starts at the center of the H-R diagram and moves horizontally to the right is known as the (red) giant branch. At low temperature and high luminosity are scattered the (red) supergiants. Finally, hot low-luminosity stars known as white dwarfs lie far away from the main sequence on the lower left. From Rolfs and Rodney 1988; courtesy of C. Rolfs.

Note, however, that as instrumentation has improved, the classes have been subdivided by adding a decimal. For example, the Sun is a star of the G2 class.

1.6 Hertzsprung-Russell Diagram

The Hertzsprung-Russell (HR) diagram, shown in figure 1.7, is used by astronomers to classify stars according to their luminosity, spectral type, color, temperature, and evolutionary stage. This is named after the Danish astronomer Einar Hertzsprung (1873–1967) and the American astronomer Henry Norris Russell (1877–1957), its creators. What makes stars appear only in certain regions of the luminosity-temperature diagram? The answer to that question lies in the fact that the HR diagram represents a snapshot, so that the points in temperature and luminosity at which stars spend a long time will be well represented and those in which they spend only a short time or are transient (on stellar timescales) will contain few stars.

Other questions that arise are, why are stars that seem to be in different stages of evolution separated on the HR diagram, and how do they evolve from one part to another distinct part of this diagram? This is one of the subjects of this book, but briefly, stars consume a succession of fuels in their cores during their lives, moving from one to the next when the first has been burned (in nuclear reactions!) to what is usually the fuel for the next stage. In so doing, they change their surface properties, so they appear in different locations on the HR diagram. For example, stars spend most of their lives, perhaps 95 %, burning the hydrogen with which they were born in their cores; this constitutes the subject matter for chapter 5. Most of the rest of their lives will be spent burning the "ashes" of hydrogen burning, beginning with helium; this is one of the subjects of chapter 6. This immediately tells us something about the locations of the stars in different parts of the HR diagram; those on the main sequence, the roughly diagonal line from upper left to lower right, which was discovered in 1912 by Hertzsprung and Russell, are by far the ones that are predominantly seen, so they must be in their hydrogen-burning phase. Along the main sequence, luminosity $\approx$ mass$^{3.5}$, in solar units. Because the luminosity increases so rapidly with mass, more massive stars burn their nuclear fuel faster than less massive ones, so they live on the main sequence for shorter times than do the less massive stars.

The 10 % of the stars in the HR diagram that do not lie along the main sequence do not follow the mass-luminosity relationship. These include the giant and supergiant stars, which are found in the upper right portion of the HR diagram. These stars are relatively cool at their surfaces, but are extremely luminous because they are large in diameter; this provides a large surface area over which to radiate their energy, thereby increasing their luminosity. The white dwarfs are at the opposite end in the lower left of the diagram. They are very small in diameter (only about 1 % of the diameter of the Sun), so even though they are hot, they are intrinsically dim.

We shall return to the HR diagram in section 9.5, when we discuss globular clusters. These objects will allow us to infer some interesting things both about the meaning of the HR diagram and about stellar evolution, as the latter is manifested in the stellar populations seen in the globular clusters. Since by chapter 9 we will have discussed all the stages of stellar evolution, revisiting the HR diagram at that point will give it much more meaning. Most notably, some of the locations on the HR diagram result from differences in the complicated stages of helium burning; these would be difficult to explain in a very meaningful way at this stage of the book.

Finally, it is important to note that selection effects, that is, the effects on apparent observations due to the ability of astronomers to see stars of a particular class, must be considered by astronomers in establishing general conclusions about stellar populations. Luminous stars are easier to observe because they can be seen from greater distances, but they are also rarer than less luminous stars. The more luminous main sequence stars reside in the upper left quadrant of the HR diagram. Thus, nearby galaxies would appear to contain a smaller fraction of their total stars in red giants and very massive main sequence stars than more distant galaxies. However, such an effect would also be expected just from the intrinsic brightness of the stars.

1.7 Some Physics Basics: Conservation Rules

There are a number of basic rules and results from physics that we will use throughout this book. Although these are generally thought by physicists to apply to microscopic systems, they will also apply to some of the macroscopic systems that we will discuss. These will be reviewed below.

Conservation of Energy

In any process, macroscopic or microscopic, energy is conserved. This applies to systems as large as massive stars as they collapse, driving the considerations of how a massive stars explodes, and as small as nuclei when they undergo interactions and, possibly, reactions with other subatomic species.

Conservation of Momentum

This will enter our considerations less frequently than conservation of energy, but it enters implicitly in many of the nuclear processes that will be discussed. It also applies to systems, again, as large as stars undergoing collapse and formation of a neutron star, as in some instances the collapse appears to have been asymmetric, resulting in a neutron star that appears to have emerged at high speed from the nebula of the star.

Conservation of Angular Momentum

As with the above two conservation rules, this can apply to collapsing stars, as the formation of a collapsed object such as a neutron star can produce a very rapidly spinning object. Of course, conservation of angular momentum also applies to nuclear reactions and to such processes as beta decay.

Charge Conservation

This will be seen, in chapter 5, to be very important to the initial nuclear reaction that occurs in hydrogen burning in not too massive stars, that is,

$$^{1}\mathrm{H} + {}^{1}\mathrm{H} \rightarrow {}^{2}\mathrm{H} + e^{+} + \nu_{e}, \tag{1.7.1}$$

(^1H refers to an ordinary hydrogen nucleus, a proton, and ^2H refers to the nucleus of heavy hydrogen, which is made up of a proton and a neutron) which cannot proceed by an ordinary nuclear strong interaction because ^2He is not stable and does not even have a resonance that could produce a long-lived, but unstable, state. In this equation, e^+ is an antielectron, that is, a positron, and ν_e is an electron neutrino. However, charge conservation also dictates that one must consider, in stellar systems, all the particles, both nuclei and electrons (and sometimes others), that can exist in a star. Indeed, if this were not the case, charge separation would quickly force the dominance of the electrostatic force over the gravitational force, something that is not observed in large-scale nature.

Lepton Conservation

The non-strongly interacting particles—electrons, muons, taus, the neutrinos, and the antiparticles of all these—are known as leptons. The number of them seems to be conserved in reactions (although there are variants of this rule). Thus, in equation 1.7.1, there are zero leptons on the left-hand side of the equation but two leptons on the right-hand side. However, of those two leptons, the e^+ is an antilepton, so it is included in that equation as (-1) lepton. It therefore cancels out the ν_e to give a net lepton number on the right-hand side of zero.

Baryon Conservation

Baryons are the strongly interacting particles that occur in nature, such as protons, neutrons, lambdas, and so on. These particles are made up of three quarks, and their antiparticles are made up of three antiquarks. They are one class of particles called hadrons, which include all of the strongly interacting particles, the other type being the mesons, such as pions, kaons, and so on. Mesons are made up of a quark and an antiquark; there are no stable mesons. Of the two classes of hadrons, baryons appear to be conserved in nuclear reactions, as is seen in the above reaction, in which there are two protons on the left-hand side, or two baryons, and a heavy hydrogen nucleus, made up of a proton and a neutron, two baryons, on the right-hand side. Mesons are not conserved in nuclear reactions. Thus the reaction

1.7.2 $$p + p \rightarrow p + n + \pi^+$$

is a perfectly acceptable reaction, as it conserves charge and baryon number. Indeed, multiple pions could have been emitted as long as their sum conserved charge.

Photons

Photons do not need to be conserved in nuclear reactions. Thus the reaction

1.7.3 $$^{2}\mathrm{H} + {}^{1}\mathrm{H} \rightarrow {}^{3}\mathrm{He} + \gamma$$

($^{1}\mathrm{H} = \mathrm{p}$; these will be used interchangeably) is an acceptable reaction, as it conserves charge and baryons and the photon, γ, that occurs on the right-hand side is actually necessary to conserve energy.

Parity Conservation

Parity refers to mirror symmetry; if you are perfectly left-right symmetric your image in the mirror will conserve parity with you. This particular symmetry is violated in β-decay.

Conservation of Isospin

This is a symmetry that involves charge but results from treating the proton and neutron as states of the same particle, the nucleon. It will not be discussed here, but we will encounter it in some of the nuclear processes we consider, and it is discussed in detail in the appendix. The Coulomb potential violates isospin conservation, so isospin is not a perfect symmetry in nuclei. It is a fairly good symmetry, however, as is discussed in the appendix.

There are some additional conservation rules that we could consider, but they will not enter our considerations in nuclear astrophysics, so they will not be discussed.

Heisenberg Uncertainty Principle

One other rule that will enter our considerations is that of the Heisenberg Uncertainty Principle, which dictates the limits to which one can simultaneously measure some of the variables associated with microscopic systems. This results from the need to describe such systems with both a wave and a particle nature. It is most often stated as

1.7.4 $$\Delta p \Delta x \geq \hbar,$$

where Δp and Δx refer, respectively, to the uncertainty in momentum and position and $\hbar$ is Planck's constant divided by 2π. However, it also applies to simultaneous measurement of the different components of the vector angular momentum as, for example,

1.7.5 $$\Delta L_x \Delta L_y \geq \hbar L_z$$

where the L_j are the three components of the angular momentum, and $i = (-1)^{1/2}$. However, we will most frequently encounter the Heisenberg Uncertainty Principle in the form

$$\Delta E \Delta t \geq \hbar, \quad (1.7.6)$$

where ΔE is the energy uncertainty associated with a nuclear state, that is, the width of a nuclear resonance, and Δt is its lifetime.

CHAPTER 1 PROBLEMS

1. How many times brighter is star no. 1 than star no. 2 if their magnitudes are given by $m_1 = 5$ and $m_2 = 15$? How far is a star from Earth if its apparent magnitude is 15 and its absolute magnitude is 4?
2. What is the speed of recession of the quasar whose spectrum is shown in figure 1.3 and which has a redshift of 3.18?
3. A nuclear resonance is found to have an energy width of $\Delta E = 0.5$ MeV. What is the "lifetime" of the quantum mechanical state represented by that resonance?
4. Which of the following reactions obeys all the necessary conservation laws, and which do not?
   a. $\bar{\nu}_e + p \rightarrow e^+ + n$ ($\bar{\nu}_e$ denotes an electron antineutrino).
   b. $\gamma + p \rightarrow \pi^+ + n$
   c. $\mu^+ + p \rightarrow e^+ + \pi^+ + n + \nu_e$

2

THE INSTRUMENTS USED TO STUDY ASTROPHYSICS

2.1 Introduction

This chapter is about some of the instruments and facilities that are used to provide astrophysical information. With each description will go some discussion of the science that it addresses, either in this or in subsequent chapters (or both). Most of these facilities provide information for nuclear astrophysics, although in a few cases the relationship might not be so apparent.

The list of instruments is not intended to be comprehensive, although an attempt has been made to include at least one example of each type of instrument that impacts nuclear astrophysics. However, there are many wonderful instruments that did not make it into this edition of the book. Certainly no slight was intended. In many cases, the information presented was extracted, sometimes without much editing, from the instrument or facility Web site. This is certainly an evolving list; the current edition is circa 2006.

The write-ups also vary considerably in what they present. This is due somewhat to the different states of maturity of the projects; where they have been in operation long enough to have acquired data this is usually presented. In some cases, the projects are yet to obtain data; in these cases, the scientific goals of the project are indicated.

2.2 The Cosmic Microwave Background: *COBE* and *WMAP*

The cosmic microwave background (CMB) is the radiant heat left over from the big bang. It was first observed—accidentally—in 1965 by Arno Penzias and Robert Wilson at the Bell Telephone Laboratories in Murray Hill, New Jersey. The properties of the radiation contain a wealth of information about physical conditions in the early universe, and a great deal of effort has gone into measuring their details. This radiation (and by extension, the early universe) is remarkably featureless; it has a blackbody spectrum that has virtually the same temperature in all directions in the sky. But there are fluctuations, and they have assumed great importance in our understanding of our universe's birth event.

In 1992, the *Cosmic Background Explorer* (*COBE*) satellite not only measured the temperature of the CMB to remarkable precision but also detected some of those fluctuations, or anisotropies, in the CMB. However the fluctuations are tiny; they represent a roughly one part in 10^5 variation on the 2.725 K background. These CMB fluctuations are related to fluctuations in the density of matter in the early universe and thus carry information about the initial conditions for the formation of cosmic structures such as galaxies, clusters, and voids. However, *COBE* had an angular resolution of 7° across the sky, which made it sensitive only to fluctuations of large size.

To refine these results, the *Wilkinson Microwave Anisotropy Probe* (*WMAP*), named after David Wilkinson, a pioneer in this field, was launched in June 2001 and has now mapped the temperature fluctuations of the CMB radiation with much higher resolution, sensitivity, and accuracy than was possible with *COBE*.

2.2.1 WMAP Mission Overview

Since *WMAP* was designed to search for fluctuations in the CMB, its science goals broadly dictated that the relative CMB temperature be measured accurately over the full sky with high angular resolution and sensitivity. Specifically, this meant mapping the relative CMB temperature with an angular resolution of better than 0.3°, a sensitivity of 20 μK per 0.3° square pixel, and systematic artifacts limited to 5 μK per pixel.

To achieve these goals, *WMAP* used differential microwave radiometers that measure temperature differences between two points on the sky. *WMAP* observes the sky from an orbit about the L2 Sun-Earth Lagrange point (see explanation below), 1.5 million km from Earth. This vantage point offers an exceptionally stable environment for observing since the observatory can always point away from the Sun, Earth, and Moon while maintaining an unobstructed view to deep space. *WMAP* scans the sky in such a way as to cover ~30 % of the sky each day, and as the L2 point follows the Earth around the Sun, *WMAP* observes the full sky every 6 months. To facilitate rejection of foreground signals from our own Galaxy, *WMAP* uses five separate frequency bands from 22 to 90 GHz.

The main features of the *WMAP* spacecraft are shown in figure 2.1. Central to its success is a pair of back-to-back microwave radiometers, shown at the

Fig. 2.1. The *Wilkinson Microwave Anisotropy Probe*. The different components are described in the text. Figure courtesy of the NASA/*WMAP* Science Team.

top of figure 2.1, with 1.4 m × 1.6 m diameter primary reflectors, which were chosen to provide the desired angular resolution. These telescopes focus the microwave radiation from two spots on the sky roughly 140° apart and feed it to 10 separate differential receivers that sit in an assembly directly underneath. This permits differential measurements of the temperature, which can be obtained with higher accuracy than can absolute temperature measurements. Large "elephant ear" radiators, one of which is seen extending out from the vertical center of figure 2.1, provide passive cooling for the sensitive amplifiers in the receiver assembly. The bottom section of the spacecraft provides the services necessary to carry out the mission, including command and data collection electronics, attitude (pointing) control and determination, power services, and a hydrazine propulsion system. The entire observatory was kept in continuous shade by a large deployable Sun shield,

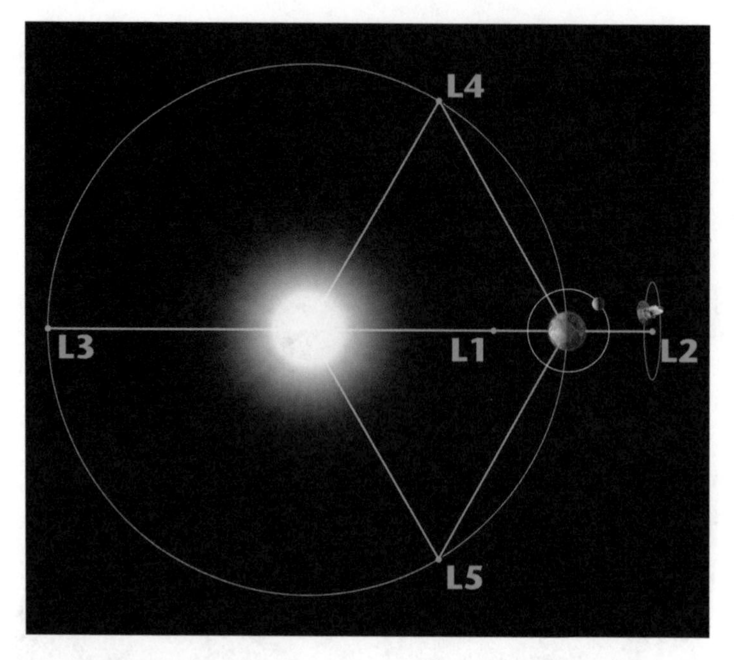

Fig. 2.2. Lagrange Points of the Earth-Sun system (not to scale). Figure courtesy of the NASA/ *WMAP* Science Team.

which is seen at the bottom of the figure and which also supports the solar panels.

2.2.2 Lagrange Points

The Italian-French mathematician Joseph-Louis Lagrange discovered five special points in the vicinity of two orbiting masses where a third, much smaller, mass can orbit at a fixed distance from the larger masses. At these Lagrange points the gravitational pull of the two large masses precisely equals the centripetal force required to rotate with them. Of the five Lagrange points, three are unstable and two are stable. The unstable Lagrange points, L1, L2, and L3 in figure 2.2, lie along the line connecting the two large masses.

The stable Lagrange points, L4 and L5, form the apex of two equilateral triangles that have the large masses at their vertices. The L1 point of the Earth-Sun system affords an uninterrupted view of the Sun, so it is the location of choice for solar observatories. The L2 point of the Earth-Sun system is home to the *WMAP* spacecraft.

The new information contained in the finer CMB fluctuations and their precise measurement by *WMAP* have shed light on several key questions in cosmology, as will be discussed in much greater detail in chapter 9.

Much of the above regarding *WMAP* was excerpted from the *WMAP* Web site at http://map.gsfc.nasa.gov/about/aboutmap_in.html.

2.3 Radio Astronomy

Radio telescopes are used to study naturally occurring radio emission from stars, galaxies, quasars, and other astronomical objects between wavelengths of about 10 m (30 MHz) and 1 mm (300 GHz). Because of the high spatial resolution that can be achieved with these telescopes, they have provided, for several decades, some of the most precise measurements achievable. Between 1 and 20 cm, the atmosphere and ionosphere introduce only minor distortions in the incoming signal (see fig. 1.5), so this region is optimal. Nonetheless, sophisticated signal processing can be used to correct for such effects, so that the effective angular resolution and image quality is really limited only by the size of the instrument. Indeed, the one shown in figure 2.3, the *Green Bank Telescope,* located in Green Bank, West Virginia, is the world's largest fully steerable single telescope; it is 100 m in diameter.

2.3.1 Principles of Operation

The sensitivity of a radio telescope, that is, its ability to measure weak sources of radio emission, depends on the area and efficiency of the antenna and the

Fig. 2.3. The *Green Bank Radio Telescope*, which has a diameter of 100 m. From http://www.nrao.edu. Image courtesy of NRAO/AUI/NSF.

sensitivity of the radio receiver used to amplify the signals. Thus, since cosmic radio sources are generally extremely weak, radio telescopes are usually large and utilize only the most sensitive radio receivers. The most familiar type of radio telescope is the radio reflector, consisting of a parabolic antenna that focuses the incoming radiation onto a small antenna referred to as the feed. In a radio telescope, the feed is typically a waveguide horn and transfers the incoming signal to the radio receiver.

The angular resolution of a radio telescope, indeed of any telescope, depends on the wavelength at which the observations are made divided by the diameter of the instrument. Thus radio telescopes must be much larger than optical telescopes to achieve the same angular resolution. The performance of a radio telescope is also limited by irregularities of the reflecting surface, the effects of wind, differential thermal deformations, and deflections due to changes in gravitational forces with antenna redirection. Since small structures can be built with greater precision than larger ones, radio telescopes designed for operation at millimeter wavelength are typically only a few tens of meters across, whereas those designed for operation at centimeter wavelengths range up to 100 m in diameter.

However, at radio wavelengths, the distortions introduced by the atmosphere are less important than at optical wavelengths (see sec. 1.3), and so the theoretical angular resolution of a radio telescope can in practice be achieved even for the largest dimensions. Also, because radio signals are easy to distribute over large distances without distortion, it is possible to build radio telescopes, when they are operated in interferometric mode, of effectively unlimited dimensions.

2.3.2 Very Large Array

The *Very Large Array*, one of the world's premier astronomical radio observatories, shown in figure 2.4, consists of 27 radio antennas arranged in a Y-shaped configuration on the Plains of San Agustin, 50 miles west of Socorro, New Mexico. Each antenna is 25 m in diameter. The data from the antennas are combined electronically to give the resolution of an antenna 36 km across, with the sensitivity of an equivalent single dish 130 m in diameter.

2.3.3 Very Long Baseline Interferometry

The highest angular resolution achieved by radio telescopes utilizes interferometry to synthesize a very large effective aperture from a number of distributed elements. Interferometer systems of essentially unlimited element separation are formed by using the technique of very long baseline interferometry,

Fig. 2.4. The *Very Large Array*, for radio astronomy. From http://www.aoc.nrao.edu/evla. Image courtesy of NRAO/AUI/NSF.

or VLBI. Data collection requires synchronization of data from different telescopes to within a few millionths of a second; this is feasible with modern instrumentation.

2.3.4 Very Long Baseline Array

The use of interferometry has lead to installations such as the *Very Large Array* and the *Very Long Baseline Array* (*VLBA*), which essentially utilizes the size of Earth as its baseline.

The *VLBA*, which was dedicated in 1993, is a system of 10 radio telescope antennas, each with a dish 25 m (82 feet) in diameter and weighing 240 tons. From Mauna Kea on the Big Island of Hawaii to St. Croix in the U.S. Virgin Islands, the *VLBA* spans more than 5000 miles, providing astronomers with the sharpest vision of any telescope on Earth or in space.

Much of the above information regarding radio astronomy was excerpted from the *National Radio Astronomy Observatory* Web site at http://www.nrao.edu/. The figures were used with the permission of the *National Radio Astronomy Observatory*.

Fig. 2.5. Artist's conception of the *Spitzer Space Telescope*. Image from http://ipac.jpl.nasa.gov/media_images/Earth_trailing2.jpg, courtesy of NASA/JPL-Caltech.

2.4 Infrared Astronomy: The *Spitzer Space Telescope*

The *Spitzer Space Telescope* (formerly, the *Space InfraRed Telescope Facility*) was launched into space on August 25, 2003. It is named after Lyman Spitzer, Jr., who made major contributions to space astronomy and played a major role in bringing the *Hubble Space Telescope* (*HST*) to fruition. During its mission, *Spitzer* will obtain images and spectra by detecting the infrared radiation from objects in space between wavelengths of 3 and 180 μm. It is essential to use an orbiting observatory for these measurements, since most of the frequency range of infrared radiation is blocked by the Earth's atmosphere and cannot be observed from the ground (see sec. 1.3). In addition, infrared backgrounds can be huge if the observatory is operated at the Earth's surface.

Consisting of a 0.85-m telescope and three cryogenically cooled science instruments, *Spitzer* is the largest dedicated infrared telescope ever launched into space. Its instruments allow observations into regions of space that are hidden from optical telescopes by vast, dense clouds of gas and dust that block radiation of optical wavelengths. Infrared light, however, can penetrate these clouds, allowing observation of regions of star formation, the centers of galaxies, and newly forming planetary systems, none of which could have been observed at optical frequencies. Infrared also brings us information about the cooler objects in space, such as smaller stars that are too dim to be detected by their visible light, extrasolar planets, and giant molecular clouds that have their primary electromagnetic emission in the infrared. Also, many molecules in space, including organic molecules, produce unique signatures in the infrared.

Because infrared radiation is primarily heat radiation, the telescope must be cooled to near absolute zero so that it can observe infrared signals from space without interference from the telescope's own heat. Also, the telescope must be protected from the heat of the Sun and the infrared radiation emitted by the Earth. To do this, *Spitzer* carries a solar shield and was launched into an Earth-trailing solar orbit. This unique orbit places *Spitzer* far enough away from the Earth to allow the telescope to cool rapidly without having to carry large amounts of coolant.

The information for this write-up and the above figure came from the *Spitzer Space Telescope* homepage at http://www.spitzer.caltech.edu/.

2.5 Optical Astronomy

2.5.1 The W. M. Keck Observatory

Observational astronomy has produced much of the information we have obtained about the universe over the past 7–8 decades. However, two recent technological developments have created a revolution in a field even as mature as this. These are adaptive optics, which greatly reduces the atmospheric distortion, discussed in chapter 1, and interferometry, which greatly improves angular resolution. Both of these techniques are now being introduced at many telescopes around the world. However, the *W. M. Keck Observatory* has been utilizing both techniques for some time. Because of this, and because the *Keck Observatory* has been at the forefront of so many modern technological developments and particularly because optical astronomical observations are so important to nuclear astrophysics, *Keck* and its capabilities will be discussed in some detail. In addition, some aspects of telescope design and operation, of importance to, but not necessarily well-known to, nuclear astrophysicists are included. The *Keck-Keck* Interferometer combines the light of the two *Keck* telescopes to obtain a 10-fold increase in angular resolution; the result is an angular resolution equivalent to that of a telescope 85 m in diameter. Adaptive optics will be described in section 2.5.2.

Fig. 2.6. The *Keck* telescopes, seen with the artificial guide star laser. Credit: Sarah Anderson, *W. M. Keck Observatory*.

The summit of Hawaii's dormant Mauna Kea volcano is the site of the *Keck Observatory*. Its basic instruments are the twin *Keck* telescopes, among the world's largest optical and infrared telescopes, which are shown in figure 2.6. Each stands eight stories tall and weighs 300 tons yet operates with nanometer precision. At the heart of each *Keck* telescope is a primary mirror 10 m in diameter, each composed of 36 hexagonal segments that work together as a single piece of reflective glass.

The *Keck I* telescope began observing in May 1993, and *Keck II* began in 1996. Mauna Kea presents an unusually desirable site for astronomy. Its 13,796-foot summit has no nearby mountain ranges to disturb the upper atmosphere or throw light-reflecting dust into the air. It is surrounded by thousands of miles of relatively thermally stable ocean. Few city lights pollute its extremely dark skies. For most of the year, the atmosphere above Mauna Kea is clear, calm, and dry.

2.5.2 The Telescopes

An altitude-azimuth design gives each 10-m *Keck* telescope the optimal balance of mass and strength. This is critically important, as a large telescope must resist the deforming forces of gravity as it tracks objects moving across the night sky.

2.5.3 The Mirrors

A telescope tracks objects, sometimes for hours, across the sky as the Earth turns. This constant movement results in deformations of the telescope structure despite all engineered precautions. Without active, computer-controlled correction of the primary mirror, scientific observations over such a time span would be impossible.

The mirror segments are precision polished. Each segment's position is kept stable by a system of extremely rigid support structures and adjustable warping harnesses, some of which are shown in figure 2.7. During observing, a computer-controlled system of sensors and actuators adjusts the position of each segment—relative to its neighbors—to an accuracy of 4 nm, about the size of a few molecules. This twice-per-second adjustment effectively counters the effects of gravity on the telescope structure.

Unfortunately, as noted in section 1.3, all Earth-based telescopes suffer from image blurring caused by the turbulent atmosphere above them. This is true of even the world's best observatory sites (high altitude, low humidity) like Mauna Kea, although to a considerably lesser extent there than in most other places. In recent years, advances in optical and computing technology have

Fig. 2.7. Some of the mechanical details of one of two *Keck* telescopes. Credit: *W. M. Keck Observatory.*

Fig. 2.8. Narrow-field image of the Galactic center taken with the Keck Laser Guide Star Adaptive Optics. Credit: *W. M. Keck Observatory*.

made it possible to greatly reduce this blurring through the use of adaptive optics (AO). At the heart of the AO system is a 6-inch-diameter deformable mirror that makes tiny changes in its shape up to 670 times per second to produce the sharpest image possible of a "guide star." This performs a nearly perfect cancellation of the atmospheric distortions, which change in times larger than this, resulting in images 10 times sharper than before the advent of AO. The successful installation of AO systems on both *Keck* telescopes has made it possible for *Keck* astronomers to study objects in far higher resolution than ever before. At some wavelengths, the *Keck* resolution approaches the diffraction limit, the fundamental limitation to resolution if the mirrors were perfect and there were no atmospheric distortion. A major advance for AO has been development of the *Keck* Laser Guide Star System, which uses a laser to excite sodium atoms that naturally exist in the atmosphere 90 km (55 miles) above the Earth's surface (see fig. 2.6), creating an "artificial guide star" that allows the *Keck* AO system to observe 70 %–80 % of the targets in the northern sky, compared with the 1 % accessible, in AO, without the laser.

2.5.4 Analysis Instrumentation

The telescope is only half of what is required to do astronomy; the instruments that are at the "astronomer's end" of the telescope are equally important. Modern astronomy is done with sensitive detection devices that convert the photons to computer images. Some of these devices are optimized for sheer photon collection, others to perform rough spectral analyses of the light they see, and still others to perform detailed high-resolution spectral analyses of the relatively bright objects they observe. Some of these details will be given in the following section on the *HST*. Different devices are also optimized for different wavelength light. These will be described briefly below.

Visible Band (0.3–1.0 μm)

- *Deep Extragalactic Imaging Multi-Object Spectrograph (DEIMOS)*. The DEIMOS is an optical spectrograph capable of gathering spectra from 130 galaxies or more in a single exposure. In "mega mask" mode, DEIMOS can take spectra of more than 1200 objects at once using a special narrowband filter.
- *Echellette Spectrograph and Imager (ESI)*. The ESI captures high-resolution spectra of very faint galaxies and quasars ranging from the blue to the infrared in a single exposure. It has produced some of the best non-AO images at the *Keck Observatory*.
- *High-Resolution Echelle Spectrometer (HIRES)*. The largest and most mechanically complex of the *Keck*'s main instruments, the HIRES resolves incoming starlight into its component colors to measure the precise intensity of each of thousands of color channels. Its spectral capabilities have resulted in, for example, the detection of planets outside our solar system.
- *Low-Resolution Imaging Spectrograph (LRIS)*. The LRIS is a faint-light instrument capable of taking spectra and images of the most distant known objects in the universe. The instrument is equipped with a red arm and a blue arm to explore stellar populations of distant galaxies, active galactic nuclei, galactic clusters, and quasars.

Near-Infrared (1–5 μm)

- *Near-Infrared Camera (NIRC)*. The NIRC for the *Keck I* telescope is so sensitive it could detect the equivalent of a single candle flame on the Moon. This makes it ideal for ultradeep studies of galactic formation and evolution, searches for proto-galaxies, and images of quasar environments.
- *NIRC-2/AO*. The second-generation NIRC works with the *Keck* AO system to produce the highest resolution ground-based images and spectroscopy in the range of 1 to 5 μm. Typical programs include mapping surface features on solar system bodies, searching for planets around other stars, and analyzing the morphology of remote galaxies.
- Near-Infrared Spectrometer (NIRSPEC). The NIRSPEC studies, for example, very high redshift radio galaxies, the motions and types of stars located near the Galactic center, the nature of brown dwarfs, the nuclear regions of dusty starburst galaxies, active galactic nuclei, and interstellar chemistry.

Mid-Infrared (5–27 μm)

- *Keck* Interferometer. The *Keck-Keck* Interferometer combines light from the two *Keck* telescopes to measure the diameters of stars, disks orbiting nearby stars, and the orbital characteristics of binary systems. It also directly detects and characterizes hot giant planets. The interferometer can reach angular resolutions to a small fraction of an arcsecond, providing the effective resolution of an equivalent telescope 85 m in diameter.

Although the *Keck Observatory* was featured in the above discussion, there are many other telescopes operating around the world. Furthermore, there are other multimirror telescopes either in operation or nearing operation, each with interferometric capabilities. This includes the *GEMINI* project in Chile, the Japanese *SUBARU* telescope in Hawaii, and the *Large Binocular Telescope* in Arizona. Future projects either in planning or construction will certainly utilize all of the capabilities discussed above in the context of the *Keck Observatory*.

This write-up and the figures were excerpted from the information given on the *Keck Observatory* Web site, http://www.keckobservatory.org/.

2.5.5 Hubble Space Telescope

The *HST* was named for Edwin Hubble. He made some major discoveries during several remarkable years in the 1920s. He was a pioneer in extragalactic

Fig. 2.9. The *Hubble Space Telescope*, floating freely in orbit, following a space repair mission. Reprinted courtesy of *STScI*.

astronomy, realizing that the Milky Way is just one of millions of galaxies in an incomparably larger setting, and then found, through the universal expansion law, the first hints that the universe began with a big bang. The *HST* was designed to operate a long-lived space-based observatory for the benefit of the international astronomical community. The obvious benefit of putting a telescope in space is that it never suffers from either the distortion due to the Earth's atmosphere (without AO a ground-based telescope can achieve about 1.0 arcsec of resolution at best [at optical wavelengths], but *HST* can do about a factor of 10 better) or the absorption by the atmosphere in some wavelengths (see chap. 1). The *HST* surely has been one of the most successful scientific instruments ever put into operation. For some of the immense number of pictures produced by the *HST*, see its Web site, http://hubblesite.org/.

HST is a 2.4-m reflecting telescope that was deployed in low-Earth orbit (600 km) by the crew of the space shuttle Discovery on April 25, 1990. At present it is hoped that it will be operated until about 2010. *HST* had to be stabilized in all three axes to ensure accurate pointing. It also had to be built in such a way that it could be serviced occasionally by astronauts. Its current suite of science instruments includes three cameras, two spectrographs, and extremely precise guidance sensors; these are discussed in somewhat more detail below.

2.5.6 Wide-Field Planetary Camera (WFPC)

The WFPC is actually four cameras. The heart of the WFPC consists of an L-shaped trio of wide-field sensors and a smaller, high-resolution ("planetary") camera tucked in the square's remaining corner. The WFPC currently on the *HST* is actually WFPC2, the second-generation WFPC.

2.5.7 Space Telescope Imaging Spectrograph (STIS)

A spectrograph analyzes the light it receives into its constituent wavelengths to determine such properties of celestial objects as chemical composition and abundances, from the characteristic emissions or absorptions of atoms and ions; temperature, from the distributions of the atomic and ionic emission or absorption lines; radial velocity, from the Doppler shifts in the emission or absorption lines; rotational velocity, from the broadening of the line shapes due to Doppler shifts in lines from different regions of the object being observed; and magnetic fields, from the line splitting or broadening. The STIS can study objects across a spectral range from the ultraviolet (UV) (115 nm) through the visible red and the near-infrared (1000 nm), using three different types of detectors, spanning a view for the elements of the detectors ranging from 25 arcsec × 25 arcsec to 50 arcsec × 50 arcsec.

The main advantage of the STIS over previous spectrographs is its capability for two-dimensional rather than one-dimensional spectroscopy. This makes it possible to record the spectra of many locations in a galaxy simultaneously rather than observing one location at a time. The STIS can also record a broad span of wavelengths in the spectrum of a star at one time. As a result, STIS is much more efficient at obtaining scientific data than were the earlier *HST* spectrographs.

2.5.8 Near-Infrared Camera and Multi-Object Spectrometer (NICMOS)

The NICMOS is an *HST* instrument that provides the capability for infrared imaging and spectroscopic observations of astronomical targets. NICMOS detects light with wavelengths between 0.8 and 2.5 μm. The arrays that make up the infrared detectors in NICMOS must operate at very cold temperatures to minimize backgrounds. NICMOS keeps its detectors cold inside a cryogenic dewar containing frozen nitrogen ice.

2.5.9 Advanced Camera for Surveys (ACS)

The ACS was installed in the March 2003 servicing mission. It actually consists of three cameras: a wide-field camera that operates from the visible to the near-infrared, a high-resolution camera, and a solar blind camera (which blocks sunlight) for stellar far-UV imaging. The primary design goal of the ACS Wide-Field Channel is to achieve a factor of 10 improvement in "discovery efficiency," compared with WFPC2. In this context, discovery efficiency is defined as the product of imaging area and instrument throughput, that is, the instrument's capability for detecting photons.

2.5.10 Fine Guidance Sensors

The Fine Guidance Sensors, in addition to being an integral part of the *HST* Pointing Control System, provide *HST* observers with the capability of precision astrometry and milliarcsecond resolution over a wide range of stellar magnitudes ($3 < V < 16.8$). Its two observing modes—position mode and transfer mode—have been used to determine the parallax and proper motion of astrometric targets to a precision of 0.2 milliarcseconds and to detect duplicity or structure around targets as close as 8 milliarcseconds (visual orbits can be determined for binaries as close as 12 milliarcseconds).

2.5.11 Some HST Science: The Hubble Ultra Deep Field Survey

Mankind's deepest, most detailed optical view of the universe was provided courtesy of the *HST*. The image, called the Hubble Ultra Deep Field (HUDF), shown in figure 2.10, was assembled from many separate exposures taken with the ACS and NICMOS during a series of nights beginning September 24, 2003, and ending January 16, 2004, and spanned 400 *HST* orbits. The HUDF image covers only a small portion of the sky, just one-tenth of the diameter of the moon, but it is considered representative of the typical distribution of galaxies in space because the universe looks essentially the same in all directions. Figure 2.10 shows a bewildering assortment of galaxies at various stages of evolution. Most are extremely "deep," that is, they are fainter than 30th magnitude (see chap. 1).

Because the most distant objects are also among the dimmest, the image allows observation of the early formation of galaxies less than one billion years old, with the NICMOS images extending perhaps to 400 million years, after the big bang (at a redshift of 12; see chap. 1). The field is far from the plane of our Galaxy in the constellation Fornax, below the constellation Orion, and so is uncluttered by nearby objects such as foreground stars. The field provides a look out of the Galaxy that allows for a clear view all the way to the horizon of the universe. The target field was, by necessity, in the continuous viewing zone of *Hubble*'s orbit, a special region where the *HST* can view the sky without being blocked by Earth or interference from the Sun or Moon. By taking multiple exposures of the same places on the sky, the *HST* was able to

Fig. 2.10. The Hubble Ultra Deep Field. This and its predecessor, the Hubble Deep Field, are arguably the most famous images produced by the *Hubble Space Telescope* and possibly in all of astronomy. Reprinted courtesy of *STScI*.

compile an exposure that was equivalent to a much longer exposure, thereby uncovering the faint objects it was able to identify. Separate images were taken in UV, blue, red, and infrared light. By combining these separate images into a single color picture, astronomers were able to infer the distances, ages, and composition of the galaxies in the Hubble Ultra Deep Field image.

All of the above information and figures were excerpted and edited from the *HST* Web site, http://hubble.nasa.gov/index.php.

2.6 X-Ray Astronomy

2.6.1 Introduction

X-ray astronomy has been an active field of research for many years, but three observatories, all launched since 1995, have provided facilities for great advances in this area of research. The first of these, the *Rossi X-ray Timing Explorer*, or *RXTE*, was launched in 1995. It was named after Bruno Rossi, a pioneer in both X-ray astronomy and space plasma physics. The second is the satellite *BeppoSAX*, named after Giuseppe (Beppo) Occhialini, a major force in Italian particle physics and astrophysics. The remaining one was launched in 2000. This was the *CHANDRA* observatory, named after the Nobel Prize–winning theoretical astrophysicist Subramanyan Chandrasekar.

Because of their high energy, X-ray photons penetrate into any mirror they might encounter. However, if they intercept their mirrors at a small grazing angle, they will be reflected and thus can be focused. Thus, *CHANDRA*'s mirrors are cylinders that have been exquisitely shaped and aligned nearly parallel to incoming X-rays; they look more like glass barrels than the familiar dish shape of optical telescopes.

2.6.2 Rossi X-ray Timing Explorer

The *RXTE* was launched on December 30, 1995, into its intended low-Earth circular orbit at an altitude of 580 km, corresponding to an orbital period of about 90 minutes. The *RXTE* features unprecedented time resolution in combination with moderate spectral resolution to explore the variability of X-ray sources. Timescales from microseconds to months are covered in an instantaneous spectral range from 2 to 250 keV. Originally designed for a required lifetime of 2–5 years, *RXTE* has passed that goal and is still performing well.

Instrumentation

The mission carries two instruments, the Proportional Counter Array (PCA) to cover the lower part of the energy range and the High Energy X-ray Timing Experiment (HEXTE) to cover the upper energy range. These instruments are equipped with collimators yielding a FWHM of 1°. In addition, *RXTE* carries an All-Sky Monitor (ASM) that scans about 80 % of the sky every orbit, allowing monitoring at timescales of 90 minutes or longer. Data from PCA and ASM are processed on board by the Experiment Data System (EDS).

The High Energy X-ray Timing Experiment

The HEXTE consists of two clusters, each containing four phoswich (phosphor sandwich: having several scintillators with dissimilar pulse shape characteristics) scintillation detectors. Each cluster can rock back and forth along mutually orthogonal directions to provide background measurements 1.5° or 3.0° away from the source every 16–128 s. The HEXTE's basic properties are as follows (a "Crab" is an observational standard based on the detection rate of photons in the appropriate wavelength range from the Crab Nebula, a particularly bright object, over a wide range of wavelengths):

- energy range: 15–250 keV,
- energy resolution: 15 % at 60 keV,
- field of view: 1° FWHM,
- collecting area: 2×800 cm^2,
- sensitivity: 1 Crab = 360 count/s per HEXTE cluster, and
- background: 50 count/s per HEXTE cluster.

Proportional Counter Array

The PCA is an array of five proportional counters with a total collecting area of 6500 cm^2. The instrumental properties are

- energy range: 2–60 keV,
- energy resolution: <18 % at 6 keV,
- spatial resolution: collimator with 1° FWHM,
- sensitivity: 0.1 mCrab, and
- background: 2 mCrab

The information for this discussion of *RXTE* comes from the *RXTE* Web site, http://heasarc.gsfc.nasa.gov/docs/xte/XTE.html. We will return to some of the important scientific contributions from *RXTE* in conjunction with the discussion of X-ray bursts in several sections of chapter 8.

2.6.3 BeppoSAX

The X-ray astronomy satellite *BeppoSAX*, was launched into a 600-km orbit at 3.9° inclination on April 30, 1996. It was deorbited on April 29, 2003, after an extremely successful mission. A major objective of *BeppoSAX* was the observation of the afterglow of gamma-ray bursts (see more detailed description in sec. 2.7.1), with the intent of providing excellent spatial localization of them. The main scientific characteristic of the *BeppoSAX* mission was the wide spectral coverage, ranging from 0.1 to over 200 keV, with a relatively large area and a good energy resolution, along with imaging capabilities (resolution of about 1′) in the range of 0.1–10 keV. The instrument complement dedicated to such purpose was composed of a medium-energy (1–10 keV) concentrator optics/spectrometer consisting of three units, a low-energy (0.1–10 keV) concentrator optics/spectrometer, a high-pressure gas scintillation proportional counter (3–120 keV), and a phoswich detector system (15–300 keV), all of which had narrow fields and pointed in the same direction (the narrow field instruments, or NFI).

The other characteristic of the mission was its capability of monitoring large regions of the sky, with a resolution of 5′ in the range 2–30 keV, to study long-term variability of sources down to 1 mCrab and to detect X-ray transient phenomena. This was realized by means of two coded mask proportional counters (wide-field cameras [WFC]) pointing in diametrically opposed directions perpendicular to the NFI. Finally, the anticoincidence scintillator shields of the PDS were used as a gamma-ray burst monitor in the range 60–600 keV. While the X-ray detection capability of *BeppoSAX* was not necessarily well matched to the spectrum of gamma-ray bursts, the gamble that gamma-ray bursts would also produce detectable X-rays, and that *BeppoSAX* would be able to provide the position resolution necessary to provide accurate spatial locations, paid off handsomely, as is discussed in chapter 6.

The primary characteristic of *BeppoSAX* was its very wide spectral coverage, with well balanced performances between the low-medium (0.1–10 keV) and medium-high (10–200 keV) energy bands. For example, the complex spectrum of a Seyfert 1 galaxy, namely, MCG-6-30–15, was observed by *BeppoSAX* NFI in 40,000 s with all the spectral components and features detected by several satellites in the past. Starting from the low-energy part there was a soft excess observed by *EXOSAT* (Pounds et al. 1986), an OVII edge around 0.8 keV observed by *ROSAT* (Nandra and Pounds 1992) and *ASCA* (Fabian et al. 1994), and an iron line at 6.4 keV and a high-energy bump above 10 keV detected by *GINGA* (Matsuoka et al. 1990). All those components were measured with good accuracy with *SAX* in a single shot for the first time. In addition, *SAX*'s capabilities provided significant contributions in several areas of X-ray astronomy:

- Compact galactic sources. Shape and variability of the continuum; narrow spectral features (iron line, cyclotron lines) as a function of the orbital and rotational phases; ultra-soft sources; discovery and study of X-ray transients.
- Active galactic nuclei. Spectral shape and dynamics of the variable continuum and of the narrow and broad components from 0.1 to 200 keV in bright objects (soft excess, warm and cold absorption and related O and Fe edges, iron line and high-energy bump, high-energy cutoff); spectral shape of objects down to 1/20 of 3C273 up to 100–200 keV; spectra of high-redshift objects up to 10 keV.

- Clusters of galaxies. Spatially resolved spectra of nearby objects and the study of temperature gradients and cooling flows; chemical composition and temperature distribution as a function of redshift.
- Gamma-ray bursts. Temporal profile with 1 ms resolution from 60 to 600 keV. X-ray counterparts of a subset with positional accuracy $< 5'$.

The information for this write-up was excerpted from the *BeppoSAX* Web site, http://www.asdc.asi.it/bepposax/. We will return to the science from this project in the discussion of gamma-ray bursts in chapter 6 and X-ray bursts in chapter 8.

2.6.4 CHANDRA

CHANDRA has provided a huge improvement in resolution of the objects it has observed over previous X-ray detections. Some of its characteristic specifications are given as

Telescope system:

- High-resolution mirror assembly: four nested pairs of grazing incidence paraboloid and hyperboloid mirrors.
- Length: each is 83.3 cm long.
- Field of view: 1° diameter.
- Angular resolution: 0.5 arcsec.
- Aspect camera: $1.4^\circ \times 1.4^\circ$ field of view.

Science instruments:

- Advanced charged coupled imaging spectrometer:
  - 10 CCD chips in two arrays provide imaging and spectroscopy; imaging resolution is 0.5 arcsec,
  - Energy range 0.2–10 keV.
- High-resolution camera:
  - Uses large field-of-view microchannel plates to give angular resolution less than 0.5 arcsec over 31×31 arcmin field of view,
  - Time resolution is 16 μsec.
- High-energy transmission grating:
  - Provides spectral resolution $\lambda/\Delta\lambda$ of 60–1000,
  - Energy range of 0.4–10 keV.
- Low-energy transmission grating:
  - Provides spectral resolution of 40–2000,
  - Energy range of 0.09–3 keV.

Fig. 2.11. Centaurus A seen at several widely different wavelengths. This series of photographs came from the *CHANDRA* space telescope, http://chandra.harvard.edu/photo/2002/0157/0157_composite.jpg. Courtesy of NASA.

A notable result that was obtained of the same object in many wavelengths of electromagnetic radiation by several observatories, one of which was *CHANDRA*, is shown in figure 2.11 for Centaurus A. This composite figure shows Centauraus A in X-ray (blue), radio (pink and green), and optical (orange and yellow). The figures present a tableau of a galaxy in turmoil. A broad band of dust and cold gas is bisected at an angle by opposing jets of high-energy particles blasting away from the supermassive black hole in the nucleus. Two large arcs of X-ray-emitting hot gas were discovered in the outskirts of the galaxy on a plane perpendicular to the jets.

The arcs of multimillion-degree gas appear to be part of a projected ring 25,000 light-years in diameter. Its size and location indicate that it may have been produced in an explosion that occurred about 10 million years ago. That explosion would have produced the high-energy jets and a galaxy-sized shock wave moving outward at speeds of a million miles per hour. The age of 10 million years for the outburst is consistent with optical and infrared observations that indicate that the rate of star formation in the galaxy increased dramatically at about that time.

Scientists have suggested that all this activity may have begun with the merger of a small spiral galaxy and Centaurus A about 100 million years ago. Such a merger could eventually trigger both the burst of star formation and the violent activity in the nucleus of the galaxy. The tremendous energy released when a galaxy becomes "active" can have a profound influence on the subsequent evolution of the galaxy and its neighbors. The mass of the central black hole can increase, the gas reservoir for the next generation of stars can be expelled, and the space between the galaxies can be enriched with heavier elements. (See the discussion of galactic chemical evolution in chap. 4.)

The above discussion was excerpted from the *CHANDRA* Web site, http://chandra.harvard.edu/.

Another X-ray observatory, *ASCA*, the *Advanced Satellite for Cosmology and Astrophysics*, a Japanese mission, was launched in 1993 and has produced a wealth of observations.

2.6.5 INTEGRAL

INTEGRAL, the *International Gamma-Ray Astrophysics Laboratory*, was launched on October 17, 2002; the launcher put *INTEGRAL* into a highly eccentric 72-hour orbit. The nominal lifetime of the observatory was 2 years but with possible extension to up to 5 years.

INTEGRAL is dedicated to the fine spectroscopy and fine imaging (angular resolution, 12 arcmin FWHM) of celestial gamma-ray sources in the energy range 15 keV to 10 MeV with concurrent source monitoring in the X-ray (3–35 keV) and optical (V-band, 550 nm) energy ranges. It is therefore a "transition instrument" between X-rays and gamma rays, which are discussed in the next section. To achieve high-energy resolution it utilizes Germanium detectors, which are operated in combination with cadmium telluride/cesium iodide imagers to achieve the fine imaging.

INTEGRAL's instrument payload consists of the following:

- Two gamma-ray instruments
  - Spectrometer (20 keV to 8 MeV). Coded aperture mask. Field of view, 16°; detector area, 500 cm^2 (Germanium array); spectral resolution, (E/dE) 500 at 1 MeV; spatial resolution, 2°.

  - Imager (15 keV to 10 MeV). Coded aperture mask. Field of view, $9^\circ \times 9^\circ$; detector area, 2600 cm^2 (CdTe array) and 3100 cm^2 (CsI array); spatial resolution, 12′.
- Joint European X-ray monitor (3–35 keV). Coded aperture mask with two high-pressure microstrip gas chambers. Field of view, 4.8°; detector area, each 500 cm^2; spatial resolution, 3′.
- Optical monitoring camera (500–850 nm). A 50-mm lens with CCD. Field of view, $5^\circ \times 5^\circ$.

INTEGRAL has been instrumental in mapping out the detailed distribution of ^{26}Al, a long-lived radioactive nuclide (half-life = 7.4×10^5 years) that might be produced in several explosive nucleosynthesis scenarios. The history of efforts to map out the Galactic distribution of this nuclide is discussed below in the context of the *Compton Gamma-Ray Observatory* (*CGRO*). However, *INTEGRAL*'s result represented a major improvement on *CGRO*'s result; it is discussed further in chapter 8.

The *INTEGRAL* write-up was excerpted from Winkler et al. (2003).

2.7 Gamma-Ray Astronomy

2.7.1 Space-Based Gamma-Ray Astronomy

As will be discussed below, one way to detect extremely high energy extraterrestrial gamma rays is by detecting the products of their interactions with the Earth's atmosphere. However, this is not practical with lower energy gamma rays. Their detection from spaceborne detectors is practical and allows the possibility of detecting the gamma rays directly, hence with intrinsically high resolution. This type of effort has been going on for several decades, initially with balloon-borne instruments, but more recently with the *CGRO*, discussed in some detail below, and *INTEGRAL*, discussed above and shown in figure 2.12. The balloon observations resulted in the initial observations of gamma rays resulting from the decay of ^{26}Al. Its half-life is sufficiently long that the ^{26}Al is expelled from the stars in which it is synthesized but short enough that an observable count rate is produced. More recent measurements with the *CGRO* have produced a remarkable map of the ^{26}Al produced in the Galaxy (see fig. 2.14). These results were then superceded by those from *INTEGRAL*, which, as noted above, are presented in chapter 8.

The *CGRO* also produced a wealth of data on gamma-ray bursts, events in which extraordinary bursts of gamma rays were emitted. Gamma-ray bursts (GRBs) are the most powerful explosions the universe has seen since the big

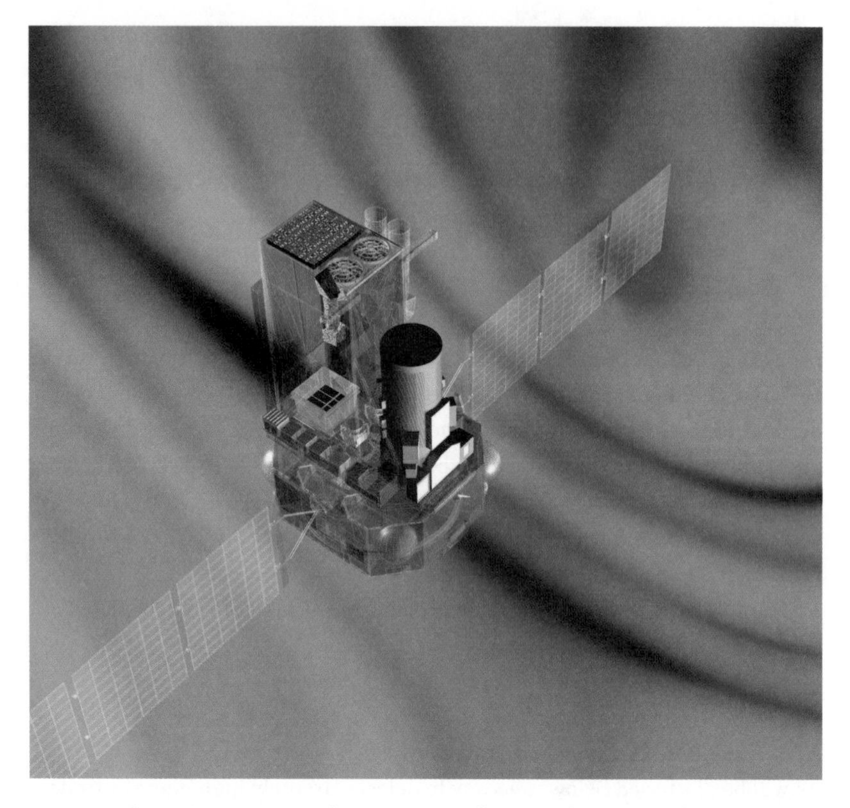

Fig. 2.12. The *INTEGRAL* X-ray and gamma-ray satellite. From http://www.dsri.dk/integral/.

bang. They occur approximately once per day and are brief but intense flashes of gamma radiation. They come from all different directions of the sky and last from a few milliseconds to a few hundred seconds. Scientists are closing in on their sources, having learned a great deal about them both observationally and theoretically. Although it was initially inferred that these events had to be galactic, just because of the prodigious energy output that would be required of them if they were extragalactic, their distribution in space ultimately was found by *CGRO* to be unrelated to the galactic plane, indicating that they were probably extragalactic. Subsequent optical observations confirmed this. Of course, this makes their energy output even more extraordinary; one of the theoretical challenges in astrophysics has been to explain the source of these GRBs. We will revisit this subject in detail in chapter 6.

2.7.2 Compton Gamma-Ray Observatory

The *CGRO*, named after Arthur Compton, who first observed "Compton scattering" of photons from electrons as well as having achieved other notable scientific objectives, was launched in 1991 and deorbited in 2000. It had four instruments that covered 6 decades of the electromagnetic spectrum from 30 keV to 30 GeV. In order of increasing spectral energy, these instruments were the Burst and Transient Source Experiment (BATSE), the Oriented Scintillation Spectrometer Experiment (OSSE), the Imaging Compton Telescope (COMPTEL), and the Energetic Gamma Ray Experiment Telescope (EGRET). Because of their special relevance to nuclear astrophysics, BATSE, COMPTEL, and EGRET will be discussed below in some detail.

2.7.3 Imaging Compton Telescope (COMPTEL)

The Imaging Compton Telescope, or COMPTEL, consisted of two detector arrays, the upper one of which used liquid scintillator and the lower NaI crystals, both of which produce a flash of light—a scintillation—when hit by a gamma ray. Detection occurs when two required successive interactions occur: an incident cosmic gamma ray is first Compton scattered in the upper detector and then is totally absorbed in the lower. The locations of the interactions and energy losses in both detectors are measured. Although the photons cannot be focused, as in the case of optical photons or X-rays, these data can be used

Fig. 2.13. An artist's conception of the *CGRO* in space. From http://cossc.gsfc.nasa.gov/cgro/. Courtesy of NASA.

to reconstruct sky images over a wide field of view with a resolution of a few degrees. COMPTEL operated in the energy range 1–30 MeV. The COMPTEL map of Galactic ^{26}Al emission is seen in figure 2.14.

2.7.4 Burst and Transient Source Experiment (BATSE)

The BATSE served as the all-sky monitor for the *CGRO*, detecting and locating strong transient GRBs as well as outbursts from other sources over the entire sky. *CGRO* had eight BATSE detectors, consisting of sodium iodide crystals coupled to photomultipliers (which produce an electrical signal from the flash of light provided by the scintillator), one facing outward from each corner of the satellite, which were sensitive to gamma-ray energies from 20 keV to over several MeV. The primary objective of BATSE was to study the phenomenon of GRBs, although the detectors also recorded data from pulsars, terrestrial gamma-ray flashes, soft gamma repeaters, black holes, and other exotic astrophysical objects.

2.7.5 Energetic Gamma Ray Experiment Telescope (EGRET)

The EGRET provided the highest energy gamma-ray window for the *CGRO*: its energy range is from 20 MeV to 30 GeV. EGRET was 10–20 times larger and more sensitive than previous detectors operating at these high energies and has made detailed observations of high-energy processes associated with diffuse gamma-ray emission, GRBs, cosmic rays, pulsars, and active galaxies known as blazars. EGRET produced images at these energies using high-voltage gas-filled spark chambers. High-energy gamma rays entering the chambers produced an electron-positron pair of particles that caused sparks along tracks that resulted in electrical signals. The path of the particles is recorded, thereby allowing the determination of the direction of the original gamma ray. The energies of the gamma rays were recorded by a NaI crystal beneath the spark chambers, providing a measure of the original total energy.

EGRET has produced the standard sky map of high-energy gamma-ray emitters; we will refer to this in subsequent sections.

The information for this write-up came from the Web site for the *CGRO*, http://cossc.gsfc.nasa.gov/cgro/.

2.7.6 The Swift Gamma-Ray Burst Mission

Once GRBs were discovered to be of cosmological origin by BATSE and *BeppoSAX*, the issue of understanding their prodigious energy output became paramount. Indeed, GRBs were such extraordinary objects that they were made the focus of an entire mission, the international *Swift Gamma-Ray*

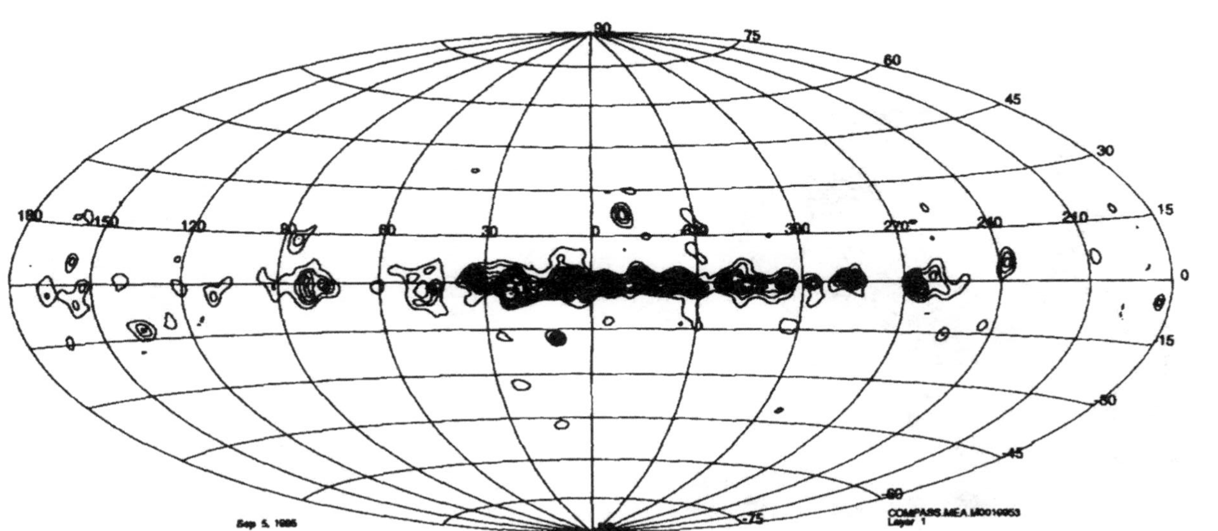

Fig. 2.14. The *COMPTEL* map of 1.809-MeV gamma rays, resulting from the decay of ^{26}Al, on the Milky Way Galaxy. A high concentration of lines indicates a region of intense gamma-ray production. Since gamma rays of that energy are the signature of the decay of ^{26}Al, they prove unequivocally that it is being produced in stars. From Oberlack et al. (1996) and Diehl et al. (1995), and the figure is from Diehl et al. (1997).

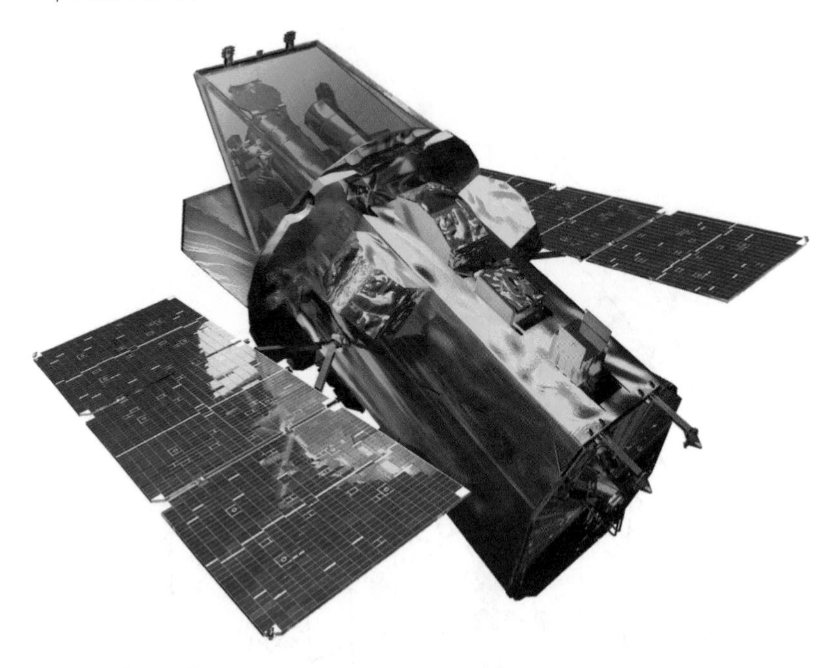

Fig. 2.15. The *Swift Gamma Ray Burst Satellite*. See text for a description of its primary components and of their functions. Courtesy of NASA. From http://heasarc.gsfc.nasa.gov/docs/swift/about_swift/factsheet.pdf.

Burst Mission. *Swift* was named for a bird of the same name because, like the swift, it can change its orientation very quickly, thereby allowing it to observe GRBs, even though their bursts last typically only tens of seconds. *Swift* was launched into a low-Earth orbit on November 20, 2004. This satellite carries three instruments to enable the most detailed observations of GRBs to date.

In *Swift* (Gehrels et al. 2004), there are three coaligned instruments, shown in figure 2.15, and known as the BAT, the XRT, and the UVOT. The XRT and UVOT are X-ray and UV/optical focusing telescopes, respectively, which produce subarcsecond positions and multiwavelength light curves for GRB afterglows. Broadband afterglow spectroscopy then produces redshifts for many of the GRBs. BAT is a wide field-of-view coded-aperture gamma-ray imager that produces arcminute GRB positions onboard within 10 seconds. The spacecraft executes a rapid autonomous slew that points the focusing telescopes at the BAT position in typically $\sim$50 s.

The positions and images derived by the various instruments are sent as soon as they are available from the spacecraft to the Gamma-Ray Coordination Network, which then broadcasts the results to the world via the Internet for rapid response by the world astronomy community for follow-up observations by other ground- and space-based telescopes. At the next satellite pass over Malindi, the more detailed data are sent to the data center where they are processed for public access within 30 minutes of the pass.

With *Swift*, a National Aeronautics and Space Administration (NASA) mission with international participation, scientists now have a tool dedicated to answering basic questions about GRBs. Its three instruments give scientists the ability to scrutinize GRBs as never before. Within seconds of detecting a burst, *Swift* relays a burst's location to ground stations, allowing both ground-based and space-based telescopes around the world the opportunity to observe the burst's afterglow. *Swift* has proved to be an extraordinarily successful mission. One of its major discoveries is of a short-time GRB, which is described in chapter 6. *Swift* enabled the identification of at least some of these types of GRBs as the possible result of collisions of neutron stars.

The above descriptions came from the *Swift* project Web site, http://swift.gsfc.nasa.gov/docs/swift/swiftsc.html.

2.7.7 High-Energy Gamma Rays

When gamma rays interact with the Earth's atmosphere they lose their energy only through the electromagnetic interaction. This produces air showers that are made up of electrons and photons produced from pair production and bremsstrahlung, which are of a very different character from the air showers produced by high-energy cosmic rays (see sec. 2.9) and which therefore allow the high-energy gamma rays to be distinguished from the more numerous high-energy cosmic rays. The differences in the character of the two types of air showers, of course, drive the design of the respective detectors of these different particles, as will be noted in several instances in this section and in the section on high-energy cosmic rays.

2.7.8 Very Energetic Radiation Imaging Telescope Array System (VERITAS)/High Energy Stereoscopic System (HESS)

Both of these observatories of very high energy gamma rays, of order 10^{11}–10^{13} eV, detect the light from the air showers produced by the incident gamma rays. This is in the form of Cherenkov light, which results from the secondary electrons moving through the atmosphere at greater than the speed at which light will travel through the atmosphere. Since this light is very directional, to 0.05° or better; the observation of such gamma rays allows determination of the direction from which they came and hence an identification of their source. The first *VERITAS* telescope is shown in figure 2.16.

Fig. 2.16. Photograph of the first *VERITAS* telescope. Reprinted courtesy of R. Ong.

Both *VERITAS* and *HESS* are made up of four telescopes that can act in "stereoscopic mode," which provides a significant improvement in both energy and directional resolution over what could be obtained from observation by a single telescope. The Cherenkov light that is detected is focused on to pixelated detectors so as to produce the directionality of the event. Timing can also be used to observe the motion of the air shower.

2.7.9 VERITAS

VERITAS consists of four 12 m in diameter, ground-based atmospheric Cherenkov telescopes designed to observe the sky in the very high energy gamma-ray regime (100 GeV to 10 TeV). It combines the power of the atmospheric Cherenkov imaging technique using a large optical reflector with the power of stereoscopic observatories using arrays of separated telescopes looking at the same shower. Each telescope will have a camera consisting of 499 pixels with a field of view of 3.5°.

The concept used in designing both *HESS* and *VERITAS* was pioneered by an earlier generation of such telescopes, which are located at the *Whipple Observatory* in Arizona. In addition to demonstrating the capabilities of this technique—in particular obtaining the necessary background rejection—the earlier instrumentation, from 1980–2000, established the field of very high energy gamma-ray astrophysics with a number of significant science results, including the detection of the Crab Nebula in high-energy gamma rays, the detection and analysis of six extragalactic sources, and important limits on a variety of high-energy phenomena ranging from pulsars and supernova remnants to primordial black holes and particle dark matter.

However, further progress in this field necessitated better background rejection to expand the capabilities at low energy, a larger effective collecting area to expand the capabilities at high energy, better angular and energy resolution commensurate with the larger effective collecting area, and improved field of view or the ability to study more than one source at a time. *VERITAS* and *HESS* both provide these improvements.

The capabilities of *VERITAS* are well matched to those of the upcoming space-based gamma-ray mission, the *Gamma-Ray Large Area Space Telescope* (*GLAST*). Together, *GLAST* and *VERITAS* will observe gamma-ray emission over an enormous energy range, from 20 MeV to greater than 10 TeV. *VERITAS* will have a relatively large collecting area and an ability to resolve very short timescale variations for bright sources, whereas *GLAST* will have a very high duty cycle and a huge field of view.

The capabilities of both *VERITAS* and *HESS* should lead to major advances in a broad range of science topics, including

1. Understanding the mechanisms of particle acceleration in active galactic nuclei (thought to be due to accretion onto supermassive black holes), pulsars, and supernova remnants. This understanding is a key to solving the mysteries of the formation of jets, the extraction of rotational energy

from spinning neutron stars, and the dynamics of shocks in supernova remnants.
2. Resolving the gamma-ray sky: unidentified sources and diffuse emission. Interstellar emissions from the Milky Way and a large number of unidentified sources are prominent features of the gamma-ray sky.
3. Determining the high-energy behavior of GRBs and transients. Variability has long been a powerful method to decipher the workings of objects in the universe on all scales.
4. Probing dark matter and the early universe. Observations of gamma rays from active galactic nuclei serve to probe supermassive black holes through jet formation and evolution studies and provide constraints on the star formation rate at early epochs through photon-photon absorption over extragalactic distances.
5. Searching for monoenergetic gamma-ray lines above 30 GeV from supersymmetric dark matter interactions; detecting decays of relics from the very early universe, such as cosmic strings or evaporating primordial black holes; or even using GRBs to detect quantum gravity effects.
6. Obtaining information about the high-energy behavior of GRBs. This requires rapid repointing to observe a burst afterglow, if it is accessible, upon a burst alert notification.
7. Obtaining high-energy and very high energy gamma-ray observations of supernova remnants. This should finally confirm or refute the long-standing hypothesis that cosmic rays with energies below 10^{14} eV are produced by shock acceleration at these sites.

HESS's four-telescope array has, in its first year of operation, acquired some spectacular data. One such data set is shown in figure 2.17; it shows some of the new very high energy gamma-ray sources that *HESS* has discovered. These include supernova remnants, pulsar wind nebulae, a microquasar, active galaxies, and a number of sources that do not easily correlate with objects known in other wavebands. Note that *HESS* is located in the southern hemisphere, in Namibia, so many of its discoveries, certainly in the context of very high energy gamma-ray sources, and those of *VERITAS*, which is located in Arizona, will be complementary.

For further information, see the *VERITAS* and *HESS* Web sites, http://veritas.sao.arizona.edu/ and http://www.mpi-hd.mpg.de/HESS.

2.7.10 GLAST

GLAST is expected to be launched in 2007. It is a high-energy gamma-ray observatory designed for making observations of the celestial gamma-ray sources

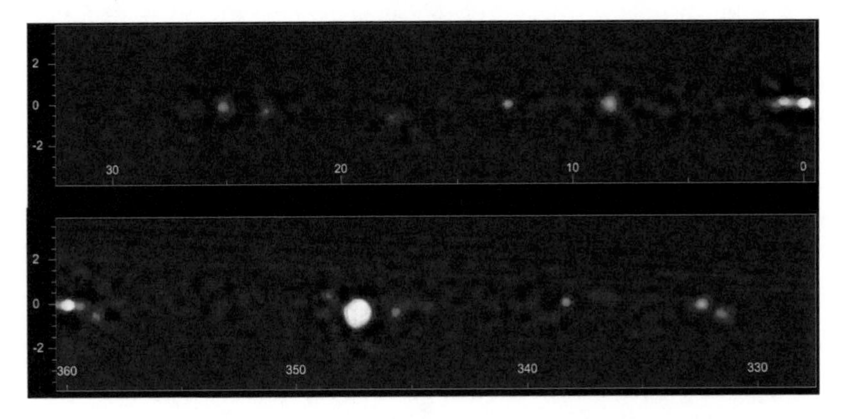

Fig. 2.17. Some of the very high energy gamma-ray sources discovered by the *HESS* telescope in its first year of operation. Only four of the objects shown are not objects newly discovered by *HESS*; the two in the upper figure near 0° (one is the Galactic center), the very bright blob in the lower figure at 347°, and the faint spot at 17°. Reprinted with permission from S. Funk.

in the energy band extending from 10 MeV to more than 100 GeV. It follows the footsteps of the *CGRO*—EGRET experiment, which was operational between 1991 and 1999.

Since *GLAST* operates above the Earth's atmosphere, the gamma rays it observes are detected directly, rather than from their interaction with the Earth's atmosphere. This is done with the Large Area Telescope (LAT). The baseline design for the *GLAST* tracker consists of a four-by-four array of tower modules. Each tower module consists of interleaved planes of silicon strip detectors (SSDs) and lead converter sheets. The SSDs are able to more precisely track the electron or positron produced from the initial gamma ray than previous types of detectors. The SSDs will have the ability to determine the location of an object in the sky to within 0.5–5 arcmin, depending on the energy. In addition, the pair conversion signature is also used to help reject the much larger background of charged cosmic rays. The high intrinsic efficiency and reliability of this technology enables straightforward event reconstruction and excellent resolution with small tails. These ease-of-use properties will maximize the mission science return for guest observers. Following the SSDs is a calorimeter, which consists of large sodium iodide detectors that, when hit by the gamma rays, produce a signal that gives the total energy of the gamma ray.

The key scientific objectives of the *GLAST* mission overlap greatly with those of *VERITAS* and *HESS*, described in the previous section.

The *GLAST* LAT has a field of view about twice as wide as EGRET's (more than 2.5 steradians) and sensitivity about 50 times that of EGRET at 100 MeV and even more at higher energies. Its 2-year limit for source detection in an all-sky survey is 1.6×10^{-9} photons cm^{-2} s^{-1} (at energies > 100 MeV). As noted above, it will be able to locate sources to positional accuracies of 30 arcsec to 5 arcmin.

Additional information can be obtained at the *GLAST* Web site, http://glast.gsfc.nasa.gov/, from which this write-up was excerpted.

2.8 Searches for Dark Matter

The motions of the stars in galaxies have demonstrated conclusively that there exists "dark matter," matter that is unseen but makes its presence known through the gravitational effects it imposes on the entities that are seen—the stars. In addition, the recent cosmological experiments, *WMAP* and the measurement of the Hubble constant (see chap. 9), identify a component of the total energy that is nonbaryonic matter. However, the nature of the dark matter has yet to be determined. One experiment to detect one of the dark matter candidates, the *Cryogenic Dark Matter Search* (*CDMS*), currently is operating in the Soudan Underground Laboratory in northern Minnesota. It searches for weakly interacting massive particles (WIMPs), perhaps the leading candidate for the dark matter particles. Their presence would be observed from the signals the WIMPs would produce when they interacted with the Ge and Si nuclei in the *CDMS* detectors. *CDMS* is now running with five "towers," stacks of six hockey puck–sized Si or Ge detectors. The WIMP signals are expected to be very small, befitting their name, so *CDMS* would detect them by searching for the signals the recoiling nuclei would produce in both ionization and phonons. Because the signals are so small, the detectors must be held at extremely low temperatures, ~50 mK, to eliminate noise. In addition, backgrounds are minimized first by siting the experiment 2000 feet underground to eliminate most of the cosmic rays that impinge on the Earth's surface, then by using intricate shielding, carefully selected materials, and a special clean room to further reduce unwanted signals.

At present, no experiment has detected a convincing WIMP signal. However, data taken with *CDMS* in 2003 with just one tower (Akerib et al. 2004; see fig. 2.18) have improved the previous best limits obtained on the WIMP scattering cross sections in another experiment, *EDELWEISS*. In addition, these *CDMS* results are strongly inconsistent with the claims of the *DAMA* experiment (as labelled in fig. 2.18) that an annual modulation in their signal, observed with (much lower resolution) sodium iodide detectors, should be interpreted as evidence for WIMP dark matter. The WIMPs would be expected to produce such an effect because of the yearly movement of the Earth

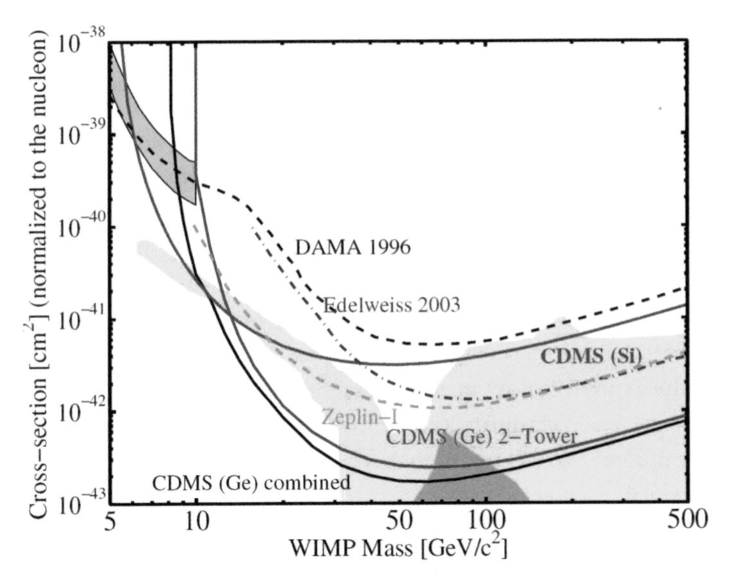

Fig. 2.18. Exclusion limits of dark matter candidates. The above figure shows the limit on the WIMP-nucleon scalar cross section from *CDMS II* at Soudan with no candidate events in 19.4 kg-day effective exposure (solid curve and, with slightly different assumptions about the data, the dashed curve). Parameter space above the curve is excluded at the 90 % confidence limit. Also shown are limits from *EDELWEISS* (*hatched line*). The *DAMA* result is as indicated in the center of the graph. The blotches are various theoretical predictions, a fairly large fraction of which are now excluded. Reprinted with permission from Akerib et al. (2004). Copyright 2004 the American Physical Society.

through the WIMP cloud that would be expected to surround the Sun. However, *CDMS*'s detectors are superior to those of *DAMA*, and *CDMS* sees no evidence for such a signal over a parameter space that has considerably greater sensitivity than that of *DAMA*. This suggests that the *DAMA* modulation may be due to instrumental seasonal variations although, in the event that the WIMP-nucleus interaction is spin dependent, the results of both experiments could be correct.

Indeed, the *CDMS* data are now impinging on predictions of theoretical models, in some cases rejecting them as possible descriptions of the WIMPs. These models assume that the WIMPs are supersymmetric particles, the result of fundamental theories of particle physics. In these theories, each particle known to modern physics has a supersymmetric counterpart, for example, an electron has a "selectron," a neutrino has a "sneutrino," and each quark has a corresponding "squark." The particle that is thought to be the strongest candidate to be the WIMP is the supersymmetric counterpart of the photon, the "photino" (see http://cdms.berkeley.edu/).

2.9 Observation of High-Energy Cosmic Rays

When high-energy cosmic rays (HECRs), nuclei from outer space, interact with the Earth's atmosphere, they lose their energy to the particles with which they interact, producing air showers. These will be hadronic showers, as the HECRs will first produce pions, which then decay to muons. Some of the muons then decay to electrons, to join with other electrons that are produced by direct scattering of the particles in the shower from atomic electrons in the atmosphere. If the primary HECR has sufficient energy, the secondary muons will often be sufficiently energetic themselves that they will penetrate far into the Earth's crust, a feature that necessitates locating very sensitive detectors far underground to eliminate as much as possible the backgrounds produced by these muons. We will return to this later in this chapter and several times in subsequent chapters.

The initial collisions of the HECRs produce showers of subatomic particles and lower-energy photons that avalanche groundward as a thin, radially expanding disk. Each air shower either dissipates in the atmosphere or, at high elevations and for sufficiently high energy cosmic rays, intercepts the ground. The various experiments described below place detectors at or below ground level to intercept some of the components of these air showers and to measure their energy, composition, and direction. Because HECRs become increasingly more rare with increasing energy, the HECR spectrum is an extremely rapidly falling distribution, as shown in figure 2.19.

One major motivation for the projects studying ultra-high-energy cosmic rays (UHECRs) is to examine those at the very highest energy. Theory suggests that the spectrum should terminate at the Greisen-Zatsepin-Kuz'min (GZK) cutoff, which is actually a turndown of the spectrum, predicted at $\sim 6 \times 10^{19}$ eV. This should result from photo-pion production in collisions of the UHECRs with the cosmic background microwave photons (see sec. 2.2 and chap. 9). The *Akeno Giant Air Shower Array* (*AGASA*), a UHECR experiment, has observed a fairly large number of events above the GZK cutoff, producing a major scientific conflict. Note that some events might be expected above the cutoff, as it would not be expected to be sharp, but would depend, for example, on the distance from Earth, hence the redshift z at which the UHECRs interacted with the CMB radiation. For example, at a z of 1, the temperature of the CMB radiation would be 6 K, not the 2.7 K observed on Earth, which would shift the energy of the GZK cutoff down in energy from that expected if the CMB radiation were at 2.7 K.

However, the *High-Resolution Fly's Eye* (*HiRes*) experiment, which is located

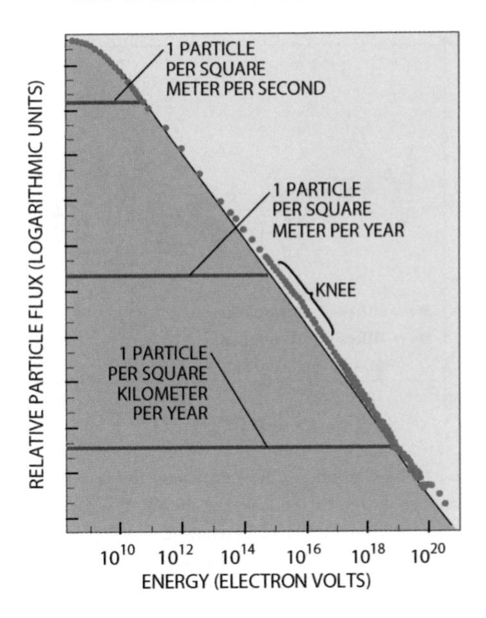

Fig. 2.19. The highest energy decades of the cosmic-ray spectrum. From *Scientific American*, January 1997, p. 47. Reprinted courtesy of Jennifer Christiansen.

in the high Utah desert and which uses a different technique to measure the HECR energy than that used by *AGASA*, does not appear to agree with the *AGASA* result, as indicated in the spectrum shown in figure 2.20. This discrepancy has spurred both a frantic data accumulation effort in the *HiRes* collaboration and the construction of *AUGER*, a detector, located in Argentina, that utilizes both the *HiRes* and the *AGASA* detection techniques to test their consistency on the same events.

2.9.1 *HiRes*

The *HiRes* experiment, located at the U.S. Army's Dugway Proving Ground in Utah, consists of two stations separated by 12.6 km, which detect UHECRs from the nitrogen fluorescence they produce as they pass through the Earth's atmosphere. The goals of the experiment are to study the spectrum, composition, and anisotropy of cosmic rays with energies from about 10^{17} eV up to beyond 10^{20} eV. The two stations at Dugway consist of a 22-mirror HiRes-I detector and a 42-mirror HiRes-II detector. Data are collected and analyzed from each of the two detectors separately (monocular) and together (stereo). The stereo data are considered much more reliable but more challenging to obtain since they require both facilities to be operating simultaneously. The stereo

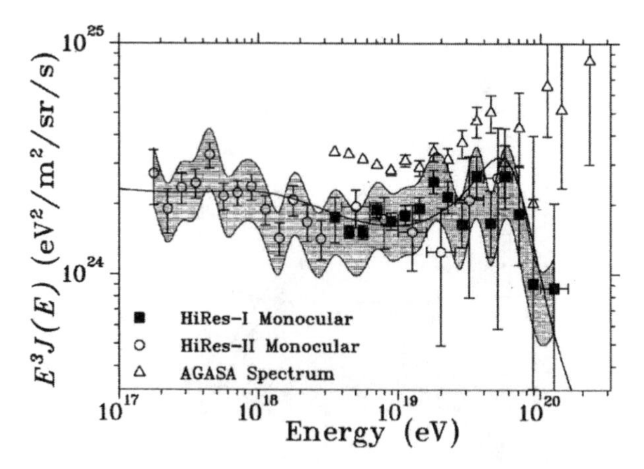

Fig. 2.20. Comparison of the *HiRes* and *AGASA* data near the possible GZK cutoff. As can be seen, the *HiRes* data strongly support the cutoff, but the *AGASA* data apparently continue well beyond it. Note, however, that the plot is somewhat misleading, as the data actually fall off as E^3. Thus a slight energy calibration difference between the two results would reconcile the discrepancy between them. Reprinted with permission from Abbasi et al. (2004). Copyright 2004, by the American Physical Society.

aspect of the experiment has maximum sensitivity to detect events above 10^{19} eV and has much higher precision for identifying the location of the event.

At the present time, there appears to be a fundamental inconsistency between the monocular *HiRes* spectrum, which exhibits a pileup/GZK cutoff at the upper end, and the spectrum from the *AGASA* air shower array experiment, which utilizes the detection of the secondary air shower muons to determine the HECR energy and shows a number of events above 10^{20} eV and no apparent cutoff in the spectrum. This is clearly seen in expanded form in figure 2.20. This conflict is crucial; if events do exist above the GZK cutoff, they will signal the advent of new physics. The final data sets from *HiRes*, both monocular and stereoscopic, are currently being analyzed.

Considerably more information about the *HiRes* project can be found on their Web site, http://hires.physics.utah.edu/.

2.9.2 The Pierre Auger Observatory

Pierre Auger can be considered to be the discoverer of the giant air showers generated by the interaction of UHECRs with the Earth's atmosphere. Thus it was appropriate to name this observatory, which is designed to intercept air showers produced by UHECRs as they reach the ground, after him. The *Pierre Auger Observatory* was designed with particular interest in the region around the GZK cutoff. The array must be large to record statistically significant numbers of rare, very high energy showers. The construction of the *Pierre Auger Observatory* began in Malargue, Argentina, during 2001. About the size of Paris (see fig. 2.21), the observatory will ultimately consist of 24 fluorescence telescopes and 1600 surface water–Cherenkov particle detector stations (see figs. 2.22 and 2.23) located about 1.5 km apart, arranged in a giant grid covering some 3000 km^2. *Auger* is a hybrid detector, utilizing both of the time-honored approaches to detecting the highest energy cosmic rays on the same events. However, as noted above, those two approaches appear not to agree as to the energy of the events they see. *Auger* should be able to resolve that discrepancy by applying both energy-measuring techniques to the same event.

As detectors are constructed and put into operation, they can begin to take data, albeit with lower event rates than they will ultimately be able to achieve. The full array should be finished in 2007, but a sufficient number of events around the GZK cutoff have already been obtained to confirm the issue between *AGASA* and *HiRes*. If the *Auger* data use the energy determination of the air fluorescence detectors, they are considerably more consistent with

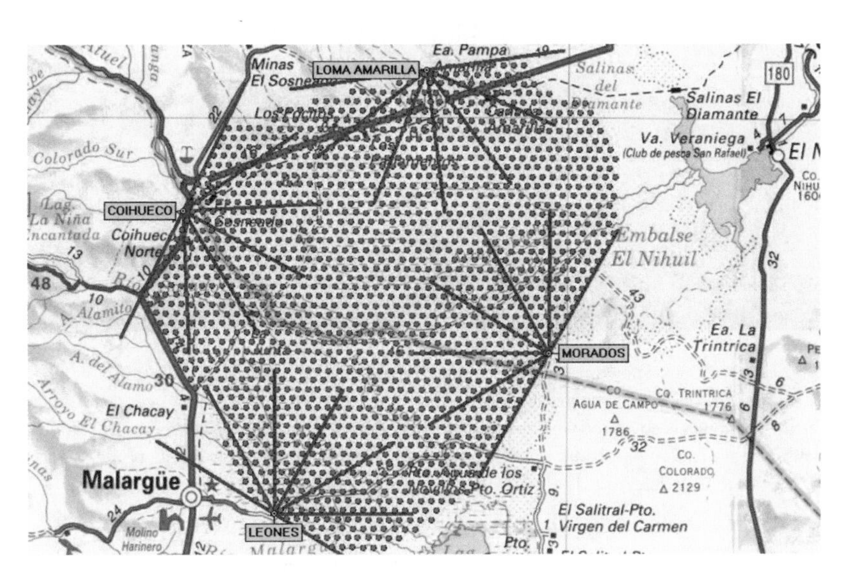

Fig. 2.21. Map of the *Auger* data stations. The points with the lines radiating from them are the fluorescence detectors, and the dots are the surface water detectors. The fluorescence detectors are aimed so as to cover the entire sky from which high-energy cosmic rays could originate that would produce the showers that would hit the surface water detectors. From http://hea.cwru.edu/auger/index.html.

Fig. 2.22 and 2.23. Shown to the left is one of the four *Auger* fluorescence detector buildings. It encloses six fluorescence telescopes, each with a 30° × 30° field of view over the surface array. The tower is equipped with antennas to communicate with the surface detector stations with global positioning system time tagging. The picture on the right is of one of the surface water Cherenkov detectors. From http://www.auger.org/photos/gallery.html.

those of *HiRes* than with those of *AGASA*, as shown in figure 2.24. However, the energies of the surface detector events appear to have energies more consistent with those from the *AGASA* data, suggesting that the physics associated with calculating the energy with one of the techniques is more complex than is being assumed. *Auger* will continue to improve its statistics around the GZK cutoff. When the issue of the existence of the cutoff is resolved with better statistics, the collaboration can begin to address other issues such as the possible existence of HECR point sources.

This write-up was excerpted from the *Auger* Web site, http://www.auger.org/index.html; considerably more information is available there.

2.10 Neutrino Astronomy

Another signature of the processes that go on in the cosmos comes to us in the form of neutrinos, the weakly interacting particles produced in weak decays but in a variety of other situations as well, for example, and importantly for astrophysics, pion decay. Neutrino astronomy has actually been going on for several decades in the solar neutrino experiment, conducted in the Homestake gold mine in South Dakota by Ray Davis, for which he won the Nobel Prize. Since its inception, there have been several other experiments, all conducted in mines around the world, aimed at detecting the neutrinos from our Sun. These experiments will all be discussed in detail in chapter 5, but some of the facilities used to detect them will be discussed briefly here.

2.10.1 Super-Kamiokande

Super-Kamiokande, or *Super-K*, shown in figure 2.25, is a 50,000-ton water Cherenkov detector located at a depth of 1 km in the Kamioka Mozumi mine

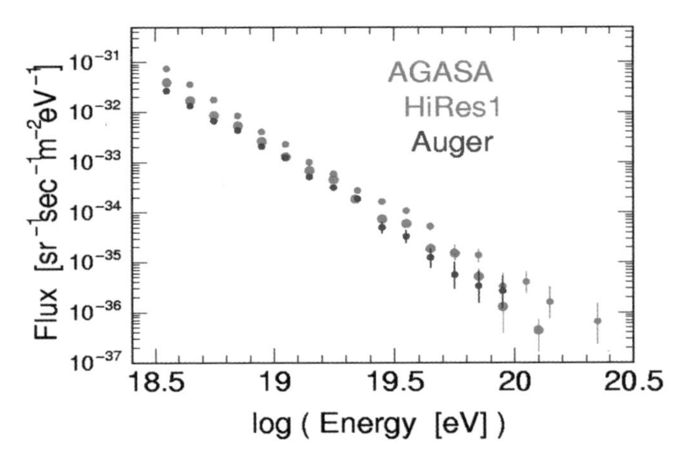

Fig. 2.24. A comparison of the *Auger* spectrum with that of *AGASA* and HiRes 1 (mono). The *Auger* points are about 25 % below those of *AGASA* and slightly lower by ~10 % than those of HiRes. Reprinted with permission from P. Mantsch.

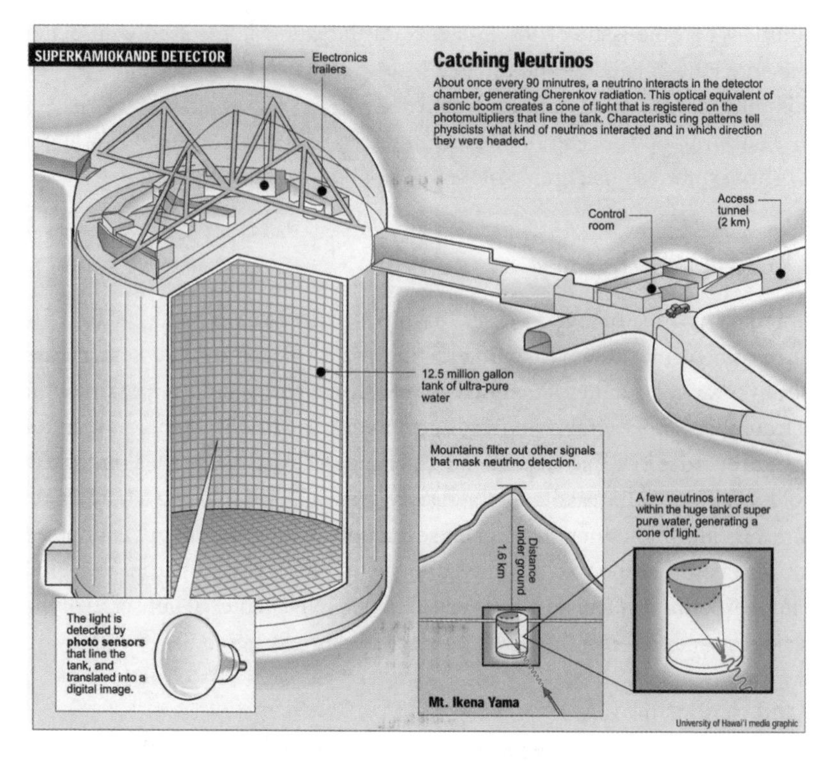

Fig. 2.25. *Super-Kamiokande* underground detector. Courtesy of Kamioka Observatory, Institute for Cosmic Ray Research, University of Tokyo.

in Japan. It was built to be used for a search for proton decay (nucleon decay in general) and observation of neutrinos (solar, atmospheric, from supernova, etc.) and cosmic rays (mostly muons: downward-going muons created by cosmic-ray particles in the atmosphere and upward-going muons formed by neutrino interactions in the Earth beneath the detector). It was also found to be an excellent detector of high-energy solar neutrinos.

Super-K is a huge tank of very clean water, 39 m in diameter and 41 m in height (see fig. 2.25). Covering the sides, top, and bottom are many (about 11,200) 20-inch photomultiplier tubes (PMTs), which are very sensitive light detectors that can detect single photons that originate in the inner volume of water.

Super-K has found evidence of the transformation of muon-type neutrinos to tau-type neutrinos (neutrino oscillations; see sec. 3.7) from atmospherically generated neutrinos. This is taken as strong evidence that neutrinos have a small but finite mass. *Super-K* has also detected solar neutrinos; this will be discussed in greater detail in chapter 5.

The information for this write-up came from the *Super-K* Web site, http://www-sk.icrr.u-tokyo.ac.jp/sk/index_e.html. For much more information, see that Web site.

2.10.2 Sudbury Neutrino Observatory

The *Sudbury Neutrino Observatory* (*SNO*), shown in figure 2.26, is located 2 km underground in a mine in Sudbury, Ontario, Canada. It uses an inner vessel of heavy water and an outer vessel that surrounds the inner one of ordinary water to measure the flux, energy, and direction of electron-neutrinos produced in the Sun. Because of its heavy water, *SNO* can also detect the other two types of neutrinos (muon-neutrinos and tau-neutrinos). *SNO* showed that the number

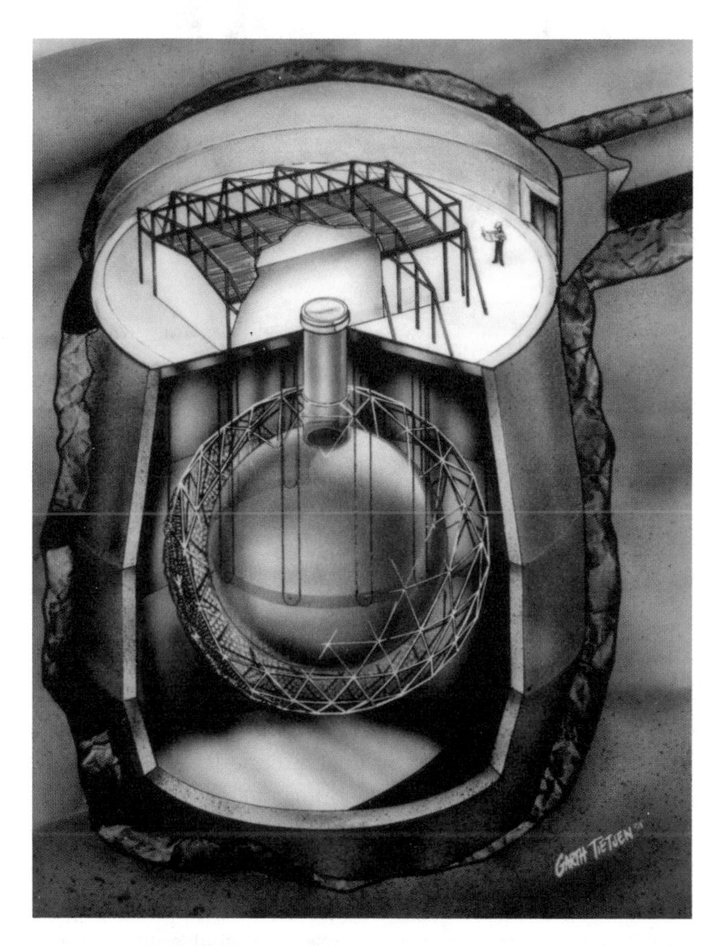

Fig. 2.26. An artist's conception of the *SNO* detector. The spherical inner vessel contains the heavy water and outer vessel contains ordinary (albeit very pure) water. Reprinted with permission from A. McDonald.

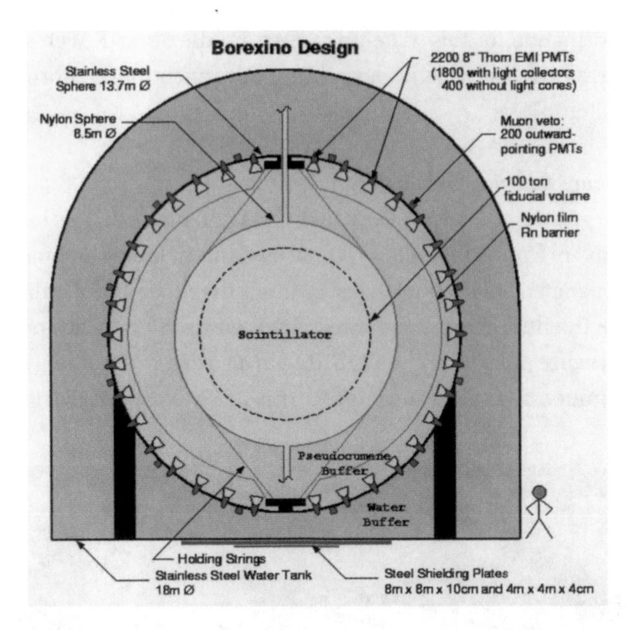

Fig. 2.27. The *Borexino* detector, with its various features labeled. Reprinted with permission from F. Calaprice.

of neutrinos emitted by the Sun was as expected from solar models, but that some of them had "oscillated" into neutrinos of a different flavor; *SNO*'s results have significantly altered our understanding of neutrinos and of the Sun. The science results that were obtained from *SNO* are discussed in detail in chapter 5, which is the chapter on hydrogen burning and the standard solar model; *SNO* has had a huge impact on our understanding of the Sun.

For much more detailed information, see chapter 5, as well as the *SNO* Web site, http://www.sno.phy.queensu.ca/.

2.10.3 Borexino

The *Borexino* project is a detector of solar neutrinos that is designed to observe the medium-energy neutrinos being emitted from the Sun. This detector is being constructed in the Gran Sasso underground laboratory in Italy, more than 1 km underground. It will ultimately utilize 300 tons of a liquid scintillator to detect the interactions between the neutrinos and electrons in the detector by observing the recoil energy of the electrons as they lose their energy in the scintillator. The objective of observing lower energy neutrinos than were observed either in *Super-K* or *SNO* imposes stringent requirements on the purity of the materials used; this has occupied a large fraction of the attention of the collaboration. However, the results from *Borexino* should produce new constraints on our understanding of the Sun, and possibly of neutrinos as well. This is discussed in greater detail in chapter 5.

A diagram of the *Borexino* detector is shown in figure 2.27. It is scheduled for completion in 2007.

For more information, see the *Borexino* Web site, http://borex.lngs.infn.it.

2.10.4 High-Energy Cosmic Neutrinos

Neutrinos are necessarily produced in potentially detectable quantities in a variety of astrophysical sites besides the Sun. For example, any site that can produce baryons at sufficient energy that they can undergo pion production will ultimately produce neutrinos from the decays of the pions into muons and neutrinos and the subsequent decays of the muons into electrons and more neutrinos. The neutrinos, like gamma rays, would point back to their source. Because the neutrinos are so difficult to detect, the neutrino telescopes that currently exist or are under construction use huge volumes of naturally occurring detection medium, for example, sea water or South Pole ice, to detect upward-going muons produced by interactions between the neutrinos and the matter below the detector (see discussion below) to indicate the presence, and the direction, of the neutrinos. This is indicated in figure 2.28 (from the *IceCube* Web site, http://www.icecube.wisc.edu/brochure/Amanda_brochure.pdf), which shows the neutrinos being produced in some cosmic accelerator and then interacting on their passage through the Earth to produce a detectable muon. Neutrinos would also be produced in copious quantities from a supernova, a stellar explosion of a massive star. However, their detection in large numbers would be limited to supernovae in our Galaxy, which apparently occur only about once every 30 years (see chap. 6), and, because their energies are tens of MeV, only those at the high-energy tail of the distributions would have any possibility of being detected, via their interaction with electrons, by the existing high-energy neutrino detectors.

The neutrinos would be indicative of the processes that occur at their very source, unlike any other astrophysical messenger, which tend to be "processed"—scattered—either by the matter through which they pass on their way to Earth or by interstellar magnetic fields. This is a result of the neutrinos' tiny cross section with matter, a result of their interacting only through the weak interaction. Unfortunately, this same fact makes them extremely difficult to detect, which necessitates very large detectors.

The neutrino interactions that will occur in all the high-energy neutrino detectors, which utilize either ice or water as the detection medium, are as

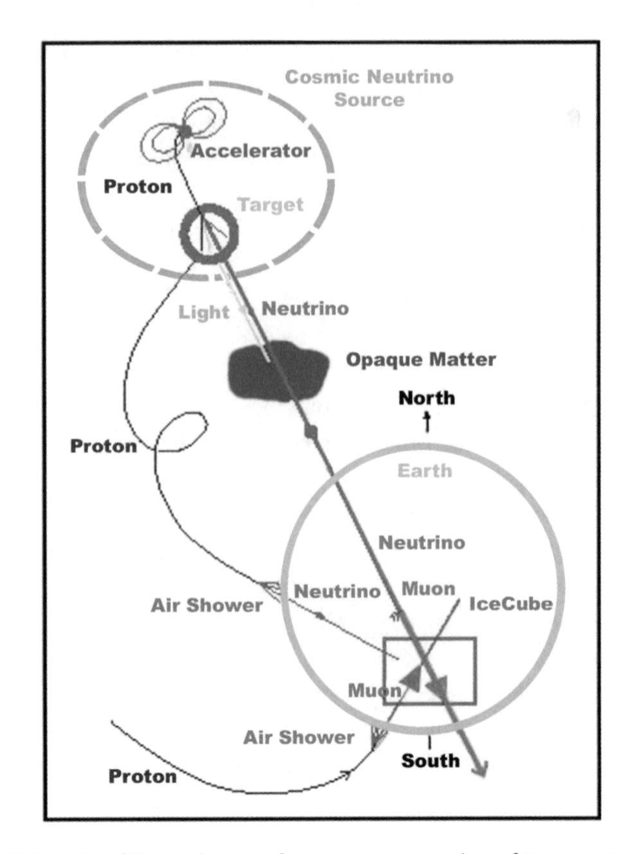

Fig. 2.28. Schematic of the production of a cosmic neutrino, then of its interaction with the Earth to produce a muon, which then can be detected by conventional means. Reprinted with permission from F. Halzen. From http://www.icecube.wisc.edu/brochure/amanda_brochure.pdf.

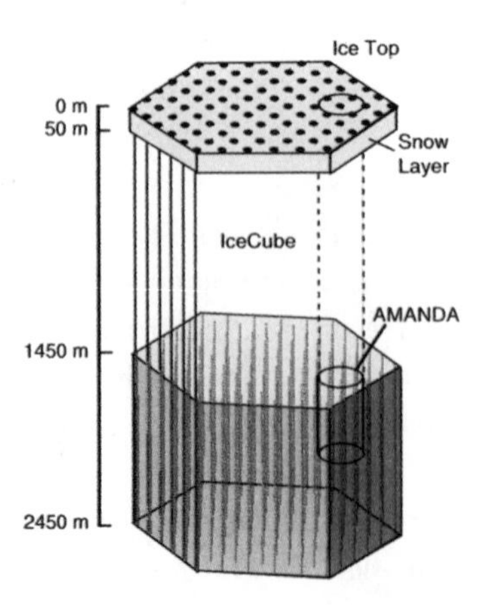

Fig. 2.29. Schematic of *IceCube* as it is configured in the ice of the South Pole. *IceTop* is a system that will identify downward-going events. Reprinted with permission from F. Halzen. From http://www.icecube.wisc.edu/brochure/amanda_brochure.pdf.

follows (for much more detail on the interactions, see the *ANTARES* Web site, http://antares.in2p3.fr/):

- charged-current ν_e interactions produce electrons, which then produce electromagnetic and hadronic showers. Unfortunately, charged-current ν_e interactions will not be distinguishable from neutral-current interactions of neutrinos of all flavors, ν_e, ν_μ, and ν_τ, with the electrons in the ice or water;
- charged-current ν_μ interactions produce $\mu^\pm$ leptons as well as a point-like hadronic shower; and
- charged-current ν_τ interactions produce $\tau^\pm$ leptons with electronic, muonic, and hadronic decay modes.

As noted above, all the high-energy neutrino detectors discussed are actually sensitive to upward-going neutrinos. These (very large) detectors have a huge number of cosmic-ray muons raining down on them at all times so the only way they can distinguish those from neutrino-induced events is to rely on the facts that most of the neutrinos that enter the Earth come out the other side and that the few that actually do interact with the rock, ice, or water to produce detectable particles will produce them in an upward-going direction. The detectors are designed to distinguish this type of event from the orders of magnitude more frequent downward-going events.

2.10.5 AMANDA and IceCube

One detector that is searching for ultra-high-energy neutrinos is *IceCube*, a detector that is located in the deep ice at the South Pole. It was designed first as a smaller version, called *AMANDA* (*Antarctic Muon and Neutrino Detector Array*). The purpose of these detectors is to observe high-energy ($\sim$1 TeV and above) neutrinos from astrophysical point sources. It looks down (see fig. 2.28), so it observes the signatures of upward-moving neutrinos as they complete their path through the Earth. If the signature is a muon, it will be extremely relativistic, so it will emit Cherenkov light when passing through the ice (the attenuation length of such light is hundreds of meters in deep ice), which allows them to be tracked by measuring the arrival times of these Cherenkov photons at the PMTs.

AMANDA and *IceCube* are made up strings (see fig. 2.29) of optical modules (OMs), each of which is made up of a PMT and passive electronics in a glass pressure vessel, on strings at depths of 1400–2400 m. *AMANDA* had an effective detection area of approximately 10,000 m^2 and an effective volume of about 10^7 m^3; *IceCube* will increase those numbers considerably. Figure 2.29 shows a schematic diagram of *IceCube*.

Results from *AMANDA-II* are summarized as follows:

1. A search using *AMANDA-II* established a flux limit for point sources.
2. There is no evidence for any neutrino point sources, although some interesting, but low-statistics, possibilities bear further investigation.
3. *AMANDA-II* has provided an extremely useful test module for the much larger *IceCube*, which began construction in 2004 and which will subsume *AMANDA*.

2.10.6 IceCube

IceCube is the second-generation version of *AMANDA*. It has been planned to consist ultimately of 80 strings of OMs (see fig. 2.30), now dubbed digital optical modules or DOMs, with a total of 4800 DOMs that are deployed in essentially the same way as was done for *AMANDA*. *IceCube* will have an effective volume of 1 km^3 of Antarctic ice. At the end of 2005, nine strings had been deployed and were working, and subsequent years will see an accelerating installation rate. Special drills have been designed to drill the deep holes in the ice into which the strings of DOMs will be placed. Because of the hostile working conditions at the South Pole, only about 3 months of the year can actually be used to install the strings, but it is hoped that the installation rate will ramp up to roughly a dozen strings per working season. This should produce a completed *IceCube* in about 2010.

For several science objectives, for example, neutrino emission from GRBs, as few as 16 *IceCube* strings will provide half the sensitivity of the full *IceCube* array. Thus, state-of-the-art science can be accomplished even during the construction of *IceCube*.

This write-up and the *IceCube* figures were excerpted from the information on the *IceCube* Web site, http://icecube.wisc.edu.

2.10.7 Astronomy with a Neutrino Telescope and Abyss Environmental RESearch (ANTARES)

ANTARES is also a large-volume detector of high-energy neutrino events. It is located in the Mediterranean Sea off the coast of Marseilles, France. It was first deployed, as a prototype, in 2003, and the first signals were received from it in March 2003. The *ANTARES* detectors are designed to look downward.

Most of the Galactic plane is visible to *ANTARES*, and the Galactic center is visible most of the sidereal day. Since the *AMANDA* and *IceCube* telescopes

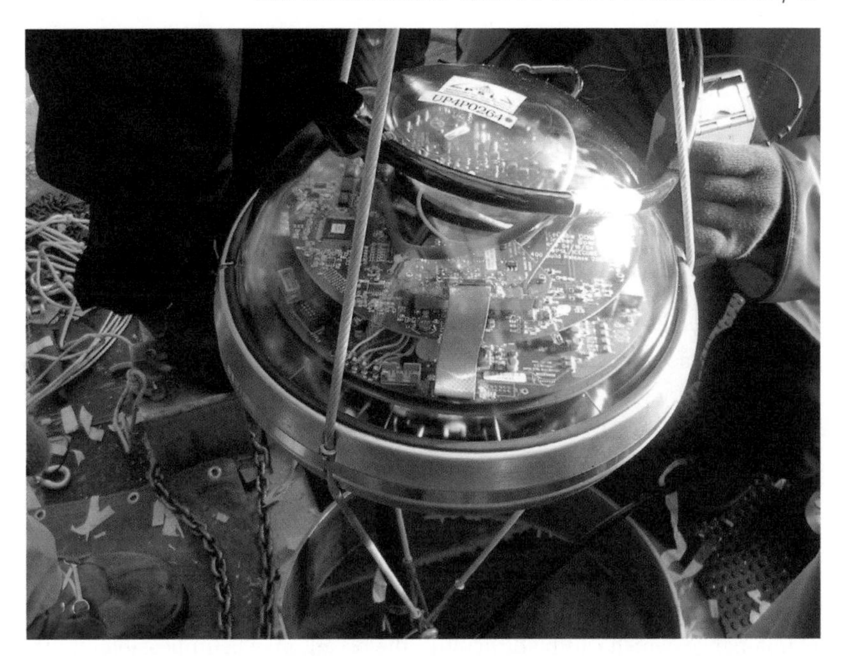

Fig. 2.30. A photograph of one of the *IceCube* DOMs, oriented essentially as it will be in *IceCube*, with the photomultiplier tube facing downward. Since these DOMs are ultimately frozen into the ice, they must be robust but, of course, the Cherenkov light from the relativistic muons must be able to penetrate their shell. In addition, the signal cables must be robust, as they too will be frozen into the ice. Figure from http://icecube.wisc.edu. Reprinted with permission from F. Halzen.

at the South pole are sensitive to positive declinations, the two detectors will have a reasonable area in common, about 1.5π sr, for cross checks. However, they will also observe from significant amounts of different space.

This above discussion was extracted from the *ANTARES* Web site, http://antares.in2p3.fr/.

2.10.8 Neutrino Extended Submarine Telescope with Oceanographic Research (NESTOR)

NESTOR also searches for neutrino-induced muons and/or showers of charged particles in a large volume of seawater. It is located at the southwest of the Peloponnese peninsula, Greece, at the seabed of the Ionian Sea. The *NESTOR* collaboration has located an 8 km × 9 km horizontal plateau at a depth of 4000 m (Tsirigotis 2004). Extensive studies of the local environment have shown that light of wavelength 460 nm has a transmission length of 55 m at that location. *NESTOR* was deployed in March 2003.

The basic detector unit is a hexagon, having a diagonal of 32 m. At the tip of each arm of the hexagonal floor is a pair of two 15-inch PMTs, one looking upward and one looking downward. The electronics for each floor are housed inside a large titanium sphere located at its center from which signals are sent to shore. Stacking 12 of these floors in the vertical, with a distance between them of 30 m creates a detector tower. The effective volume of a *NESTOR* tower in reconstructing through-going muons of energy >10 TeV is greater than 20,000 m^3.

Recent results from the *NESTOR* collaboration can be found at Aggouras et al. (2005). This discussion was excerpted from the *NESTOR* Web site, http://www.nestor.org.gr/.

The sites selected for *NESTOR* and *ANTARES* (the Mediterranean) are certainly more hospitable than that of *IceCube* (the Antarctic, where the "working season" usually involves days that have about the same level of hostility as a bad winter day in the most extreme northern climes inhabited by humans). In addition, water and ice present different issues. For example, coating on the light-sensing modules and currents may present problems in water, whereas maintenance, which can be difficult in the ocean, may be impossible in ice. Furthermore, the scattering length of light in ice is different from that in water. Time will tell which approach turns out to produce the best results. However, high-energy neutrinos will provide a promising new window on the universe.

2.11 Gravity Wave Detection

Detection of gravity waves has long been a scientific dream, but with the *Laser Interferometer Gravitational-Wave Observatory* (*LIGO*), this is becoming a reality. *LIGO* will be used for research into the nature of gravity, and it will open up an entirely new window onto the universe. It will thus be a scientific tool both for physics and for astronomy. *LIGO* consists of two widely separated installations within the United States, which are operated in unison as a single observatory. When it reaches maturity, this observatory will be open for use by the national community and will become part of a planned worldwide network of gravitational-wave observatories. Construction of the facilities was completed in 1999. Initial operation of the detectors began in 2001, and the first data run took place in 2002. Data acquisition has continued since that time, with gradual improvements taking place as time has gone on. Although *LIGO* has not yet seen any real events, it has developed a technology that will be extended to *Advanced LIGO*, for which scientists are currently gearing up

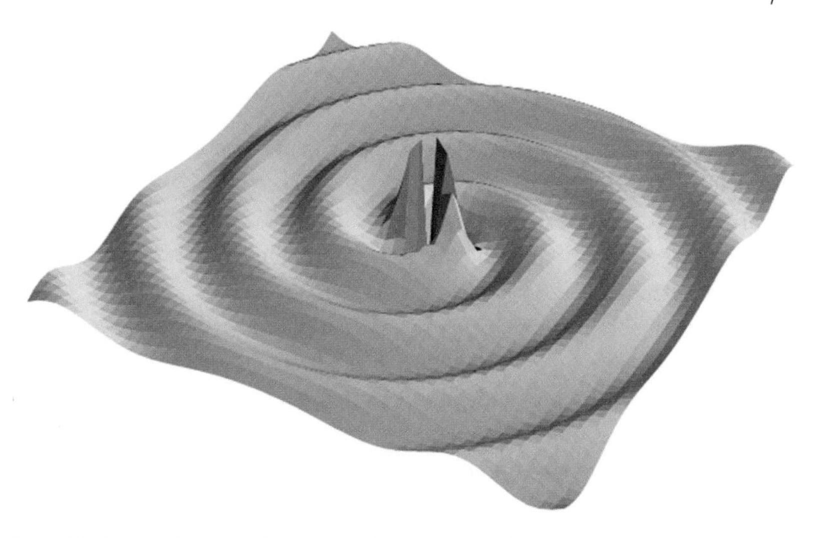

Fig. 2.31. Artist's drawing of gravitational waves, as ripples in space time. Courtesy of the *LIGO* collaboration.

and which will have a much greater sensitivity to events in nature that can produce gravity waves. *Advanced LIGO* will have a noise level that is 1/10 that of *LIGO* over all frequencies and an even greater reduction at the important lower frequencies.

Gravitational waves are ripples in the fabric of space and time (see fig. 2.31) produced by violent cosmic events, for example, by the collision of two black holes or by the cores of supernova explosions. Gravitational waves are emitted by accelerating masses much as electromagnetic waves are produced by accelerating charges. Albert Einstein predicted the existence of these gravitational waves in 1916 in his general theory of relativity, but only since the 1990s has technology developed sufficiently to permit even the hope of detecting them. Although gravitational waves have not yet been detected directly, their influence on a binary pulsar (two neutron stars orbiting each other) has been measured accurately and is in good agreement with the predictions. Joseph Taylor and Russel Hulse were awarded the 1993 Nobel Prize in Physics for their discovery of this binary pulsar.

2.11.1 LIGO's Scientific Goals?

General relativity describes gravity as a manifestation of the curvature of space-time. *LIGO* will permit scientists to test this description for rapidly changing, dynamical gravity (the space-time ripples of the gravitational waves) and also for the extremely strong, dynamical gravity of two black holes as they collide.

More specifically, *LIGO* has the possibility to make advances in both physics and astronomy, including

- Verifying, through direct measurement, general relativity's prediction that gravitational waves exist.
- Testing general relativity's prediction that these waves propagate at the same speed as light and that the graviton (the fundamental particle that accompanies these waves) has zero rest mass.
- Testing general relativity's prediction that the forces the waves exert on matter are perpendicular to the waves' direction of travel and stretch matter along one perpendicular direction while squeezing it along the other. This is a manifestation of the spin of the graviton; thus *LIGO* will test general relativity's prediction that the spin of the graviton is twice that of the photon.
- Firmly verifying that black holes exist and testing general relativity's predictions for the violently pulsating space-time curvature accompanying the collision of two black holes. This will produce the most stringent test ever of Einstein's general relativity theory.
- Observing the spiraling together and coalescence of pairs of neutron stars and in some cases the implosion of the coalesced star to form a black hole.
- Observing the swallowing of a neutron star by a black hole and the collisions and coalescences of black holes. This motivation has recently received a new burst of hope from the observations of the characteristics of short-time GRBs, which suggest that they are the result of neutron star–black hole collisions.
- Observing the birth of a neutron star in a supernova explosion and the pulsation and spin of this newborn neutron star.

The way in which such a detector works, as sketched in figure 2.32, is as follows. The larger the gravitational-wave detector, the more sensitive it will be. Detecting the very weak waves that are predicted requires two installations, each with a vacuum pipe 4 feet in diameter arranged in the shape of an L with 4-km arms. Since gravitational waves penetrate the Earth unimpeded, these installations need not be exposed to the sky and are entirely protected by a concrete cover. At the vertex of the L and at the end of each of its arms are test masses that hang from wires and are fitted with mirrors. The main building at the vertex serves as the control center and houses vacuum equipment, lasers, and computers. Shifts in the interference patterns of the light from ultrastable

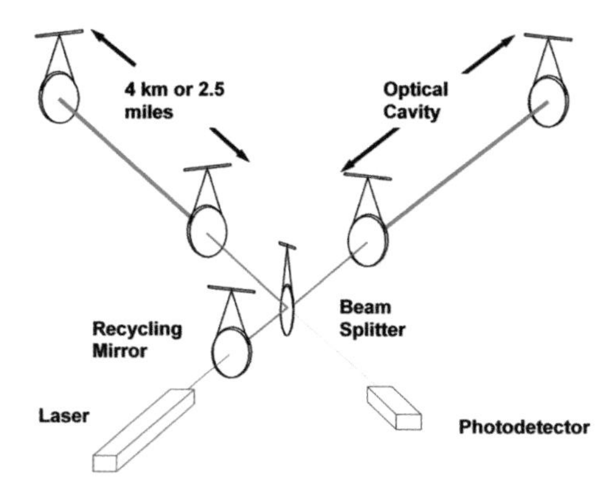

Fig. 2.32. Schematic drawing of the *LIGO* gravitational wave detector. Courtesy of the *LIGO* collaboration.

laser beams traversing the vacuum pipes measure the tiny motions of the test masses, that is, the effect of gravitational waves on those masses.

When the gravitational waves enter the *LIGO* detector they will decrease the distance between the test masses in one arm of the L and increase it in the other. These changes are minute: just 10^{-16} cm, or 1×10^{-8} of the diameter of a hydrogen atom over the 4-km length of the arm. Although these are unimaginably small distances, the tiny changes can be detected by isolating the test masses from all other disturbances, such as seismic vibrations of the Earth and gas molecules in the air, and by bouncing high-power laser light beams back and forth between the test masses in each arm and then interfering the two arms' beams with each other. The tiny changes in test mass distances throw the two arms' laser beams out of phase with each other, thereby disturbing their interference and revealing the form of the passing gravitational wave.

However, at least two detectors located at widely separated sites are essential for the unequivocal detection of gravitational waves. Local phenomena such as microearthquakes, acoustic noise, and laser fluctuations can cause a disturbance at one site, simulating a gravitational-wave event, but such disturbances are unlikely to happen simultaneously at widely separated sites. The *LIGO* sites are near Livingston, Louisiana, and at Hanford, Washington. The sites, which are separated by nearly 2000 miles, are both flat and large enough to accommodate the 4-km interferometer arms. Both are also far enough from urban development to ensure that they are seismically and acoustically quiet.

Other gravitational-wave detectors around the world are in various stages of development. These include *VIRGO* (Europe), *GEO 600* (Europe), *TAMA300*

(Japan), and *AIGO* (Australia). Descriptions of these facilities can be found at http://wwwcascina.virgo.infn.it/, http://www.geo600.uni-hannover.de/, http://tamago.mtk.nao.ac.jp/, and http://www.gravity.uwa.edu.au.

This write-up and the accompanying figures were only slightly modified from those found on the *LIGO* Web site, http://www.ligo.caltech.edu/. They were used courtesy of the *LIGO* Laboratory.

2.12 Particle Accelerators of Special Relevance to Nuclear Astrophysics

2.12.1 The Gesellschaft fur Schwerionenforschung (GSI) Darmstadt Facility

A major advance made in nuclear physics in the late 1900s is the storage-cooler ring built at the *GSI* at Darmstadt, Germany. This device is shown schematically in figure 2.33, which shows the heavy-ion synchrotron SIS, the fragment separator FRS, and the storage ring ESR. Primary ions accelerated in the synchrotron to an energy of several hundred MeV per atomic mass unit are then directed to a postacceleration target at the entrance of the fragment separator, where they interact with its nuclei to produce exotic nuclei by in-flight projectile fragmentation. In the separator fragments within a small band of "magnetic rigidity," the magnetic field times the radius of curvature $= B\rho$, are selected and transported to the storage ring.

Initially, the exotic beams stored in the ESR are "hot," that is, their relative momentum spread $\Delta p/p$ is rather high (typically 10^{-2}), and their angular divergence and beam size are large. Thus the beam must be cooled in order to be stored by one of three means: stochastic, electron, or laser cooling. Since much of the data from this facility were obtained through the use of electron cooling, that technique will be the focus of the rest of the discussion.

With this facility, exotic ions within the same band of magnetic rigidity $B\rho$ but dispersed over a wide range of mass, are produced and simultaneously injected into the storage ring. If, however, specially shaped degraders are placed in the first dispersive focal plane of the separator, one single nuclear species can be singled out for injection into the storage ring, owing to the nuclear-charge-dependent energy loss ΔE in the degrader (Geissel et al. 1992). Most of the fragments generated by this technique will be at the proton-rich side of the stable nuclides. However, if fission of relativistic uranium projectiles is used initially to produce the exotic nuclei, neutron-rich fragments can be preferentially produced. A unique feature of this facility is the capability for production and storage of highly charged exotic ions—mainly bare or hydrogen-like ions—due to the projectile fragmentation being performed at high energies.

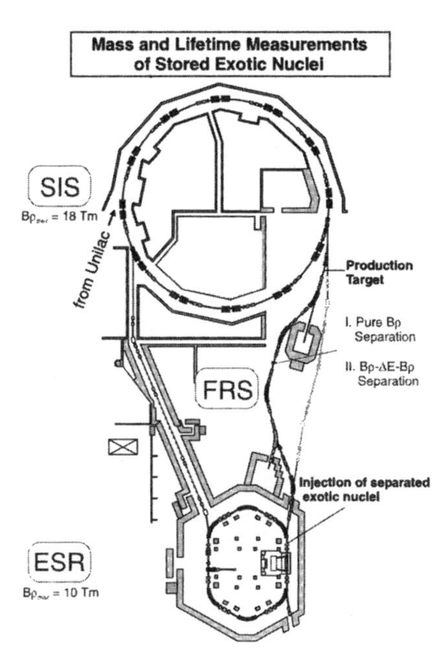

Fig. 2.33. The combination of the heavy ion synchrotron SIS, the fragment separator FRS, and the storage-cooler ring ESR for production, storage, and cooling of exotic highly charged ions. From Bosch (2003). Credit: *Journal of Physics B: Atomic, Molecular, and Optical Physics.*

As noted above, much of the data obtained with this facility utilized electron cooling. In this cooling mode, first proposed by Budker (1967), an intense electron beam is directed to be co-moving with the heavy ions over part of their trajectory. Through Coulomb interactions, the ions quickly assume both the transverse and longitudinal momentum of the cold electrons, cooling to a typical $\Delta p/p$ of 10^{-5} to 10^{-7}. The ions cooled in this way can be stored in the low vacuum of the storage ring for times of the order of hours. The limitation is that short-lived ions are difficult to store, as the time required for cooling can be many seconds. (A modification of this technique, "isochronous mass spectrometry," can extend the range of observable lifetimes down to much shorter values; as low as a few microseconds.) The resulting beams move around the ring with constant velocity, that is, the velocity of the electrons that did the cooling, and the different species of ions are detected by Schottky mass spectrometry (Bosch 2003).

In Schottky mass spectrometry, the ions, all of which have the same velocity to a negligible uncertainty, pass through a pair of plates—Schottky noise pickups—with each revolution of the storage ring. In so doing, they induce a

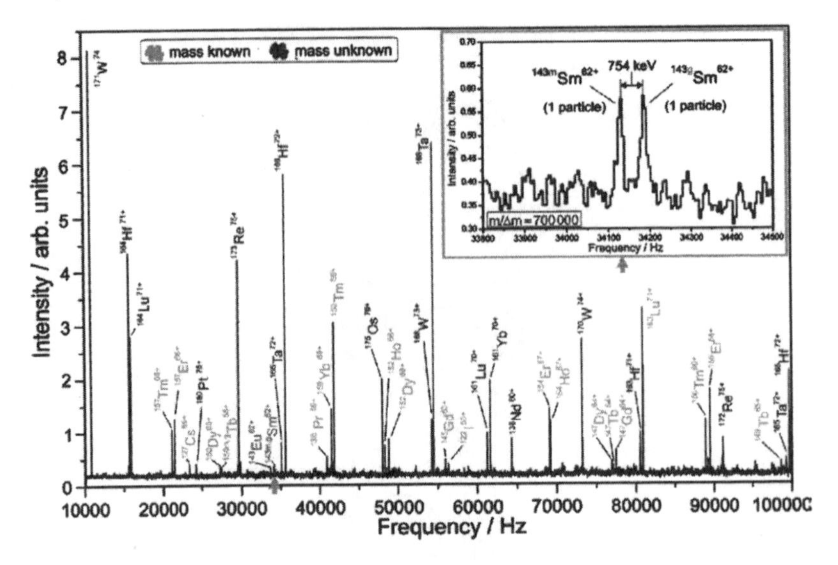

Fig. 2.34. Schottky spectrum of fragments from a primary ^{209}Bi beam, stored and electron cooled in the ESR at *GSI*. The main spectrum shows the difference of the 30th harmonic of the revolution frequencies of the many stored ion species and of a local oscillator operating at about 60 MHz. It covers roughly the full acceptance of the ESR. The inset shows the ability of the system to resolve the ground and isomeric states of ^{143}Sm, each populated by only a single ion. From Bosch (2003). Credit: *Journal of Physics B: Atomic, Molecular, and Optical Physics.*

signal proportional to the square of their charge on the plates. These signals are sampled, Fourier transformed, and mixed with a local oscillator to convert the original revolution frequencies into a useful range. A spectrum of ions is seen in figure 2.34, which shows about 50 well-resolved simultaneously recorded ionic species. The inset shows the high resolving power of the storage ring: shown there are the ground state and isomeric state of ^{143}Sm, only 754 keV apart in mass units of mc^2. Also shown is the capability of the detection system to observe single ions. Accurately determined masses are crucial to many of the considerations of nuclear astrophysics, as will be discussed in several later chapters.

An interesting example of one of the capabilities of this facility is provided by a half-life measurement, that of ^{168}Ta (a neutron-poor nucleus). With this technique, the decay of the intensity of the stored species of interest can be measured from the Schottky spectrum. This is shown for this case on the left-hand side of figure 2.35. Ions are lost through two mechanisms, either β decay (actually positron emission in this case) or scattering from the gas atoms in the storage ring. The latter loss mechanism is small for an ion with a lifetime as short as that of ^{168}Ta but in any event can be measured by noting the attrition in other species that are either stable or much longer lived. The right-hand side of this figure shows the decay curve measured with this technique.

Perhaps the most important lifetime measurement made with this facility was that of ^{187}Re, a cosmochronometer that has been used to infer the age of the universe (Bosch et al. 1996). This lifetime measurement, and its implications, will be revisited in chapter 7.

2.12.2 The National Superconducting Cyclotron Laboratory (NSCL) at Michigan State University

The *NSCL* is designed to produce primary beams of heavy ions of energies of tens of MeV per atomic mass unit, from which one can, by a nuclear reaction, produce secondary beams of short-lived nuclei. It is continuously being upgraded to provide higher energy and higher intensity beams of heavy ions. The result of this is more intense beams of nuclei farther from stability, radioactive nuclear beams (RNBs), many of which are of great astrophysical

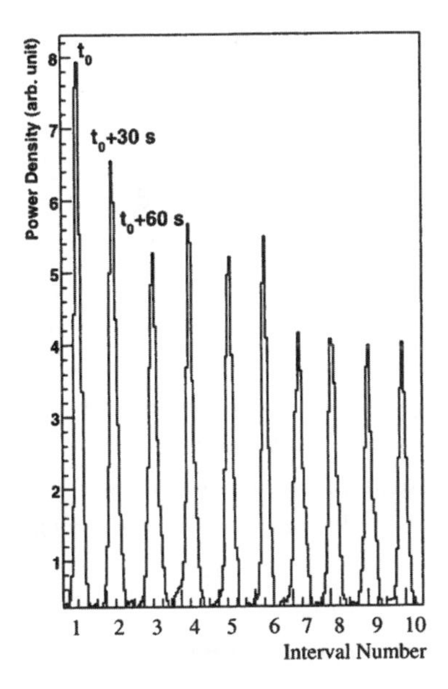

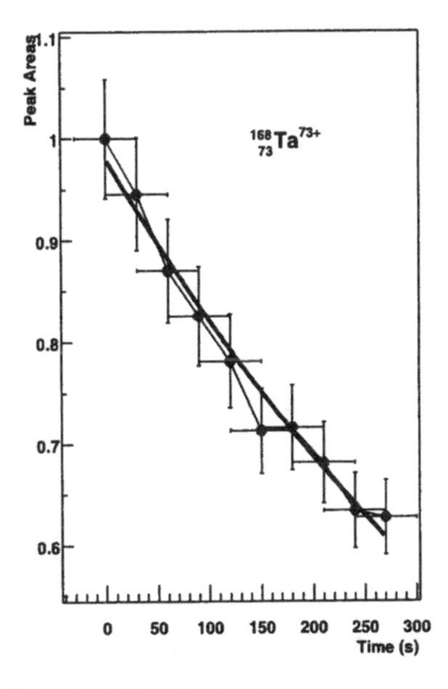

Fig. 2.35. Lifetime measurement of bare $^{168}Ta^{73+}$. The left-hand side shows the Schottky lines of the parent $^{168}Ta^{73+}$ nucleus in intervals of 30 s, and the right-hand side shows the corresponding decay curve. The extracted half-life in the emitter rest frame (β^+-decay only) amounts, after small corrections for losses in the ring due to charge-changing processes, to $T_{1/2} = 5.2$ m, whereas the half-life of neutral ^{168}Ta is only 2.0 m (β^+-decay + orbital electron capture decay). From Bosch (2003). Credit: *Journal of Physics B: Atomic, Molecular, and Optical Physics.*

interest. This is especially true of the processes of nucleosynthesis occurring at high temperatures and on short timescales, most notably, the r-process (see chap. 7) and the rp-process (see chap. 8).

This facility is currently being used extensively to study the properties of unstable nuclei—their half-lives and decay modes—and in some cases to study reactions on them. The *NSCL* has pioneered many of the techniques currently being used in both nuclear physics and nuclear astrophysics studies of nuclei far from stability. Several of the results it has produced will be quoted in subsequent chapters.

Figure 2.36 indicates the basic configuration of the facility used to produce the beams of radioactive nuclei. A small cyclotron, the K500, injects a preaccelerated beam of heavy ions into a larger cyclotron, the K1200. The preacceleration phase achieves a sufficiently energetic beam that it can be passed through a stripper foil to increase its charge state, thereby permitting acceleration to a higher energy in the K1200 acceleration phase than would otherwise have been possible. Intense primary beams of heavy nuclei are extracted from the second cyclotron and directed to a production target, at which is produced a plethora of nuclei, many of which are far from stability. Following this is a mass analysis system, a complex system of magnets and an absorber that selects out the unstable nucleus or nuclei of interest, ^{78}Ni in the example shown. This nucleus is very far beyond the neutron-rich side of stability but is of special interest because it has shell closures in both protons and neutrons. Several types of detectors can be used to identify the nuclei of interest and measure their decay properties.

Considerably more information about the NSCL can be found on its Web site, http://www.nscl.msu.edu/.

Another RNB facility of the high-energy fragmentation type that has produced some excellent physics results and to which reference is made in subsequent chapters is SPIRAL, Système de Production d'Ions Radioactifs Accélérés en Ligne, which is located at *GANIL*, the *Grand Accélérateur National d'Ions Lourds*, in France.

2.12.3 The Center for Nuclear Study Radioactive Ion Beam Separator (CRIB)

A low-energy in-flight type RNB separator, *CRIB*, has been installed for nuclear physics and nuclear astrophysics research by the Center for Nuclear Study, at the University of Tokyo, in the RIKEN Accelerator Research Facility. This is one of several accelerators around the world that can first produce large quantities of unstable nuclei and then collect them into a beam on which nuclear reactions can be studied. *CRIB* consists of a double achromatic system

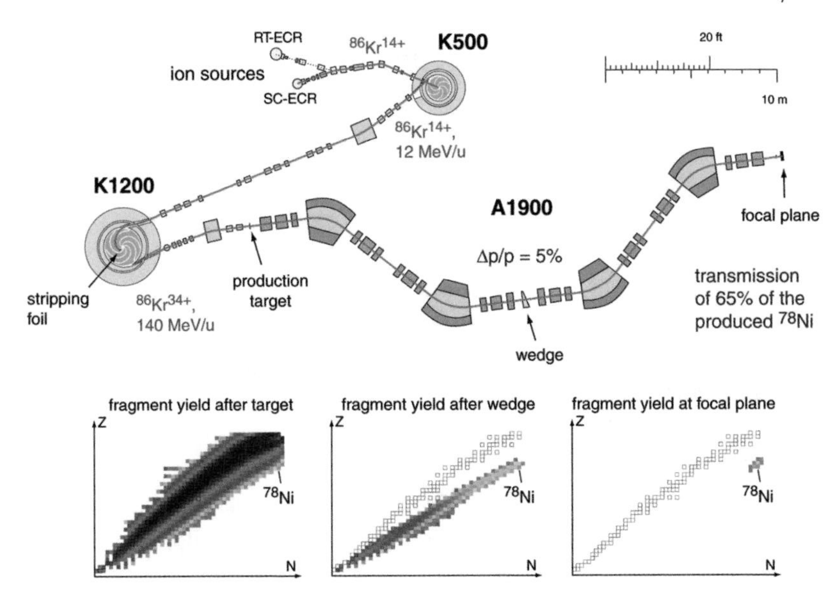

Fig. 2.36. The *National Superconducting Cyclotron Facility* at Michigan State University. Shown are the two cyclotrons, the K500 and the K1200, followed by the mass analysis system. The inserts at the bottom show how the mass analysis system selects the species of interest at each stage; the stable nuclides are shown for reference. Reprinted with permission from A. Stoltz.

and a Wien filter (a combination of electrostatic and magnetic fields set to transmit only ions of a specific charge to mass ratio), as indicated in the schematic shown in figure 2.37. It is capable of providing RNBs at 5–10 MeV per nucleon. Since *CRIB* was developed at the end of 2000, many proton-rich RNBs were successfully produced via the (p,n), (d,t), and (^{3}He,n) reactions in "inverse kinematics," the configuration indicated in figure 2.37.

The studies on which the *CRIB* has focused primarily involve proton-rich unstable nuclei, which play an important role in explosive hydrogen burning, primarily in the rp-process (see chap. 8). Although a great deal of experimental effort is needed to understand this process, only a few reactions have thus far been studied directly with RNBs. *CRIB* was constructed for this purpose, and nuclear astrophysics experiments using low-energy RNBs are being conducted. *CRIB* uses a windowless gas target in addition to the magnetic double-achromatic system. The Wien filter is successfully operating to improve both the purity and the emittance of RNBs of interest. A series of resonant scattering experiments have been made with the RNBs from *CRIB*, which includes ^{21}Na+p, ^{22}Mg+p, ^{23}Mg+p, ^{25}Al+p, and ^{26}Si+p, in addition to

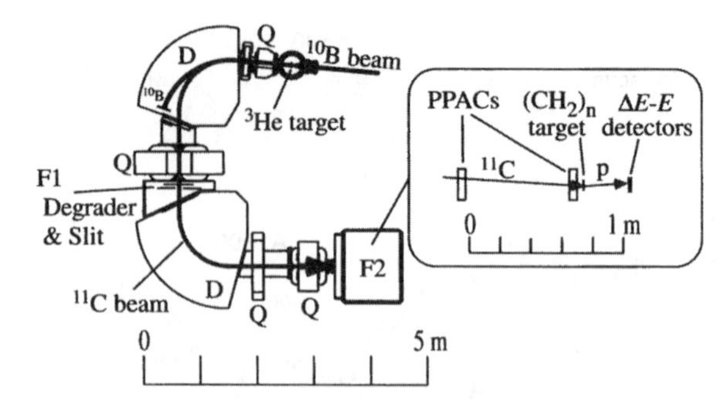

Fig. 2.37. Schematic view of the experimental setup of the *CRIB* ion optics and the characteristics of the double achromatic system. From Yanagisawa et al. (2005). Copyright 2005, with permission from Elsevier.

earlier studies of ^{11}C+p and ^{12}N+p. In most cases new states were observed. The reaction ^{14}O(α,p)^{17}F, which is critical for understanding the breakout process of the hot CN cycle (see chap. 8), is also being investigated by the thick-target method using an intense ^{14}O beam and a cooled He gas target.

This reaction and the concomitant data are discussed further in chapter 8. An upgrade of the AVF cyclotron is planned; it will optimize the accelerator facility for intense RNBs of about 10^8 particles per second with *CRIB*.

Much of the preceding discussion and the above figure were excerpted from the *CRIB* Web site, http://www.cns.s.u-tokyo.ac.jp/ann03/online/pdfs/a20_notani.pdf.

2.12.4 The Isotope Separation and Acceleration Facility (ISAC) at TRIUMF

ISAC, located in British Columbia, Canada, is the current generation of RNB facilities at TRIUMF. During the past 15 years, a group at TRIUMF built a facility that could create and isolate beams of many interesting, short-lived nuclei of intermediate mass. The facility was called *TISOL* (*TRIUMF Isotope Separator Online*). *TISOL* created the desired radioactive isotopes by bombarding a suitable target with 800 MeV protons from TRIUMF's cyclotron. It then separated these radioisotopes from others that were formed simultaneously, often at considerably greater intensity, and provided them to experimenters as a low-energy beam of particles. This facility was used to produce a low-energy beam of ^{16}N nuclei, from which a measurement was made of one of the components of the ^{12}C(α,γ)^{16}O reaction cross section (see chap. 6), which is critically important to nuclear astrophysics. The *ISAC* facility is the next phase of this project. It utilizes the very intense, intermediate-energy proton beam from the TRIUMF cyclotron on some target to produce the species of interest. These are extracted and injected into a new isotope separator online facility. Then these unstable nuclei are accelerated to the desired energies via a linear accelerator, that is, to those energies at which nuclear reactions occur in stellar environments, to perform the reaction studies.

A picture of the facility is shown in figure 2.38. The facility allows measurement of cross sections of reactions that are important to understanding what nuclear reactions occur when stars explode or when they accrete matter in X-ray bursts and concomitant rp-process scenarios (see chap. 8). While scientists understand the nuclear reactions occurring within many stellar environments, a clear picture of the very hot environments of novae, supernovae, or X-ray bursts requires this new information. These hot environments lead to reactions involving short-lived radioactive species. TRIUMF's *ISAC* enables study of the very reactions that occur in those explosive conditions.

One experiment that has been performed with *ISAC* is a measurement of the cross section for the ^{21}Na(p,γ) reaction (a ^{21}Na RNB incident on an H

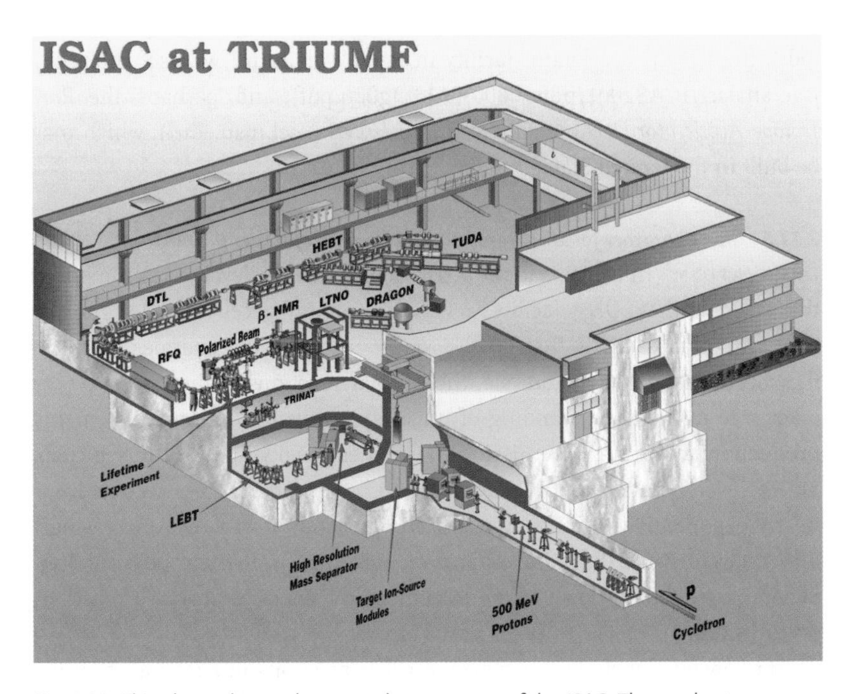

Fig. 2.38. This shows the accelerator and target room of the *ISAC*. The accelerator components, labeled RFQ and DTL, are at the back (*top*) of the picture, and the reaction component analyzer DRAGON, the Detector of Recoils and Gammas of Nuclear Reactions, is in the foreground. From http://www.triumf.ca/isac/is-20jan2003.jpg. Reprinted with permission from L. Buchmann.

target, with the emitted gamma ray being detected), which is of importance to hot hydrogen-burning scenarios such as those that might occur in the accretion of matter onto a white dwarf or a neutron star. This experiment and the motivation for it are discussed in detail in chapter 8.

ISAC significantly increases the types of nuclear beams available at TRIUMF. These species can be used for a wide range of studies over and above those mentioned here; the range of applications can include materials science, industrial uses, atomic physics, surface science, nuclear structure explorations of very deformed nuclei, and many more.

Much of the discussion of *ISAC* came from its Web site, http://www.triumf.ca/isac/isac_home.html.

2.12.5 Future Accelerators of Astrophysical Significance

It is expected that more powerful accelerators will be built in the future to replace the current generation of RNB facilities, and some are already under construction. This list includes an upgraded facility at *GSI* Darmstadt (see, e.g., http://www.ep1.rub.de/~panda/db/papersDB/henning_walter_proc_leap03.pdf), the RIKEN accelerator facility in Japan (see http://wwwsoc.mii.ac.jp/jsac/analsci/ICAS2001/pdfs/0600/0613_3g02n.pdf), and, perhaps, the *Rare Isotope Accelerator* (*RIA*) facility (see http://www.nscl.msu.edu/), which may be built in the United States.

2.12.6 The Laboratory for Underground Nuclear Astrophysics (LUNA)

MOTIVATION. In contrast to the RNB facilities being developed throughout the world is the *LUNA* facility, a very low energy accelerator designed to study nuclear reactions that involve stable nuclei at the energies at which they occur in stars. For example, the nuclear reactions of the pp-chain play a key role in the understanding of nucleosynthesis of the elements, energy production, and neutrino emission in stars, especially in our Sun (see chap. 5). However, because of the Coulomb barrier, the reaction cross section drops nearly exponentially with decreasing energy. Thus it becomes increasingly difficult to measure the cross sections or, equivalently, the astrophysical $S(E)$-factors (see chap. 3) down to the extremely low energies at which reactions occur in stars.

The low-energy studies of thermonuclear reactions in a laboratory at the Earth's surface are greatly hampered by the effects of cosmic rays interacting with the detectors. Passive shielding around the detectors provides a reduction in gammas and neutrons from the environment, but it produces a concomitant increase in gammas and neutrons due to the cosmic-ray interactions in the

Fig. 2.39. Beam analysis system and detection chamber of *LUNA*. Reprinted with permission from C. Rolfs.

shielding itself. Thus a 4π active shield could only partially reduce the problem of cosmic-ray activation, even if it were possible to make it perfectly effective in identifying the cosmic rays. The *LUNA* collaboration has solved this problem by installing an accelerator facility in a laboratory deep underground.

The world's first underground accelerator facility has been installed at the

Laboratori Nazionali del Gran Sasso in Italy by the *LUNA* collaboration. The initial aim of the project was to measure the cross section of the reaction $^{3}He(^{3}He,2p)^{4}He$ over the full range of the solar Gamow peak (see sec. 3.9), which represents the effective energy range in the Sun. In this energy region, the cross section is as low as 8 pb ($= 8 \times 10^{-36}$ cm^2) at $E = 25$ keV, and about 20 fb at $E = 17$ keV. Additional measurements have been made to define the effects of electron shielding in calculating the very low energy cross sections; these are discussed in detail in chapter 4. Because some of the techniques used in measuring very low energy cross sections are unusual, some of these details will be given.

THE ACCELERATOR. Measurements of cross sections at extremely low energies require extremely good energy stability over the long periods of time required to make the measurements. Thus, the *LUNA* source and accelerator voltage are designed to ensure this capability. As shown in figure 2.39, the beam transport is provided by a double-focusing 90° analyzing magnet and an electromagnetic steerer. The shim angles of the analyzing magnet are adjustable between 25° and 40° to allow focusing and defocusing of the beam. The beam intensity is measured by a beam calorimeter. The typical beam current in the target area down to a beam energy of 40 keV is as high as 400 μA of ^{3}He.

THE GAS TARGET SYSTEM. The target is provided by a windowless target chamber at a pressure of 0.5 mbar of ^{3}He, and that allows for recirculation for the target gas as well as for the ^{3}He out of the ion source, a process that also involves continuous cleaning of the gas.

THE DETECTOR SETUP. Four ΔE-E detector telescopes (which have first a thin detector and then a thick one to provide both energy measurement and particle identification) are placed in a rectangular target chamber around the beam at distances of 2.5 and 3.5 cm from its axis. Each telescope consists of transmission surface barrier silicon detectors with a 0.25-μm thick Al layer deposited on both sides of the detectors. The ΔE and E detectors both have an active area of 2500 mm^2; the ΔE (and E) detector has a thickness of 140 μm (and 1000 μm). A mylar foil and an Al foil (each of 1.5-μm thickness) are placed in front of each telescope. They stop the intense elastic scattering yield and shield the detectors from beam-induced light. The detectors are maintained permanently at low temperature (about −20°C) using a liquid recirculating cooling system (see Greife et al. 1994; Arpesella et al. 1995) to minimize noise.

We will return to the discussion of *LUNA* and to some of the extraordinary data it has obtained in both chapter 4 and chapter 5. The information for this write-up was excerpted from the *LUNA* Web site, http://www.lngs.infn.it/site/exppro/luna/luna.html.

It is anticipated that a second-generation underground accelerator laboratory may be built in the *Deep Underground Science and Engineering Laboratory* (*DUSEL*), which may be built in the United States.

2.13 Gamma-Ray "Accelerator" Facilities

We will see in chapter 3 that nuclear reactions initiated by fairly energetic gamma rays are important to our considerations of nuclear astrophysics. Although the facilities to perform this type of measurement are still in their infancy, several either do exist or are being planned. Because their work has been limited to date, they will not be discussed in detail. The names of some of these are *DANILAC*, *HIgS*, *Sansoken*, and *Forshungszentrum Rossendorf.* It is anticipated that they may appear as significant additions to a future edition of this textbook.

2.14 Underground Laboratories

While it might seem odd to think of performing any research associated with astronomy deep underground, many experiments require the low background environments that can be achieved therein. Indeed, as mentioned above, *Super-Kamiokande* is located in the Kamioka mine in Japan, *Borexino* and *LUNA* in the Laboratori Nazionali del Gran Sasso in Italy, and *CDMS* in the Soudan mine in Minnesota. The reason for siting these low count rate experiments underground is simply that siting them at the Earth's surface would permit a cosmic-ray background rate that would swamp out any signals from desired real events. Of course, siting experiments under the sea, as has been done with *ANTARES* or *NESTOR*, or the South Pole ice, as has been done with *AMANDA* and *IceCube*, accomplishes the same shielding effect.

The motivation for siting experiments deep underground is indicated in figure 2.40, which shows the cosmic-ray background rates at a variety of sites around the world. It can be seen that going thousands of feet underground vastly diminishes the cosmic-ray flux and hence the backgrounds from these events. As can be seen from that graph, going from the 2000-foot depth of the *Waste Isolation Pilot Plant* to the 6800-foot depth of the Sudbury mine in Canada reduces the cosmic-ray background by 3 orders of magnitude, an important factor in extremely low event rate experiments. Note, however, that this background does not ever really go to zero; at some level cosmic neutrino-induced events become the dominant background.

The detector *SNO*, designed to detect both the solar neutrinos that may have changed flavors on their trip from their point of creation in the Sun to the Earth and those that have not, is sited in the Sudbury mine in Canada. It is currently the deepest mine, at 6800 feet below the surface, in which physics experiments are housed. Indeed, the *SNO* experiment could not have been performed at a shallow site.

Indicated on figure 2.40 is the Deep Underground Laboratory. This refers to an initiative by the United States to develop a *Deep Underground Science and Engineering Laboratory*, or *DUSEL*. If built, this facility would have the capability to perform not only the neutrino and dark matter experiments suggested in the different sections of this chapter but many other types of experiments, including geoscience (see http://www.earthlab.org/plan.html), mining engineering, and geobiology. This laboratory would have sufficient size to allow construction of a huge detector for observing proton decay events and for doing long baseline experiments on neutrino beams generated by accelerators 1000 miles away.

However, the deepest existing underground laboratory at present is the *Sudbury Neutrino Observatory Laboratory* (*SNOLab*), at a depth of 6800 feet. This has been a fairly small laboratory, but it is now being expanded to a size of 45,000 square feet. While this is still not a large laboratory, it will certainly permit several wonderful new experiments in underground science.

The 2 km of over-burden at the site provides 6010 meters water equivalent of shielding from cosmic rays and offers a uniquely low background environment for the next generation of experiments exploring the frontiers of particle physics and astrophysics. Fields of potential research projects for *SNOLab* include

- low-energy solar neutrinos;
- neutrinoless double beta decay;
- dark matter searches; and
- detection of neutrinos from supernova explosions.

Project Status

Construction of the surface facility has been completed. The design of the underground excavations is complete, and a detailed design of the laboratory infrastructure is underway. Excavation is expected to be completed in 2006.

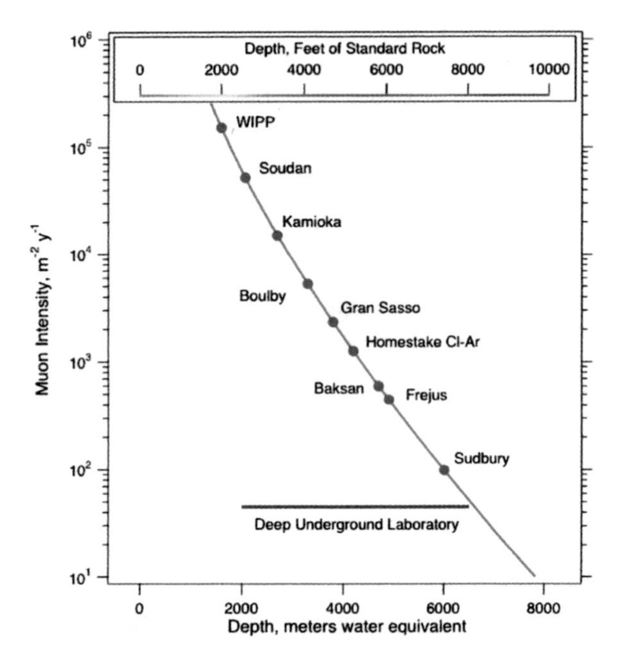

Fig. 2.40. Intensity of muons produced from high-energy cosmic rays interacting with nuclei in the Earth's atmosphere as a function of depth. Indicated on the graph are various existing underground laboratories and the proposed *Deep Underground Science and Engineering Laboratory*. Reprinted with permission from R. G. H. Robertson.

Space for experiments is expected to become available in early 2007. Excavation will be done in two phases. The first phase will include a large new experimental hall, a network of smaller laboratory areas, and the associated support spaces. A second excavation phase will permit the addition of another large experimental cavity if need be and a funding permit at a future date. The first phase will include a large new experimental hall, 60 feet × 50 feet × 50 feet (seen at the top in fig. 2.41), a network of smaller laboratories, and the associated support spaces (seen in the center of fig. 2.41). A possible cryo-pit to permit use of heavy cryogens such as argon and xenon will be built, as indicated in the upper right corner in figure 2.41.

2.15 Laser Fusion Facilities

A new type of facility is currently under construction that might provide extraordinary new capabilities in studies of nuclear astrophysics and astrophysics and more generally in studies of high-energy density physics. This is

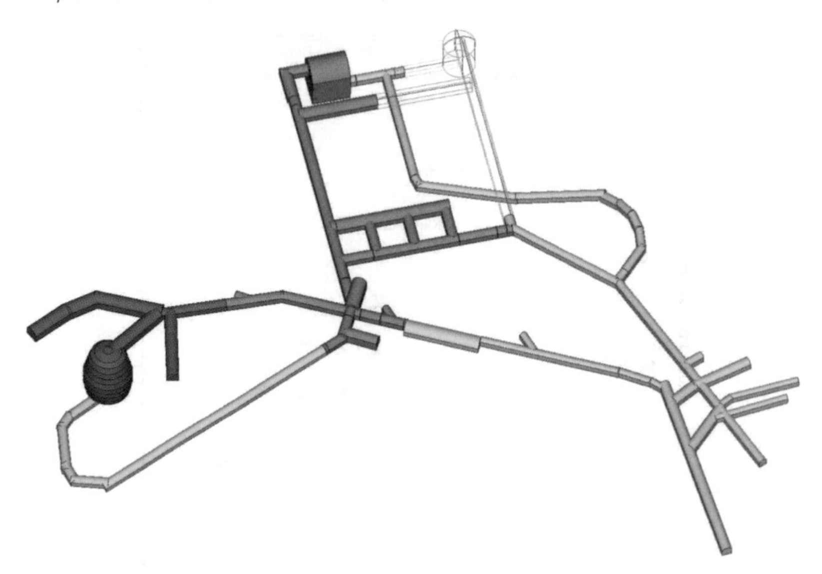

Fig. 2.41. This shows a view from above of the layout of the new *SNOLab* at about the 6800-foot level. The different components of the existing and planned laboratory are discussed in the text. From http://www.snolab.ca. Reprinted with permission from A. McDonald.

a facility that uses converging high-intensity laser beams to implode pellets containing deuterium and tritium sufficiently that they would undergo the nuclear reaction

$$^{2}\mathrm{H} + {}^{3}\mathrm{H} \rightarrow {}^{4}\mathrm{He} + \mathrm{n}.$$

This is one of the reactions that, in big bang nucleosynthesis, produces much of the ^{4}He observed in the universe (see chap. 9). It also can occur at relatively low temperatures, so it is a possibility for generating energy in the future (with much further development!). The facility that will produce these reactions is being built at the Lawrence Livermore National Laboratory. It is known as the National Ignition Facility, or NIF; it will converge 192 beams of laser light, initially produced by a single laser, amplified many times, and then brought to bear on a "hohlraum," a small cylinder that produces the X-rays that heat a millimeter-size pellet that contains, among other things, ^{2}H and ^{3}H. The total amount of energy delivered to the pellet will be 1.8 MJ, with a peak power of 500 TW, at a wavelength of 351 nm, producing conditions that, for about 20 picoseconds, achieve a matter temperature of $> 10^{8}$ K, a radiation temperature of 3.5×10^{6} K and a density of $> 10^{3}$ g cm^{-3}. These are conditions that sound much like those of stars. Specifically, some of the instabilities that occur in type Ia supernovae will also occur in the NIF implosion process, making them available for laboratory study. The facility will also produce, for the same 20 picoseconds, 10^{26} neutrons cm^{-3}, a number that is well above what is thought to be achieved in a core-collapse supernova and which is thought to drive the r-process of nucleosynthesis (see chap. 7). This facility fills a building that is the size of three football fields; this size is necessary to provide the amplification of the laser beams required to achieve the conditions sought. The facility is expected to be fully operational in 2010.

For the short time during which the stellar conditions exist in the imploded pellet, one might think of studying nuclear reactions under the conditions in which they actually occur in stars, that is, on fully ionized nuclei. This is the first time that nuclear astrophysicists have ever had this opportunity. In addition, it might be possible to study the sorts of instabilities that occur in supernovae and to study the hydrodynamics and plasma physics relevant to astrophysics. Of course, this will require nuclear astrophysicists to develop radically new approaches to performing measurements, as the conditions inside the chamber in which the pellet undergoes thermonuclear reactions are unlike those in which normal nuclear physics experiments are carried out.

Use of this facility, and of others around the world that are being developed in Japan, France, and Russia, may make possible a completely new generation of experiments in nuclear astrophysics.

CHAPTER 2 PROBLEMS

1. Calculate the energy of the GZK cutoff.
2. Assuming that the cross section for pion photoproduction from protons interacting with the 2.7 K CMB photons is about 1 μb, calculate the typical range of protons above the GZK cutoff before they have such an interaction.
3. Assume the first diffraction minimum is given by $\theta_{min} = \lambda/D$, where D is the diameter of the telescope and λ is the wavelength of the electromagnetic radiation being observed. Assuming the θ_{min} for $\lambda = 500$ nm and an 8 m telescope, determine how large D would have to be for a radio telescope to provide the same angular resolution. How large would D have to be for an X-ray telescope of typical wavelength?
4. You are working as a grants program manager for one of the funding agencies. You have two proposals: one for a space-based high-energy gamma-ray detector, which looks down at the atmosphere, and the other for a ground-based detector such as *VERITAS* or *HESS*. Because of the expenses of launch and space station maintenance, the space-based system is considerably more expensive than the

ground-based system, but its efficiency is higher since it can look at a much larger portion of the atmosphere. Consider only the construction costs, and assume that the space-based system costs $2 billion (which includes some prorated coverage of the launch system) and that the ground-based system costs $20 million. Assuming that the only criterion involved in making your decision as to which project to fund is to minimize the number of dollars per event detected, estimate the detection efficiencies of the two systems and decide which system to support.

3

NUCLEAR BASICS OF NUCLEAR ASTROPHYSICS

3.1 Nuclear Masses, Q-values, Isotopic Abundances, and Nuclear Nomenclature

Although the interior of a star provides the cauldron in which nuclei are synthesized, the properties of the individual nuclei also play a role in determining the abundances of the nuclides that stars produce. The nuclear masses determine nuclear stability, decay lifetimes, and decay modes, and, in addition, their participation in various astrophysical processes. For example, the masses of the light nuclei determine what reactions can occur at the (usually low, at least by nuclear standards) energies at which astrophysical reactions take place as well as the amount of energy given off in each nuclear burning process. This in turn determines, at least at some level, the temperature in the region of the star in which the process operates and hence the structure of the star. As another example, the masses of the heavy neutron-rich nuclides determine the pathway through the nuclei of the periodic table through which the r-process, one of the processes by which heavy nuclei are synthesized, proceeds. This ultimately determines the abundances of the nuclides synthesized in that process.

Many mass formulae have been developed (see, e.g., Haustein 1988; Moller, Nix, and Kratz 1997) to provide a generalized description of the known masses and to extrapolate to unknown masses, but all contain the basic features of nuclei that have been known for many decades. The mass of a nucleus having Z protons and N neutrons, and mass number $A = N + Z$ is given as the sum of the masses of its proton and neutron constituents (in energy units) minus its binding energy, BE. However, mass formulae are usually given as atomic masses, that is,

$$M(Z, N) = ZM_{\rm H} + NM_{\rm n} - {\rm BE}, \quad (3.1.1)$$

where M_H is the mass of the hydrogen atom and M_n is the mass of the neutron. Atomic masses are based on atomic mass units, u, which choose the mass of the ^{12}C atom to be 12.000 u. One u is equal (in energy units) to 931.494 MeV (Tuli 2000). In this system, the mass of the hydrogen atom is 1.007825 u and that of the neutron 1.008665 u. Basically, the nucleus is characterized reasonably well (Segre 1977) as a liquid drop of uniform density. The textbook form of the binding energy BE is then given as a sum of the most basic physical characteristics of nuclei: their number of baryons, surface energy, Coulomb energy, symmetry energy, and pairing energy. Indeed, the simplest form of the nuclear binding energy (see fig. 3.1) is most illuminating as to the properties of the nucleus it represents. In this expression the binding energy BE (defined here as a positive number) is given as

3.1.2 $$\mathrm{BE} = C_1 A - C_2 A^{2/3} - C_3 Z^2/A^{1/3} - C_4(A/2 - Z)^2/A + \delta_{NZ}.$$

The first term represents the fact that the binding energy per nucleon is found to be remarkably constant for stable nuclei throughout most of the periodic table (see fig. 3.1), a manifestation of the (short-range) nuclear force, which is attractive and is responsible for holding nuclei together. One result of this, or at least a correlated fact, is that the nuclear volume increases approximately as A, so that the nuclear radius varies approximately as $r = R_o A^{1/3}$. The second term represents the fact that nucleons at the nuclear surface will not be surrounded by other nucleons that can exert an attractive force on them, so they will have a reduced binding energy from those in the nuclear center. This term is proportional to the size of the nuclear surface, so is proportional to $A^{2/3}$. The third term represents the Coulomb energy of the nucleus, which, since the Coulomb force between the protons is repulsive (and since it is long range, is the reason that nuclear sizes have an ultimate limit), will have the opposite sign as the first term. There is also a nuclear symmetry energy, a result of the Pauli principle, so that the fourth term will also reduce the binding energy per nucleon but will tend to be small when the numbers of protons and neutrons are nearly equal (which, of course, they are not for stable heavy nuclei). Were it not for the Coulomb force, this term would tend to keep the proton and neutron numbers equal throughout the entire periodic table, as that would maximize the binding energy in that situation. The final term is the pairing energy, a result of the fact that nuclei in which the neutrons and protons are paired tend to have increased binding energy over those in which they are not. Thus this term increases the BE when the numbers of protons and neutrons are even.

The above mass formula represents the masses of stable nuclei reasonably

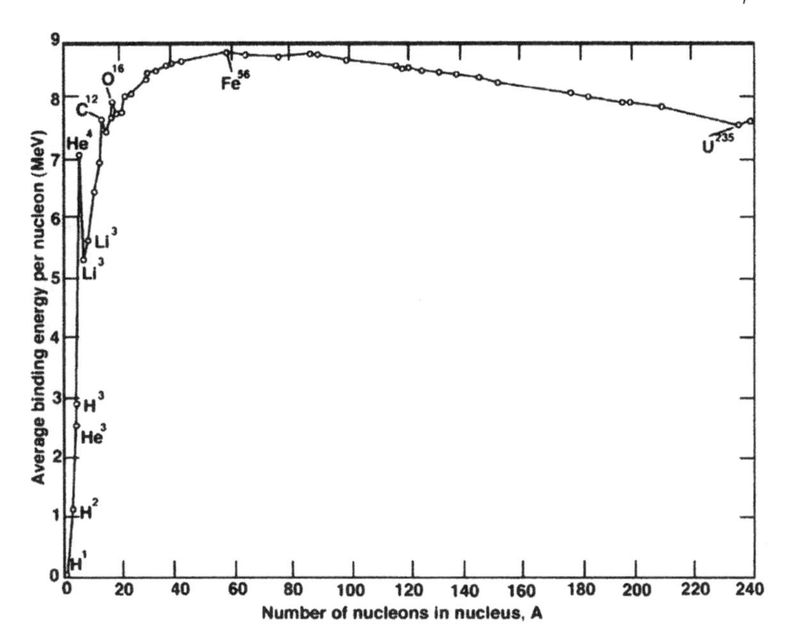

Fig. 3.1. The binding energy per nucleon for stable nuclei. Note that the nuclei with the largest binding energies are in the region of iron. The eventual decrease of the binding energy curve at high-mass numbers tells us that nucleons are more tightly bound when they are assembled into two middle-mass nuclides rather than into a single high-mass nuclide. Thus, energy can be released by the nuclear fission, or splitting, of a single massive nucleus into two smaller fragments. The rise of the binding energy curve at low-mass numbers, on the other hand, tells us that energy will be released if two nuclides of small mass number combine to form a single heavier-mass nuclide via nuclear fusion. From Wikipedia, at http://en.wikipedia.org/wiki/Binding_energy.

well with the constants (Segre 1977): $C_1 = 15.67$ MeV, $C_2 = 17.23$ MeV, $C_3 = 0.697$ MeV, $C_4 = 93.15$ MeV, and $\delta_{NZ} = 0$ if $(N, Z) =$ (even, odd) or (odd, even), $-12A^{-1/2}$ if $(N, Z) =$ (odd, odd), and $+ 12A^{-1/2}$ if $(N, Z) =$ (even, even), where, for example, (even, odd) refers to a nucleus that has an even number of neutrons and an odd number of protons.

Of particular relevance to astrophysics are the gross features of the nuclear masses. Figure 3.1 shows that the binding energy per nucleon increases up to the iron-nickel region, then slowly decreases as A increases further. This results from the fact that the ratio of internal nucleons to surface nucleons increases with A, so the surface energy term decreases in importance as A increases. However, around mass 40, the Coulomb energy begins to overcome the symmetry energy term, and $N > Z$ thereafter for stable nuclei. These two terms continue their nuclear balancing act, with the long range Coulomb force gaining importance compared with the short-range nuclear force as Z increases. Ultimately the Coulomb force wins out, and the heaviest stable nucleus is ^{208}Pb. Of course, there are long-lived nuclei all the way up to the uranium-plutonium region and beyond. The binding energy per nucleon plays a very important role in determining which processes of nucleosynthesis can produce which nuclei. Very briefly, fusion processes can produce new nuclei as long as they are exothermic, that is, up to the iron-nickel nuclei (although, as we will see, other factors, for example, the Coulomb barrier, also influence this). Beyond that mass region, other processes take over, most notably (because of the Coulomb barrier), neutron capture processes.

The binding energy per nucleon also influences the abundances of the nuclides. Since all nuclei from ^{12}C up to the heaviest nuclei are produced in the hot cauldron of a star, those that are less susceptible to destruction in the accompanying high temperature photon bath will be more likely to survive and ultimately to be expelled into the interstellar medium so as to be counted in the abundances of the nuclei. As noted in section 1.1, most significant here is the stability of the "alpha-nuclei," those nuclei that have even numbers of both protons and neutrons, for example, ^{4}He, ^{12}C, ^{16}O, ^{20}Ne, ^{24}Mg, ^{28}Si, ^{32}S, ^{36}Ar, and ^{40}Ca. These effects can be seen in figures 1.1, 1.2, and 3.1, in which these nuclei are seen to be especially abundant. This is predicted by the pairing term, the last one, in the mass formula. Additional effects on the binding energy, for example, shell closures (see next section) will also have important effects.

One also sees the effect of the nuclear pairing force on the heavy nuclides, where the even Z nuclei always have more stable isotopes than the adjacent odd Z nuclei. Another interesting feature is that the neutron-rich heavy nuclei are generally more abundant than the neutron-poor ones. This is a direct consequence of the processes by which those nuclides are synthesized, as will be discussed in subsequent chapters.

Of course, modern nuclear mass formulae take into account many more nuclear effects than those indicated in equation 3.1.2, such as deformation and isospin; the interested reader is referred to Moller, Nix, and Kratz (1997) and Haustein (1994) for more detail on this subject. These effects are important for the stable nuclei, but they become very important in extrapolating to the nuclei through which some of the most extreme processes of nucleosynthesis pass. Many techniques have been used over the past several decades in measuring atomic mass, including magnetic analysis, that is, mass spectrometry. However, the obvious need for masses of nuclei far from stability has motivated several new techniques; these are discussed in detail in

Lunney, Pearson, and Thibault (2003). One technique utilized magnetic analysis, but of beams of nuclei produced at the Isotope Separator Online (ISOLDE) (see, e.g., Kluge 1986; Ravn et al. 1978) at CERN, which had the capability of producing many of the far-from-stability nuclides needed to describe the r-process of nucleosynthesis (see chap. 7). The nuclides at ISOLDE are produced by bombarding targets selected for their capability of providing the rare nuclides of interest with a high-energy proton beam. Chemical selectivity is also used to enhance the beams of the nuclides of interest with respect to isobars because they often have masses so similar that they are difficult to separate with typical magnetic spectrometers. While ISOLDE has produced beams of some very important nuclei, especially neutron-rich nuclei near the $A = 130$ r-process (see chap. 7) progenitor mass peak (e.g., ^{130}Cd and ^{132}Sn), this facility also often requires heroic efforts in developing the techniques of chemical selectivity that are needed to provide the beams of interest and will always have difficulty providing some beams simply because of the chemical selectivity required. An additional limitation results from the time required to stop the energetic proton beam in a target and extract from rest the rare nuclei that are produced, roughly several tenths of a second. Thus, measurement of half-lives as low as 0.1 s is extremely difficult. The mass resolution achievable with ISOLDE, however, is quite good—typically 200 keV/c^2.

Another technique that has been developed is a Penning trap, in which ions of the rare nuclei produced, for example, by bombarding a target with a high-energy proton or heavy-ion beam, are trapped by a complicated set of electric and magnetic fields (Brown and Gabrielse 1986). The trapped ions are driven resonantly at their cyclotron frequency by radiofrequency power (see, e.g., Bollen and Schwarz 2003). The ions in the Penning trap are subjected to an excitation frequency for a prescribed exposure and then are ejected from the trap. The energy gained from the excitation is determined by the ion's time of flight to a microchannel plate. The spectrum of time of flight versus frequency exhibits a minimum at the resonant cyclotron frequency, thereby determining the mass. In the past, the source for such a device was one that produced very low energy ions, for example, an ISOL type source. Thus it suffered from the same limitations as ISOLDE, both in chemical selectivity and ultimate half-life of the ions being considered. However, the "gas stopper" has profoundly affected that limitation (see, e.g., the Web site for the Canadian Penning Trap: http://joseph_vazsg.tripod.com); it can slow high-energy heavy nuclei in a noble gas, for example, helium, to energies that are tractable for the Penning trap. Thus the Penning trap might be able to utilize recoiling ions at some point in the near future. That will greatly reduce both the dependence on chemical selectivity and the half-life limitation. While the ability of the Penning trap approach does depend on the comparison between the half-life and the time the ion is stored in the trap, this technique does permit mass measurements for nuclides with half-lives below 100 ms and for somewhat longer-lived nuclides, with an accuracy well below 100 keV/c^2.

A third technique for mass measurement has been developed using a heavy-ion storage ring at *GSI Darmstadt* (see, e.g., Bosch 2003). In this facility heavy ions, for example, uranium, bombard a target, producing a cocktail of exotic nuclides. These are reaccelerated to a fairly uniform energy and then stored in a storage ring and cooled in a region of high-density electrons. The frequencies of the stored ions are measured as they circulate and are then related to some reference beam allowing mass resolution of about 50 keV/c^2. This facility circumvents the chemical selectivity necessary for the ISOL technique, as the rare ions are produced in high-energy heavy-ion collisions. However, it does not circumvent the half-life limitation of the ISOL-type facilities, as the cooling time of the circulating ions is typically a few seconds, making measurement of half-lives of less than a second difficult. It should be noted, however, that many species of ions can be measured simultaneously, as can be seen in the example shown in figure 2.34.

Despite the effort that has been put into producing beams of these rare nuclides using radioactive nuclear beam (RNB) facilities, the number of nuclides needed to describe nucleosynthesis still vastly exceeds the number that have been studied. Even with the next-generation RNB facilities it will not be possible to make all the nuclei that are necessary to understand the most extreme processes of nucleosynthesis. Thus refinement of the mass formulae is very important to astrophysics.

Reaction *Q*-values are calculated from the difference in masses of the initial and final state particles as, for the reaction $a_i + A_i \rightarrow a_f + A_f$,

3.1.3
$$Q = [M(a_i) + M(A_i) - M(a_f) - M(A_f)]c^2.$$

Since the masses are generally given as rest-mass energies in the tables, their combinations automatically result in energies.

The standard nuclear physics notation for denoting a reaction such as $a_i + A_i \rightarrow a_f + A_f$ is $A_i(a_i, a_f)A_f$. This usually denotes that A_i is the target nucleus, a_i is the incident projectile, and is usually the lighter nucleus or particle in the entrance channel (although this is not the case in inverse kinematics, which is often used in experiments involving short-lived nuclei), a_f is the outgoing projectile that is detected, and A_f is the residual nucleus, which is usually the heavier exit-channel nucleus or particle and is usually not

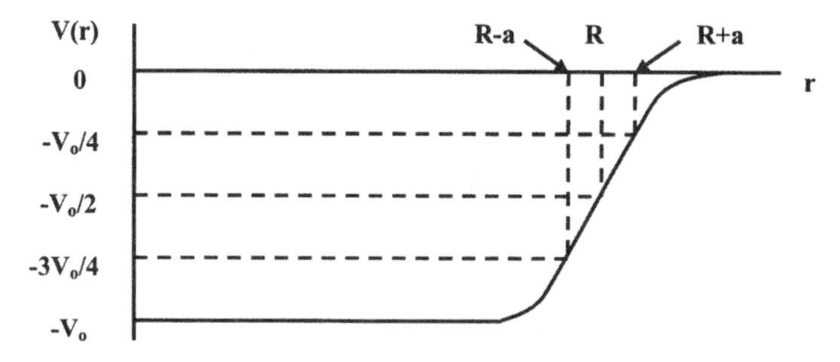

Fig. 3.2. The Woods Saxon potential well. A typical depth for the nuclear potential well seen by a proton or a neutron is 50 MeV.

detected (although it is in some cases). Variations on this notation will also be used, in ways that will either be obvious or defined, below.

3.2 The Nuclear Potential: Scattering and the Shell Model

The protons and neutrons of which nuclei are comprised are in quantum states described by the Schrodinger equation. The potential well for a nucleon in a nucleus reflects the fact that the strong interaction is short range, so that the potential seen by the individual nucleons, which is the average over the interactions with all the nucleons of the nucleus, results in a relatively simple form. Indeed, the nucleons in a large nucleus will only see the nuclear forces from other nearby nucleons; they will be outside the range of nucleons that are more distant. Thus, at least for the nuclei relatively close to stability, the matter distribution is usually well represented by a Woods-Saxon form. This same form is also used to represent the nucleon-nucleus (or α-particle-nucleus) elastic scattering potential, albeit with slightly larger radii to account for the nonzero range of the nucleon-nucleon interaction. The potential is given by

3.2.1 $$V(r) = V_o\{1 + \exp[(r - R)/a]\}^{-1}$$

and is shown in figure 3.2. This potential goes to zero at distances well removed from the nucleus. It has a halfway radius of R, usually taken to be $R = R_o A^{1/3}$ (A is the mass number of the nucleus), where R_o is usually taken to be around 1.2 fm for proton-nucleus interactions and 1.3 fm for α-nucleus interactions. The size of the "diffuseness" a is usually around 0.5–0.8 fm; this characterizes the range over which the nuclear potential goes from large to small. (At the halfway radius minus a it is about 3/4 of its value at $r = 0$, and at the halfway radius plus a it is about 1/4 of that value.) The actual nucleon-nucleus scattering potential is considerably more complex than just one term; it must include, in addition to the basic potential term, terms to characterize absorption (in inelastic scattering and nuclear reactions) and spin-orbit effects. A Coulomb potential must also be added in the case of charged-particle scattering; indeed, it is often the dominant term in reactions at the energies relevant to nuclear astrophysics. Some terms are surface peaked; it is usually convenient to represent them as derivatives of the above form.

The solutions to the Schrodinger equation with this potential, but without the absorption terms to characterize scattering, are the single-particle shell-model states (see, e.g., Segre 1977). These states will have the usual radial component that depends on a "principal quantum number n" and an "orbital angular momentum quantum number ℓ," a component, a Legendre polynomial, $Y_{\ell m}$, that depends on the angular variables θ and φ and quantum numbers ℓ and its "projection m" and a component χ_{sm_s} that depends on the spin variables:

3.2.2 $$\psi_{n\ell m_s}(r, \theta, \varphi) = R_{n\ell}(r) Y_{\ell m}(\theta, \varphi) \chi_{sm_s}.$$

The resulting states are grouped into bands of orbitals, which are separated from other bands by large energy gaps. Since the Schrodinger equation can be separated into a part involving only angular variables and a part involving only the radial variable, each state will be characterized by principle and orbital angular momentum quantum numbers and, since protons and neutrons are spin 1/2 particles, a total angular momentum quantum number, j, that results from the vector addition of the orbital angular momentum and the spin. The lowest lying state is found to be the $1s_{1/2}$ state, followed by a shell containing the $1p_{3/2}$ and $1p_{1/2}$ orbits, followed by another shell containing the $1d_{5/2}$, $2s_{1/2}$, and $1d_{3/2}$ states, followed by another shell containing the $1f_{7/2}$, $2p_{3/2}$, $1f_{5/2}$, and $2p_{1/2}$ states, and so on. In addition to the large energy gaps between the shells, the last shell also is found to have a relatively large energy gap between the $1f_{7/2}$ and the $2p_{3/2}$ orbitals. Standard spectroscopic notation is used here, that is, (s, p, d, f, . . .) refer to values of the orbital angular momentum of (0, 1, 2, 3, . . .) in units of $\hbar$. The leading integer indicates the principal quantum number, and the subscript indicates the total angular momentum. The large energy separations between the shells lead to the nuclear "magic numbers" at neutron or proton numbers of 2, 8, 20, 28, (40), 50, 82, and 126. According to the Pauli principle, each state of total angular momentum j can contain $2j + 1$ protons and $2j + 1$ neutrons, so the lowest mass doubly magic nucleus is that with protons and neutrons that fill the lowest lying ($1s_{1/2}$) orbit: ^{4}He.

Table 3.1 Sequence of Shell-Model Orbitals

$n\ell_j$	Number of N/P	$\Sigma\ N/P$	Nuclei
$1s_{1/2}$	2	2	^{4}He
$1p_{3/2}$	4	6	
$1p_{1/2}$	2	8	^{16}O
$1d_{5/2}$	6	14	
$2s_{1/2}$	2	16	
$1d_{3/2}$	4	20	^{40}Ca
$1f_{7/2}$	8	28	
$2p_{3/2}$	4	32	
$1f_{5/2}$	6	38	
$2p_{1/2}$	2	40	Zr
$1g_{9/2}$	10	50	^{90}Zr, Sn
$1g_{7/2}$	8	58	
$2d_{5/2}$	6	64	
$2d_{3/2}$	4	70	
$3s_{1/2}$	2	66	
$1h_{11/2}$	12	82	^{132}Sn, Pb
$1h_{9/2}$	10	92	
$2f_{7/2}$	8	100	
$2f_{5/2}$	6	110	
$3p_{3/2}$	4	104	
$3p_{1/2}$	2	112	
$1i_{13/2}$	14	126	^{208}Pb
$1i_{11/2}$	12	138	
$2g_{9/2}$	10	148	
$2g_{7/2}$	8	156	

The next lowest mass doubly magic nucleus fills the ($1s_{1/2}$) and ($1p_{3/2}$, $1p_{1/2}$) shells, so it has eight protons and eight neutrons: ^{16}O. The next lowest doubly magic nucleus fills those shells plus the ($1d_{5/2}$, $2s_{1/2}$, $1d_{3/2}$) shell, so it has 20 protons and 20 neutrons: ^{40}Ca. This procedure can be continued, although the energy gaps become more complicated beyond the ones described here, and the stable nuclei begin to deviate from $N = Z$. Table 3.1 indicates the ordering and grouping of the shell model orbitals (see, e.g., Segre 1977; de-Shalit and Talmi 1963).

In nuclei, the state with the larger of the two possible total angular momenta for any total orbital angular momentum always is the more tightly bound. This is the result of the nuclear spin-orbit interaction, which acts in the opposite direction from the atomic spin-orbit force. In some cases, it splits the two total angular momentum states so much that they end up in different shells; thus spin-dependent forces are also much more important to the nucleons in nuclei than they are to electrons in atoms.

The shell model can provide extremely useful information just from the very basic information given here. For example, ^{16}O is a doubly magic nucleus, so it would be expected to have a core of eight protons and eight neutrons that would require a significant amount of energy to excite (i.e., to promote nucleons from the $1p_{3/2}$ or $1p_{1/2}$ orbitals into higher lying 1d or 2s orbitals, or even more highly excited orbitals). The ordering indicated above for the orbitals, even the shells, is approximately correct, so the next orbital above the doubly magic ^{16}O would be $1d_{5/2}$. Thus one might expect the next neutron after those in ^{16}O to fill that orbital, producing a ground state in ^{17}O having a spin of 5/2 and a positive parity, denoted as $J^{\pi} = 5/2^{+}$. Since the next orbital is $s_{1/2}$, the first excited state might be expected to have a spin of 1/2 and a positive parity. Both are correct. Of course, the states are not pure single-particle states; "configuration mixing" will distort the simple picture presented here. Nonetheless, the simplest picture of the shell model can often produce reasonably accurate nuclear physics observations or conclusions.

In regions between the closed shells, nuclei can develop collectivity of many of the nucleons within the nucleus. This will affect the nuclear energetics as well as the shapes of the nuclei involved. One interesting question that has arisen as nuclear astrophysicists have begun to study nuclei far from stability is whether or not the magic numbers that apply to the stable or nearly stable nuclei will apply to very neutron- or proton-rich nuclei; indeed, this appears not to be the case, at least for some of the magic numbers. These are questions that are under intense theoretical and, for lighter nuclei, experimental scrutiny at present, but their answers will be known for sure for heavier nuclei only when nuclear physicists can produce and study these far-from-stability nuclei.

While extensive studies have not been made for heavy nuclei that are far from stability, some do exist for lighter nuclei that will serve to illustrate the point. If one observes the excitation energy of the first excited states as one moves through a series of isotopes or isotones (nuclei with the same neutron number but varying proton number), one can observe the evolution of the gap in binding energy of the orbital being filled. This has been done for several isotonic sequences, some of which appear to exhibit "shell quenching," that is, the apparent disappearance of the closed shells that are well known for stable nuclides. Several suggestions have been offered for the cause of the effect. In

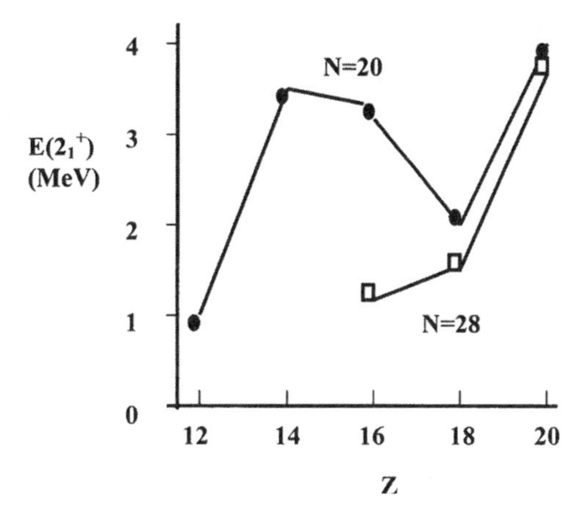

Fig. 3.3. Excitation energies of the first excited 2+ states the $N = 20$ and $N = 28$ isotonic sequences. Results are from Cottle and Kemper (1998).

the case of the $N = 20$ shell closure, "intruder states," that is, the $f_{7/2}$ orbital that would normally lie far above the highest filled orbital, the $d_{3/2}$ orbital, dip in energy to a value that is much closer to that of the $d_{3/2}$ orbital (see, e.g., Werner et al. 1996). Regardless of the explanation, the result is unmistakable and can be observed in a variety of signatures of the shell effects that have been studied. These include energies of excited states (discussed further below; see Cottle and Kemper 1998), inelastic scattering excitation strengths of excited states (Cottle and Kemper 1998; Chiste et al. 2001), and masses (Sarazin et al., 2000), among others.

Perhaps the simplest result to study in this context is the excitation energies of excited states in the isotonic sequences. In nuclei in which shell closures are well verified, excitations of nucleons from a filled shell to the next vacant shell would require appreciably more energy than those for which a closed shell does not exist. The excitation energies of the lowest lying $J^\pi = 2^+$ states (for the ground states, $J^\pi = 0^+$, as is the case for all even-even ground states in the periodic table) for the nuclides in the $N = 20$ and $N = 28$ isotonic sequences can be seen in figure 3.3 as a function of the proton number Z of the nuclides. For the $N = 20$ isotones it can be seen that this energy is large for $Z \geq 14$ (Si) but drops to <1MeV for ^{32}Mg, a value that is typical of non-closed-shell nuclei. The $N = 28$ shell closure is also obvious for ^{48}Ca but is much less so for ^{46}Ar and lighter $N = 28$ nuclei. Note that most of the nuclei represented in figure 3.4 are short-lived radioactive nuclides. Thus nearly all these studies of these effects have required the use of RNBs.

Nuclear transfer reactions with incident projectile energies of tens of MeV have been used for decades to determine spectroscopic information about nuclear states. For example, consider a reaction in which an ^{16}O target is bombarded with a deuteron, an ^{2}H nucleus, which deposits its neutron with the target nucleus producing ^{17}O. The outgoing proton is then observed; energy conservation requires that the energy of each proton will bear a direct correspondence to the state in which it leaves the residual ^{17}O nucleus. Since this single-particle stripping reaction will be a very simple one if it can populate states in the residual nucleus by having the deuteron simply drop off into an available orbit, one would expect to see the single-particle states strongly populated, while more complicated states, perhaps involving excitations of the ^{16}O core to accommodate the configuration mixing that might exist in the wave function of that state, would be weakly populated. This is indeed the case; the ground and first excited states of ^{17}O, which appear to look, respectively, like a single neutron in a $d_{5/2}$ and $s_{1/2}$ orbital outside an ^{16}O core, are strongly populated, whereas many of the states at higher excitation are either populated weakly or not observed at all in this reaction. However, fragmentation of the single-particle strength, for example, the $1d_{5/2}$ strength, is also observed; it is divided up among the states having $J^{\pi} = 5/2^{+}$. Nuclear physicists characterize the strength of the transition to each of these states by a spectroscopic factor, S which is proportional to the probability of the single-particle component in each state's wave function. Specifically, if the center-of-mass energy is sufficiently high that the reaction is a direct reaction (see sec. 3.8) and the reaction mechanism is a simple one-step process involving the transfer of a single nucleon from a light incident projectile to the target nucleus, one can then write

$$(d\sigma/d\Omega)_{\mathrm{exp}} = N\{(2J_f+1)/[2J_i+1)(2j+1)]\}C^2 S(d\sigma/d\Omega)_{\mathrm{th}}, \quad (3.2.3)$$

where N is an overall normalization factor that permits use of a simplifying assumption (the zero range approximation) in the reaction code that is usually used for such calculations and that produces the theoretical prediction of the cross section $(d\sigma/d\Omega)_{\mathrm{th}}$, J_i and J_f are the initial and final state spins, and j is the transferred total angular momentum. The quantity S is the single-particle spectroscopic factor, and C is an isospin Clebsch-Gordan coefficient (see the discussion of isospin and Clebsch-Gordan coefficients in the Appendix, and see Schiffer 1971 for an isospin-specific discussion), which depends on the particular reaction being studied. The factor $C = (t_i t_{3i} T_i T_{3i}; T T_3)$, where t_i and T_i are the isospin of the incident projectile and target, t_{3i} and T_{3i} are their projections, and T and T_3 are the total isospin and projection. This is given for completeness; the formalism associated with direct reactions is a separate topic in itself and will not be dealt with in this book.

Note that it would be difficult to populate a low-lying negative parity state in this reaction. The lowest lying unfilled negative-parity single-particle orbital is the $1f_{7/2}$ orbital, which occurs in the next highest lying shell. One could consider promoting particles from the 1p orbits into 1d or 2s orbits, which would produce a negative parity state. However, such a reaction would be complicated, as it would involve both transfer of the neutron from the incident deuteron and excitation of core nucleons. Although such processes do occur, they do not usually lead to strongly populated states in simple transfer reactions.

Suppose instead that one performed a single-particle pickup reaction in which an ^{18}O nucleus was bombarded with a deuteron, it picked up one of the neutrons in the target nucleus, and a triton, an ^{3}H nucleus, was observed. To zeroth order, the two neutrons in the ^{18}O ground state that are outside the ^{16}O closed shells would be in the $1d_{5/2}$ orbit, so removal of one of those neutrons would produce the ^{17}O ground state, which should be strongly excited. The first excited state, containing a $2s_{1/2}$ neutron coupled to an ^{16}O core, would be difficult to excite in this reaction, so this simple picture would not predict it to be strongly populated. However, any ^{17}O states that had strong components in their wave functions of two neutrons in the $1d_{5/2}$ orbit and a neutron vacancy—hole—in the $1p_{1/2}$ or $1p_{3/2}$ orbits could be strongly populated in this reaction. These would be negative parity states; they would be weakly populated in the stripping reaction described above, as any such reaction would have to both pick up a neutron and inelastically excite the core.

More than one particle can also be transferred in reactions; such studies tell us about other types of configurations in the wave functions of the states they populate, for example, effects of clustering into substructures such as α-particles (^{4}He nuclei). The combined information from decades of nuclear structure studies, however, has given us an excellent picture of the structure of many nuclides. Nuclear astrophysicists have used the information derived from these spectroscopic studies, specifically, the energies and spectroscopic strengths of the states observed, to calculate the properties of resonances that influence the rates at which nuclear reactions occur in stars. This is especially valuable in situations in which it is not possible to measure the cross sections for the reactions of interest directly. This will be discussed further in section 3.8.

Other types of reactions have become important to astrophysics. One such class of reactions is Coulomb breakup reactions. These are especially important to experiments involving short-lived nuclei; in all such reaction studies, the nucleus that would normally be the target nucleus but that is too short-lived to be made into a target becomes a beam, and the usual incident projectile becomes the target. However, in many cases beams of these nuclei are still too weak to permit study of the reaction of interest; this is especially true of studies of proton or neutron radiative capture reactions, which are of frequent interest to nuclear astrophysics. In some of these cases, beams of the nucleus that will be the reaction product—the residual nucleus—might be capable of being produced, permitting study of the inverse reaction. In such cases, that beam may be brought to bear on some heavy nuclear target, for example, lead, and the reaction can be studied in which the beam nuclei are broken up by the intense photon field that results from the beam nuclei interacting with the lead nucleus through the Coulomb force (Baur and Rebel 1996). Although this would appear to represent the inverse of the reaction that is desired, and thus would provide definitive information, technically it really only can be used to study the proton capture that goes from the ground state of the target nucleus to the ground state of the residual nucleus (the excited states are different in the forward and inverse reactions). The beauty of the Coulomb breakup technique is that the photon field produced by the interaction of a nucleus with a heavy nucleus is so intense that the probability of the reaction occurring is greatly enhanced. This is an obvious advantage when the intensity of the RNB is low. A classic case of this type of experiment is the use of Coulomb breakup of ^{14}O to ^{13}N + ^{1}H to infer the cross section for ^{13}N(p, γ)^{14}O (Motobayashi et al. 1991; Decrock et al. 1991; Kiener et al. 1993). In this particular case, the nonexistence of proton-bound excited states of ^{14}O greatly simplifies the equivalence relationship between the two reactions. But even in more complex situations, much valuable information has been obtained from Coulomb breakup reactions; we will return to this type of experiment in subsequent chapters.

Another class of reaction studies that has been used to provide information for nuclear astrophysics is the asymptotic normalization coefficient (ANC) approach (Mukhamedzhanov and Timofeyuk 1990). In the ANC approach, low-energy direct-transfer reactions are used to measure the long-range tails of the distributions of the transferred particles, which are the quantities that one needs to calculate the cross section of interest if the reactions are at very low energies. Since these studies involve direct reactions, the ANC approach actually measures the direct component of the reaction of interest but not the resonant component (see secs. 3.7 and 3.8.). Since the resonant components often dominate the yield, this might seem to provide information of secondary importance. However, the direct component of capture reactions is often important and can be extremely difficult to measure at low energies. Thus the ANC method can provide very important information for nuclear astrophysics.

Perhaps the best way to illustrate the ANC technique is by an example. One reaction that is of great importance to the detection of solar neutrinos, and to understanding the standard solar model (see chap. 5), is the $^7Be(p, \gamma)^8B$ reaction. It is dominated at the energies at which it occurs in the Sun by the direct reaction mechanism, so the ANC approach can provide precisely the information required. In calculating the cross section for this reaction at very low energies, one needs to know the tail of the distribution of the captured proton (this book will not deal with this, but see Austern 1970 or Satchler 1983 for details of direct nuclear transfer reactions). But this is precisely what is measured in transfer reactions between interacting low-energy heavy ions. In this study (Azhari et al. 1999), the reaction studied was $^{10}B(^7Be,^8B)^9Be$ at an energy of 12 MeV per nucleon. In this case, a proton is transferred from the target nucleus (^{10}B) to the beam nucleus (^7Be), creating the nucleus (^8B) that is of interest. The magnitude of the cross section is determined by the amplitude of the wave function of the transferred proton. However, that is the same amplitude to which the $^7Be(p, \gamma)^8B$ reaction is sensitive; thus this transfer reaction allows calculation of the cross section for this astrophysically important reaction.

The results from ANC measurements are in good agreement with those from direct measurements of the proton capture cross sections in cases in which that comparison can be made. This certainly gives us confidence in the ANC method of determining the direct reaction component of cross sections of importance to nuclear astrophysics.

3.3 Nuclear Collective Models

Although many models have been devised to describe nuclear phenomena, two stand out as having withstood the tests of time and are often used in describing nuclear phenomena. These are the vibrational and the rotational nuclear models. The first describes the excitation energies of the nuclear levels in terms of a vibrating system; in this case one would expect that the energies would, by analogy with atomic levels, grow as $(\text{spin})^2$. This is observed to a reasonable approximation in many nuclei. The other extreme is the rotational nuclear model; in this case the nucleus is treated as a rigid rotor, and the energies of successive states are given by spin(spin + 1), as would be expected for a rotating quantum mechanical system. Such systems are also observed in many nuclei, especially those far from closed shells.

Both models are the result of collective effects of the nucleons in the nuclei involved. This collectivity can have important effects not only in producing the observed energy spectra, but also in enhancing, for example, inelastic scattering for nuclei that have vibrational model structures, in many cases by 1–2 orders of magnitude over what would be expected from the single-particle shell model. Thus clearly these collective models do describe phenomena that are simply not contained in the simple shell model. Of course, modern shell-model codes can include the many quantum mechanical configurations needed to describe the collectivity that drives the collective models.

In the vibrational model in an even proton number–even neutron number nucleus that is well between closed shells, the ground state will have spin-parity of $J^\pi = 0^+$. The low-lying excited states can be formed by exciting particles from the lowest lying orbitals, creating "holes," to higher lying orbitals. These two orbitals will usually have the same parity, so the low-lying excited states will have positive parity (coupling a positive parity particle to a positive parity hole or a negative parity particle to a negative parity hole will both produce a positive parity level). Notable exceptions to these are observed in the doubly magic nuclei such as ^{16}O, ^{40}Ca, and ^{208}Pb. In these nuclei, the ground states will still be $J^\pi = 0^+$, but simple particle hole excitations have to change major shells, which will produce particles and holes of opposite parity, so will necessarily result in first-excited states having negative parity, for example, 3^-. The minimum energy configurations in ^{16}O (see the above Table 3.1 of the ordering of shell-model orbitals) will have protons or neutrons promoted from $1p_{1/2}$ orbits to $1d_{5/2}$ orbits; coupling each particle-hole pair thus created will create a negative parity wave function. In ^{40}Ca, the excitations will result from promoting $1d_{3/2}$ protons and neutrons to $1f_{7/2}$ orbits, again creating a negative parity wave function. The situation is more complicated for ^{208}Pb, even aside from the fact that the protons and neutrons are in different orbitals, so it cannot be so readily guessed by looking at the ordering of the shell-model orbitals.

One useful model for characterizing wave functions in many nuclei is the particle-core coupling model, a hydrid of the vibrational model and the shell model. Since we will refer to this model in subsequent sections to illustrate some important points in nuclear astrophysics, we will discuss it in some detail. This model has been especially useful in characterizing states in nuclei that are not more than a few nucleons away from a core nucleus that is well characterized by strong vibrational states, especially those near closed shells. An example is provided by ^{41}Ca, in which the low-lying states are built on the collective states of ^{40}Ca ($J^\pi = 0^+, 3^-$, and 5^-) coupled to neutrons in $f_{7/2}$ or $p_{3/2}$

orbitals. As noted above, the ^{40}Ca collective states are formed by promoting particles from the $d_{3/2}$ orbital, the last one filled in the ^{40}Ca ground state, to the $f_{7/2}$ orbital, the lowest lying unfilled orbital in ^{40}Ca. Confirmation of the veracity of this model, and a striking observation of the effects of the Pauli principle, are provided by ^{40}Ca(d,p)^{41}Ca and ^{41}Ca(α,α′)^{41}Ca reaction data (α denotes a ^{4}He nucleus, d an ^{2}H nucleus, and p an ^{1}H nucleus; these notations will be used interchangeably). The spins of the states observed in the low-lying spectrum go up to $J^{\pi} = 17/2^{+}$, which suggests that the two highest spin states of the sequence, the $15/2^{+}$ and $17/2^{+}$ states, can be represented formally by coupling a ^{40}Ca(5^{-}) core to an $f_{7/2}$ neutron, that is,

3.3.1
$$|J^{\pi}\rangle = \sum U(3\ 7/2\ 5\ 5; 8\ J)|n(f_{7/2})\rangle|^{40}\mathrm{Ca}(5^{-})\rangle.$$

Here the "*U* coefficients" $U(3\ 7/2\ 5\ 5;\ 8\ J)$ perform the required angular momentum coupling (see, e.g., Edmonds 1963 or deShalit and Talmi 1963 for definitions and discussion of these coupling coefficients). For these coefficients to be nonzero, the orbital angular momenta of the $f_{7/2}$ neutron, 3, and of the ^{40}Ca(5^{-}) core state, 5, must add vectorially to that of the 15/2 or 17/2 states, $L = 8$. In addition, the total angular momenta of the $f_{7/2}$ neutron, 7/2, and of the ^{40}Ca(5^{-}) state, 5, must also add vectorially to that of either of the two ^{41}Ca states, either $J = 15/2$ or 17/2. One can represent these wave functions pictorially, as indicated below. In this pictograph, the $ket|n(f_{7/2})\rangle$ represents the $f_{7/2}$ orbit neutron that is coupled to the ^{40}Ca 3^{-} and 5^{-} states. The two components in the brackets represent, above, the particle (x)–hole (o) excitation in the core neutrons (*N*), and below the particle-hole excitation in the core protons (*P*). Consideration of the structure of the core becomes very important to understanding the results of these reactions, as is discussed in the next paragraph.

3.3.2
$$|15/2^{+}\rangle \text{ or } |17/2^{+}\rangle = U(3\ 7/2\ 5\ 5; 8J)|n(f_{7/2})\rangle \begin{pmatrix} f_{7/2}\text{-x-o-o-o-o-o-o-o-} & \text{-o-o-o-o-o-o-o-o-} \\ d_{3/2}\text{-o-x-x-x-} & \text{-x-x-x-x-} \\ N & P \end{pmatrix}$$
$$+\, U(3\ 7/2\ 5\ 5; 8J)|n(f_{7/2})\rangle \begin{pmatrix} \text{-o-o-o-o-o-o-o-o-} & \text{-x-o-o-o-o-o-o-o-} \\ \text{-x-x-x-x-} & \text{-o-x-x-x-} \end{pmatrix}$$

In the above pictographs, the particle-hole excitation in the core is represented by a particle in the $f_{7/2}$ orbital and a hole in the otherwise filled $d_{3/2}$ orbital.

A remarkable feature appears in the ^{41}Ca(α, α′) results; the strength of excitation of the $17/2^{+}$ state is about half what would be expected from the strength of excitation of the ^{40}Ca(5^{-}) state. However, the shell-model orbitals that would be involved in creating the 5^{-} state must result from promotion of the protons from the last orbital in ^{40}Ca to be filled, which are $d_{3/2}$, to those that are just unfilled, which are $f_{7/2}$, and similarly for the neutrons. In order to form the wave function for the $17/2^+$ state the $f_{7/2}$ nucleons that form the 5^- state must be aligned with the extra $f_{7/2}$ neutron. This is not a problem for the protons, but doing so for the neutrons would violate the Pauli principle; the first term in the pictorial wave function is therefore prohibited. This produces the observed reduction in strength of the inelastic excitation (Goode and Boyd 1976). This simple picture is confirmed by the ^{39}K(α, d)^{41}Ca reaction, where the $17/2^+$ state in ^{41}Ca is much more strongly excited than would be expected from the simple particle-core coupling model, in which the Pauli principle is not taken into account (Goode and Boyd 1976). This is because the entire $17/2^+$ wave function looks much like a deuteron with both nucleons in the $f_{7/2}$ orbits, coupled to the ^{39}K ground state; it is therefore much more sensitive to this deuteron transfer reaction than would predicted by the simple particle-core coupling model.

3.4 Nuclear Electromagnetic Transitions

There will be numerous situations in nuclear astrophysics in which it will be important to understand the origin and nomenclature for electromagnetic transitions in nuclei. In dealing with electromagnetic transitions in nuclei, as in atoms, the transitions will be characterized by the multipolarity λ of the transition operator and the initial and final state wave functions. From the theory of electromagnetism, the electric and magnetic multipole operators are (Preston 1962)

$$Q_{\lambda\mu} = \sum_{i}\left[e_i r_i^{\lambda} Y^*_{\lambda\mu}(\Omega_i) - i g_{si}\mu_o k(\lambda+1)^{-1}\bar{\sigma}_i \times \bar{r}_i \cdot \nabla(r^{\lambda}Y^*_{\lambda\mu})_i\right], \tag{3.4.1}$$

where the second term is usually dropped (as will be done here), and

$$M_{\lambda\mu} = \mu_o \sum_{i}[g_{si}\bar{s}_i + 2g_{li}\bar{\ell}_i/(\lambda+1)] \cdot \nabla(r^{\lambda}Y^*_{\lambda\mu})_i, \tag{3.4.2}$$

where μ_o is the nuclear magneton $= e\hbar/2m_p c$ (m_p is the mass of the proton), the gs are the gyromagnetic ratios, the σ's are the Pauli spin matrices, s is the nucleon spin, and the sums are over the i nucleons of the nucleus.

From time-dependent perturbation theory, the transition rate will be given by

$$R^{EM}_{if} = (2\pi/\hbar)|\langle f|H_{int}|i\rangle|^2\rho_E, \tag{3.4.3}$$

where ρ_E is the number of final states per unit energy interval. The initial state will be the wave function of the nuclear excited state, and the final state will be the product of the lower excitation state and the outgoing photon having wave number k. The outgoing waves can be represented by Bessel functions of order λ, and, in the special case in which $kR_{\text{nuc}} \ll 1$ (which is usually the case),

$$j_\lambda(kr) \approx (kr)^\lambda/(2\lambda + 1)!!, \tag{3.4.4}$$

where $(2\lambda + 1)!! = 1{\cdot}3{\cdot}5{\cdot}\ldots\ (2\lambda + 1)$. The selection rules are readily determined by the matrix elements in equation 3.4.3, as the wave functions that represent the initial and final states will have a definite parity, that is,

$$\psi(\bar{r}) = \pm\psi(-\bar{r}) \tag{3.4.5}$$

with this parity being established by the spherical harmonic in the wave function. For example, it is $(-1)^\lambda$ for ψ_i, if the spherical harmonic is of order λ. Thus, for a dipole transition, the transition amplitude will involve an integral

$$\int Y_{\lambda_f\mu_f}(\theta_f, \phi_f)\, Y_{1(1,0,-1)}\, Y_{\lambda_i\mu_i}(\theta_i, \phi_i)\, d\Omega \tag{3.4.6}$$

where the subscript (1, 0, −1) refers to the three possible projections of the dipole operator. Assuming integer spins, this integral will vanish unless

$$\mu_f - \mu_i = 0, \pm 1, \tag{3.4.7}$$

and

$$\lambda_f = \lambda_i \pm 1. \tag{3.4.8}$$

For the more general case, the parity of the electric multipole operator of multipolarity λ is $(-1)^\lambda$, so, for noninteger spins, a similar argument can be developed to show that for electric multipole transitions to occur at all, the parities of the initial and final states are related by

$$\pi_i \pi_f = (-1)^\lambda. \tag{3.4.9}$$

Thus, for an electric dipole transition, the initial and final states must have opposite parity, for a electric quadrupole transition they must have the same parity, and so on.

The parity of the magnetic multipole operator of multipolarity λ is $(-1)^{\lambda+1}$. Thus for magnetic multipole transitions to occur the parities of the initial and final states are related by

$$\pi_i \pi_f = (-1)^{\lambda+1} \tag{3.4.10}$$

Angular momentum coupling is obviously an important feature of electromagnetic transitions; this is summarized in the Wigner-Eckart theorem, which couples the initial state and final state total angular momentum with the multipolarity of the transition resulting from the general transition operator $T_{\lambda\mu}$, that is,

3.4.11 $$\langle J_f M_f | T_{\lambda\mu} | J_i M_i \rangle = (J_i \lambda M_i \mu | J_f M_f)(J_f || T_\lambda || J_i),$$

where the first term on the right-hand side is just a Clebsch-Gordan coefficient (see the appendix), and the second term is "the reduced matrix element of T_λ" and is independent of the orientations M_i, M_f, and μ. (see, e.g., deShalit and Talmi 1963 or Edmonds 1963) The selection rules for the total angular momenta of the initial and final state wave functions are contained explicitly in the Clebsch-Gordan coefficient, so that

3.4.12 $$|J_f - J_i| \leq \lambda \leq J_f + J_i.$$

This applies to either electric or magnetic transitions of order λ and is certainly consistent with what would be inferred from simple vector addition.

The transition rate for emission of a photon with wave number $k = E/\hbar c$ with an angular momentum characterized by $\lambda\mu$ and of either the electric or magnetic type can be written, after all the normalizations are done properly (Preston 1962), as

3.4.13 $$R_{if}(\sigma\lambda\mu) = [8\pi(\lambda + 1)/\lambda(2\lambda + 1)!!][k^{2\lambda+1}/\hbar]|\langle f | O_{\lambda\mu} | i \rangle|^2,$$

replacing the transition operator of equation 3.4.11 by $O_{\lambda\mu}$, which stands for either $Q_{\lambda\mu}$ or $M_{\lambda\mu}$, (see eqs. 3.4.1 and 3.4.2); furthermore, the superscript *EM* (see eq. 3.4.3) has been replaced by the $(\sigma\ \lambda\ \mu)$ notation. The term σ is either *E* or *M*; it designates whether the expression is for an electric or magnetic transition. Usually, the initial state magnetic substates M_i are averaged over, and those of the final states M_f are summed over, resulting in the reduced transition probability $R_{if}(\sigma\lambda)$

3.4.14 $$B(\sigma\lambda, J_i \rightarrow J_f) = (2J_i + 1)^{-1} \sum_{M_i M_f} |\langle f | O_{\lambda\mu} | i \rangle|^2,$$

3.4.15 $$R_{if}(\sigma\lambda) = [8\pi(\lambda + 1)/\lambda(2\lambda + 1)!!](k^{2\lambda+1}/\hbar)\, B(\sigma\lambda\, J_i \rightarrow J_f).$$

The actual values for the electromagnetic transition rates will depend on the specific form of the nuclear wave function. For example, nuclei having wave functions that are well characterized by single-particle shell-model states, as one might expect to be the case for nuclei just one nucleon above or below a closed shell, will have very different electric and magnetic multipole transition rates from those of nuclei having many nucleons outside of a closed shell. However, if it is assumed that just one nucleon undergoes a change of state, the integrals can be evaluated for the general case. The relevant expressions are given in Segre (1977). For an electric multipole transition rate of multipolarity λ, the expression is

3.4.16 $$R(E\lambda) = (e^2/\hbar c)[(\lambda+1)/\lambda]\omega/[(2\lambda+1)!!]^2(kR)^{2\lambda}(2J_f+1)S|I(E\lambda)|^2,$$

where

3.4.17 $$I(E\lambda) = \int_0^\infty R_i(r)(r/R)^\lambda R_f(r)^* r^2 dr.$$

In these expressions, S is an angular momentum coupling factor that is of order unity but obviously must be nonzero for the transition to occur. The asterisk denotes the complex conjugate and $\omega =$ (transition energy)/$\hbar$. The integrals are over the nucleus, so the variable r is of the order of the nuclear radius.

For a magnetic multipole transition rate of multipolarity λ, with just one nucleon undergoing a change of state, the expression is

3.4.18 $$R(M\lambda) = (e^2/\hbar c)[(\lambda+1)/\lambda]\omega/[(2\lambda+1)!!]^2(kR)^{2\lambda}(2J_f+1)S(\hbar/m_p cR)^2|I(M\lambda)|^2,$$

where

3.4.19 $$I(M\lambda) = [\mu_p\lambda - \lambda/(\lambda+1)]\int_0^\infty R_i(r)(r/R)^{\lambda-1}R_f(r)^* r^2 dr.$$

where μ_p is the magnetic moment of the proton = 2.79 nuclear magnetons, one nuclear magneton $= e\hbar/2m_p c$, and m_p is the mass of the proton.

If the angular momentum factors S in equations 3.4.16 and 3.4.18 are set to unity and assuming that the nuclear radius is given by $R = 1.4A^{1/3}$ fm, rough values for the above transition rates can be obtained (see Segre 1977). The resulting formulae are

3.4.20

$$R(E\lambda) = 4(\lambda+1)/\{\lambda[(2\lambda+1)!!]^2\}[3/(3+\lambda)]^2(E_\lambda/140)^{2\lambda+1}A^{2\lambda/3}m_p c^2/\hbar,$$

3.4.21

$$R(M\lambda) = 0.088(\lambda+1)/\{\lambda[(2\lambda+1)!!]^2\}[3/(2+\lambda)]^2(E_\lambda/140)^{2\lambda+1}A^{(2\lambda-2)/3}[\mu_p\lambda/2 - \lambda/(\lambda+1)^2]m_p c^2/\hbar.$$

It is possible, using these expressions (and following Preston 1962), to obtain a rough estimate of the value of the matrix elements in equation 3.4.14 by setting the spherical harmonics to one. Then

3.4.22 $$\langle f|Q_{\lambda\mu}|i\rangle \approx Ze R^\lambda,$$

Table 3.2 Properties of Electromagnetic Transitions

Type of transition	Parity Change?	Energy dependence
E1	Yes	E^3
E2	No	E^5
M1	No	E^3
E3	Yes	E^7
M2	Yes	E^5

3.4.23
$$\langle f | M_{\lambda\mu} | i \rangle \approx A(e\hbar/2m_p c)\, R^{\lambda-1}.$$

Although this approximation is somewhat crude, as all the factors involving the multipolarity have been set to unity, it does allow some useful observations from these expressions. The ratio of the magnetic and electric transitions of the same multipolarity is $(\hbar/m_p c)/R$, which is typically of order 0.1. When squared, as required in equations 3.4.16 and 3.4.18, this ratio becomes roughly 0.01. Thus electric transitions will generally dominate over magnetic transitions of the same multipolarity. However, the parities of electric and magnetic transitions of the same order are opposite, so these would not compete anyway. Of greater interest is the relative magnitude of multipoles differing in order by unity. These will be transitions of the same parity, that is, they will be able to connect the same states, so they can compete in transitions in which they can both satisfy the angular momentum requirements. Thus, for example, an E2 and an M1 transition, both of which can cause a transition from a $3/2^+$ level to a $1/2^+$ level, can be shown to have a ratio that is roughly unity, when it is taken into account that large enhancements to E2 transitions occur as a result of the collectivity that often occurs in states in nuclei in between the closed shells. In general, an electric transition of multipolarity λ will compete with a magnetic transition of multipolarity $\lambda - 1$ if the total angular momenta of the initial and final states are such that both may participate in the decay.

It is useful to end this section with a summary of some of the dependences of the different transition rates on the energy and parity of the initial and final nuclear states. These are given in table 3.2 above.

We will return to the application of electromagnetic transitions in section 3.8.

3.5 β-Decay

β-decays are crucial to many of the processes of nuclear astrophysics. For example, for stars in their initial burning stage, hydrogen burning, four protons are fused through a complex set of reactions to make a ^{4}He nucleus. Obviously, this requires two protons to be converted to two neutrons through β-decays. In addition, the heavy nuclides are synthesized by successive neutron captures, so at some stage neutrons must have converted to protons to produce stable heavy nuclides. We will return to these and other processes in subsequent chapters.

The energetics of β-decay are indicated for one system in figure 3.4, which shows the several β-decay possibilities associated with mass 14 u nuclei. A neutron-rich nucleus can β-decay via the process

3.5.1 $$(N, Z) \rightarrow (N-1, Z+1) + e^- + \bar{\nu}_e,$$

assuming $M(N, Z) > M(N-1, Z+1)$. Here $\bar{\nu}_e$ denotes an electron antineutrino. Proton-rich nuclei can, in principle, undergo either positron emission or electron capture, that is,

3.5.2 $$(N, Z) \rightarrow (N+1, Z-1) + e^+ + \nu_e,$$

if

$$M(N, Z) > M(N+1, Z-1) + 2m_e c^2,$$

and

3.5.3 $$(N, Z) + e^- \rightarrow (N+1, Z-1) + \nu_e,$$

if $M(N, Z) > M(N+1, Z-1)$. However, assuming atomic masses are used, the threshold for positron emission exceeds that for electron capture by twice the rest-mass energy of the electron, 1.022 MeV. Thus many nuclei that are forbidden to decay by positron emission can decay by electron capture, However, the energy dependence of the two processes usually dictates that, once the threshold for positron emission is exceeded by a fraction of an MeV that process will dominate over electron capture.

Some of the details of β-decay are worth considering, as they involve some of the basic information with which nuclear astrophysicists must deal. In what follows we will refer to electrons, but the derivation could apply equally well to positron decay. We will develop an approach that utilizes time-dependent perturbation theory rather than the more elegant approach that utilizes Dirac spinors; the results are equivalent. The initial state is that of the mother nucleus ψ_i and the final state a product of the wave functions of the daughter nucleus ψ_f, the electron ψ_e, and the antineutrino $\psi_{\bar{\nu}}$. Using time-dependent perturbation theory to calculate the rate λ at which β-decay occurs then requires evaluation of the expression

3.5.4 $$d\lambda(E) = (2\pi/\hbar)|H_{fi}|^2 dN/dE,$$

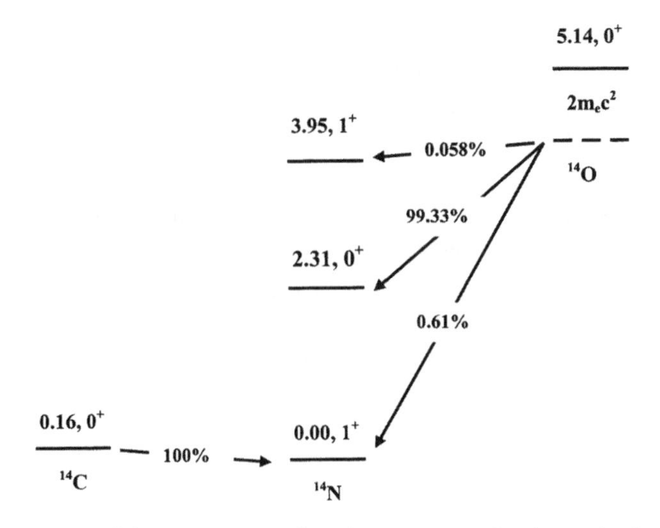

Fig. 3.4. Energetics of electron or positron decay for mass 14 u nuclei. The Q-value for positron decay of ^{14}O is sufficiently large that it can go to three states in ^{14}N. Note also that the energy available for positron decay is reduced by twice the rest-mass energy of the electron from the atomic masses. All energies are given in MeV.

where

3.5.5 $$H_{fi} = \langle \psi_f \psi_e \psi_{\bar{\nu}} | H | \psi_i \rangle,$$

where H is the operator that governs the β-decay process, and dN/dE is the density of final states. This expression gives the rate for β-decay to a state having specific electron and antineutrino energy; the actual transition rate involves an integral over all possible combinations of energy sharing of the two light particles, the electron and the antineutrino, emitted in this process. The recoiling daughter nucleus will have a small amount of energy; this is usually not important to β-decay considerations.

The factor dN/dE is given by the usual phase space expressions, as can be determined from the standard normalization-in-a-box formalism. For a single particle in the box, the density of states dN can be written

3.5.6 $$dN = 4\pi n(ndn),$$

where $n_x = p_x L/h =$ the quantum number along the x-axis, p_x is the momentum along that axis, and $n = (n_x^2 + n_y^2 + n_z^2)^{1/2}$. In the present case, however, such factors are required for both the antineutrino and the electron. In addition, their energies are not independent, since they share the total kinetic energy available. Thus

3.5.7 $$dN = (4\pi)^2(L/h)^6 p_e^2 dp_e p_{\bar{\nu}}^2 dp_{\bar{\nu}} \delta(K_{\bar{\nu}} + K_e - K_T),$$

where the *K*s represent the kinetic energies of the electron and antineutrino, K_T is the total kinetic energy available for them to share, the *p*s are the corresponding momenta, and L is the dimension of the box over which the normalization is performed. Because of the δ-function, one can integrate over $dp_{\bar{\nu}}$. The energies can be written in terms of the total energy of the electron W expressed in units of the rest-mass energy of the electron,

3.5.8 $$W = (K_e + mc^2)/mc^2.$$

W_o is the maximum value of W, that is, it is the sum of the end-point energy of the electron distribution and the rest-mass energy of the electron (assuming the mass of the antineutrino is negligible). This gives

3.5.9 $$dN = (4\pi)^2(L/h)^6 m^5 c^4 W(W^2 - 1)^{1/2}(W_o - W)^2 dW.$$

3.5.10 $$= (4\pi)^2(L/h)^6 m^5 c^4 g(W) dW,$$

which defines $g(W)$. We ultimately wish to determine the total β-decay rate over all possible electron energies. If we ignore the effect of the Coulomb force on the electron for the moment, $g(W)$ will be the only factor that involves the electron energies, at least for the simple transitions we will consider. Thus we can integrate it from one, its minimum value, to W_o, its maximum value, to give

3.5.11 $$f(W_o) = \int g(W) dW = (W_o^2 - 1)^{1/2}[W_o^4/30 - 3W_o^2/20 - 2/15] + (W_o/4)\ln[W_o + (W_o^2 - 1)^{1/2}].$$

Note that, for large electron kinetic energies, the leading term in this expression varies as W_o^5, with a strong dependence on energy.

The Coulomb force of the nucleus will, of course, distort the electron energy distribution. The effects of this are represented by the function $\rho(Z, R, W)$ of nuclear charge and radius and of electron energy. This function is conveniently included in a redefinition of the β-decay operator H in terms of a dimensionless operator M, the function ρ, and g, which is the magnitude of the strength of the interaction

3.5.12 $$\langle |H_{fi}|^2 \rangle = g^2 \rho(Z, R, W) |M_{fi}|^2,$$

where the brackets indicate that an average has been taken over spin and isospin. The function ρ is close to 1.0 for very light nuclei or very energetic electrons but otherwise must be accounted for (see, e.g., deShalit and Feshbach 1974). We will return to the evaluation of the matrix elements M_{fi} shortly.

The more general expression for the phase space factor now becomes

3.5.13 $$f(Z, R, W_o) = \int \rho(Z, R, W)W(W_o - W)(W^2 - 1)^{1/2} dW.$$

This now gives the expression for the rate for β-decay over all combinations of energy of the electron and antineutrino as

3.5.14 $$\lambda = (2\pi/\hbar)g^2|M_{fi}|^2(4\pi)^2(L/h)^6 m^5 c^4 f(Z, r, W_o).$$

Returning now to the evaluation of the matrix element H_{fi}, we often find that the nuclear wave functions ψ_i and ψ_f are generally complicated. For now, however, we will consider light nuclei, for which the nuclear wave functions are less complicated and the calculation of the matrix elements is correspondingly simplified. The volume over which the integral in the matrix element is to be performed is that of the nuclei involved, so it is small. This allows expansions of the electron and antineutrino wave functions, both of which are given by outgoing waves, that is,

3.5.15 $$\psi_{e,\bar{\nu}} = L^{-3/2} \exp(i\bar{k}_{e,\bar{\nu}} \cdot \bar{r}_{e,\bar{\nu}}),$$

where, for example, $\bar{k}_e$ represents the (vector) wave number of the electron = $\bar{p}_e/\hbar$, and L is the size of the box over which the waves are normalized and the above phase space calculations were done. Because the wavelengths of the electron and neutrino are so much larger than the size of the nucleus, over which the integral for H_{fi} is to be performed, ψ_e and $\psi_{\bar{\nu}}$ can be approximated to be their zeroth-order term, or just $L^{-3/2}$ for the cases we are now considering, that of "allowed" transitions. For such cases, the transitions occur only between protons and neutrons of the same orbital angular momenta, so the overlap between wave functions is large. For heavier nuclei, this will normally not be the case. Then higher order terms of ψ_e and $\psi_{\bar{\nu}}$ must be taken into account to describe transitions that would be forbidden in the present analysis; these allow for the possibility of orbital angular momentum changes. Note that the factors of L from the above phase space considerations just cancel out those that normalize the wave functions of the antineutrino and electron.

The half-life is given from the above considerations as

3.5.16 $$t_{1/2} = (\ln 2)/\lambda.$$

The β-decay strengths are usually written as so-called $ft_{1/2}$ values, where, now including all the considerations above,

3.5.17 $$ft_{1/2} = (\ln 2)[(2\pi^3\hbar^7)][g^2|M_{fi}|^2 m^5 c^4]^{-1}$$

for reasons discussed below. Further evaluation of this expression requires separation of M, the β-decay operator, into its space- and spin-dependent parts:

3.5.18 $$|M_{fi}| = |M_{fi}^{\text{space}}||M_{fi}^{\text{spin}}|.$$

Evaluation of M_{fi}^{space} involves integration over space of the products of the initial and final nuclear wave functions, together with whatever additional factors have resulted from expansion of the electron and antineutrino wave functions. Since there are no such additional factors in the allowed transitions we are considering, M_{fi}^{space} is simply the overlap integral between the initial and final nuclear wave functions. However, expressing β-decays in terms of their $ft_{1/2}$ values conveniently removes the energy dependence from the righthand side and thus gives an expression that depends only on spin factors that are of order unity and the spatial integral. In the general case, this will be over the wave functions and the spatial dependences generated by the expansion of the electron and antineutrino wave functions, which in many cases will be required to allow the transition to occur at all. Thus, the $ft_{1/2}$ values provide a way of categorizing transitions by the order of the expansion of the electron and antineutrino wave functions, which determines their "order of forbiddenness."

For allowed transitions, the operators that determine M_{fi}^{spin} can be written in a more convenient form. Since β-decay involves conversions between protons and neutrons (or, technically, between up and down quarks), that aspect of M_{fi}^{spin} can be represented by isospin creation and annihilation operators. In the present context, these are defined so as to obey the same commutation rules as do the Pauli spin matrices, but they act on "charge space" instead of "spin space." For our present purposes we will forego a formal discussion of isospin and simply adopt operational definitions (but see the appendix for a more extensive discussion). Thus $t_-|\text{neutron}\rangle = |\text{proton}\rangle$, $t_+|\text{proton}\rangle = |\text{neutron}\rangle$, $t_3|\text{neutron}\rangle = (+1/2)|\text{neutron}\rangle$ and $t_3|\text{proton}\rangle = (-1/2)|\text{proton}\rangle$. For nuclei, $T_3|\text{nucleus}\rangle = (N - Z)|\text{nucleus}\rangle$. $T = T_3$ for the ground state but can attain higher values in excited states (although T_3 will still be the same for all states in that nucleus). For example, in figure 3.4, the ^{14}C ground state, ^{14}O ground state, and ^{14}N first excited state, all with spin-parity assignments of 0^+, are members of a $T = 1$ isospin triplet of states having $T_3 = +1$ in ^{14}C, $T_3 = 0$ in ^{14}N, and $T_3 = -1$ in ^{14}O. These states have essentially identical nuclear structure, differing only in the projection of their isospin quantum number. Note, however, that the opposite sign convention is sometimes also used for isospin! A good discussion of isospin is provided by the textbook by Park (1992).

β-decay can also involve spin changes so, including the possibility for both spin and isospin changes, we can write (following deShalit and Feshbach 1974)

3.5.19 $$\left|M_{fi}^{\text{spin}}\right|^2 = |C_F|^2|M_{\text{F}}|^2 + |C_{\text{GT}}|^2|M_{\text{GT}}|^2,$$

where

3.5.20 $$|M_{\text{F}}|^2 = (2J_i + 1)^{-1}\sum_{f,i}|\langle f|\sum_k \tau_{(\pm)}(k)|i\rangle|^2$$

3.5.21 $$|M_{\text{GT}}|^2 = (2J_i + 1)^{-1}\sum_{f,i}|\langle f|\sum_k \tau_{(\pm)}(k)\bar{\sigma}(k)|i\rangle|^2$$

where the sum over the index k is over the different nucleons of the decaying nucleus. The vector $\bar{\sigma}$ is the Pauli spin operator; it allows for spin changes. The subscripts F and GT refer to Fermi and Gamow-Teller transitions, where the selection rules for each are

$$\text{Fermi}: \Delta J = 0, \text{ no change in parity;}$$

3.5.22 $$\Delta T = 0, \Delta T_3 = \pm 1.$$

$$\text{Gamow}-\text{Teller}: \Delta J = 0, \pm 1 (0 \to 0 \text{ not allowed}), \text{ no change in parity;}$$

3.5.23 $$\Delta T = 0, \pm 1 (0 \to 0 \text{ not allowed}), \Delta T_3 = \pm 1.$$

In the above rules, $\Delta T_3 = +1$ (-1) implies electron (positron) emission.

A β-decay from a 0^+ to a 0^+ state would be a pure Fermi decay, for example, that occurring from the ^{14}O ground state to the 0^+ first excited state in ^{14}N since both are $T = 1$ states. The weak branch of the ^{14}O decay to the 1^+ ^{14}N $(T = 0)$ ground state, however, is a pure Gamow-Teller decay. Many decays are mixtures of the two modes, for example, the β-decay of the neutron.

Bound-State Beta Decay

One curious phenomenon that exists for about two dozen nuclei in the periodic table is bound-state beta decay (Takahashi and Yokoi 1983; Takahashi et al. 1987), in which the half-life of a nucleus in its atomic state can be very different from what it would be if the atom is fully or nearly fully stripped of its electrons. In this event, the electron emitted from the nucleus may not be emitted into the continuum, as is usually the case in β-decay, but rather into one of the inner-shell electronic states. For heavy nuclei this can result in a large shift in the energetics of the decay. The nuclei that can undergo bound-state β-decay are listed in table 3.3, along with their atomic *Q*-values (some of which are negative), their log *ft* values (those in parentheses are uncertain), their atomic decay rates λ_n, their decay rates into the continuum if the atom is fully stripped λ_c, and their rates for bound-state β-decay λ_b. Where multiple values are given for a single isotope, the decay can occur to several states in the residual nucleus. In one case, $^{110}Ag^m$, the decay is from an isomeric state. For the nuclides/states for which the *Q*-value is negative, decay cannot occur from the atom. Note that in some cases the possibility of decay to an excited state can greatly speed up the decay rate. This is particularly important for ^{187}Re; this produces a special effect in nucleocosmochronology (see chap. 7), the determination of the age of the universe from the long-lived nuclides that were produced in the first-generation stars.

Testing the predictions of bound-state β-decay theory requires producing the ions in their fully stripped state and then letting them decay; this necessitates use of a storage ring. As discussed in chapter 2 and in section 3.1, such a facility has been built at *GSI Darmstadt* (see, e.g., Bosch et al. 2003). Measurements of half-lives for several bound-state β-decay ions were measured. Most notably, the huge reduction in half-life for ^{187}Re predicted (Takahashi et al. 1987) for the fully ionized state, over that of the atom, 14 years to 43.5×10^9 years, was qualitatively confirmed. In this case, the huge reduction in half-life results from the fact that the shift in energy, because of the stripping of the electrons, allows decay to an excited state in ^{187}Os to which decay is energetically forbidden in the ^{187}Re atom but to which decay is greatly enhanced over that to the ^{187}Os ground state, that is, the level of forbiddenness is considerably less for the decay to the ^{187}Os excited state than it is to the ground state. The actual measured half-life of fully stripped ^{187}Re was found (Bosch et al. 1996) to be 32.9 years.

In that experiment, the fully stripped ^{187}Re ions were stored in the GSI storage ring. Attrition of the ^{187}Re ions results from two mechanisms, one being scattering on the gas in the storage ring, and the other being (bound-state) β-decay to ^{187}Os. The first mechanism can be monitored by observing some of the other ions that are stored in the storage ring (see fig. 3.5). The second was observed by the fact that, when a ^{187}Re ion does decay, its nuclear charge will change, although its ionic charge will remain the same in bound-state β-decay. Indeed, since the resulting ^{187}Os ions will have a slightly different mass than the fully stripped ^{187}Re present, the stored ^{187}Os ions can be observed to build up in the storage ring as the ^{187}Re decays. Ultimately, a small amount of stripper gas was inserted in the storage ring to strip off

Table 3.3 Atoms/Ions That Can Undergo Bound State β-Decay

Parent	Q (keV)	log ft	λ_n (s^{-1})	λ_c (s^{-1})	λ_b (s^{-1})
^{3}H	18.62	3.06	1.8e−9	1.8e−9	1.8e−11
^{14}C	156.5	9.04	3.8e−12	3.8e−12	6.0e−14
^{32}Si	225.0	8.02	2.2e−10	2.1e−10	1.4e−11
^{33}P	248.5	5.02	3.2e−7	3.1e−7	2.1e−8
^{35}S	166.8	5.00	9.2e−8	9.0e−8	1.2e−8
^{45}Ca	256.5	5.98	4.9e−8	4.8e−8	5.9e−9
^{63}Ni	65.9	6.64	2.2e−10	1.9e−10	3.0e−10
^{66}Ni	227.0	4.12	3.5e−6	3.4e−6	1.0e−6
^{93}Zr	60.1	(10.0)	1.4e−14	1.1e−14	1.3e−14
^{95}Nb	159.8	5.09	2.3e−7	2.1e−7	2.4e−7
^{106}Ru	39.4	4.30	2.2e−8	1.2e−8	2.1e−7
^{107}Pd	33.2	9.93	3.4e−15	1.8e−15	8.2e−15
$^{110}Ag^{m}$	83.7	5.37	2.2e−8	1.7e−8	7.8e−8
^{151}Sm	76.3	7.53	2.3e−10	1.5e−10	2.3e−9
	54.8	9.13	2.1e−12	1.1e−12	3.8e−11
^{155}Eu	252.7	8.78	5.8e−10	5.8e−10	9.0e−10
	192.7	8.57	4.0e−10	3.5e−10	9.2e−10
	166.2	7.91	1.2e−9	9.7e−10	3.3e−9
	147.4	7.47	2.2e−9	1.8e−9	7.6e−9
	134.7	8.73	8.9e−11	7.2e−11	3.6e−9
	106.6	8.94	2.7e−11	2.0e−11	1.6e−10
^{163}Dy	−2.8	(5.0)	0	0	1.6e−7
^{171}Tm	96.4	6.32	1.1e−8	7.6e−9	1.0e−7
	29.7	6.45	2.3e−10	4.0e−11	2.3e−8
^{187}Re	2.64	(11.0)	5.1e−19	0	1.4e−14
	−7.11	(7.5)	0	0	1.6e−9
^{191}Os	141.3	5.32	5.2e−7	4.0e−7	3.4e−6
^{193}Ir	−56.3	(7.5)	0	0	1.6e−10
	−57.9	(7.4)	0	0	1.7e−10
	−76.5	(7.5)	0	0	8.8e−12
^{205}Tl	−53.5	(12.0)	0	0	7.0e−17
	−55.8	(5.4)	0	0	6.6e−8
^{210}Pb	63.1	7.84	1.9e−10	7.7e−11	7.5e−9
	16.6	5.46	8.0e−10	0	8.3e−7
^{228}Ra	39.0	(6.5)	2.3e−9	1.8e−10	2.3e−7
	14.7	(5.0)	1.5e−9	0	4.9e−6
^{227}Ac	44.1	7.09	5.4e−10	9.8e−11	7.2e−8
	34.8	6.97	3.5e−10	2.9e−11	8.3e−8
	19.6	6.75	1.0e−10	0	1.1e−7
^{241}Pu	20.8	5.79	1.5e−9	0	1.9e−6

SOURCE: Takahashi et al. (1987).

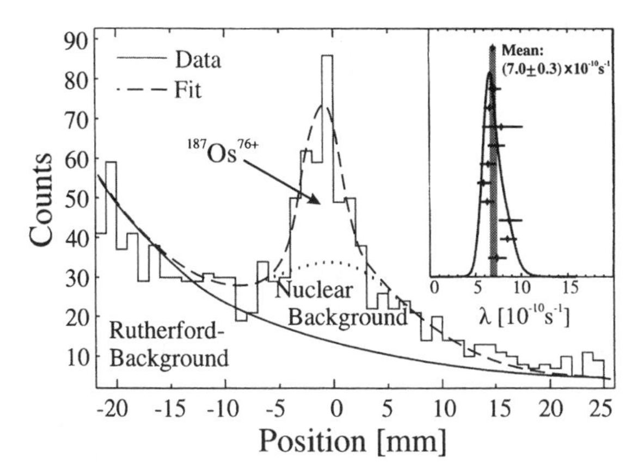

Fig. 3.5. Results of the experiment to measure the half-life of ^{187}Re using the *GSI Darmstadt* storage ring. Shown is the position spectrum of ions deflected by the first dipole magnet behind the gas jet target. The peak is due to $^{187}Os^{76+}$ nuclei from the β-decay of $^{187}Re^{75+}$. The background from elastic scattering (*solid line*) and from nuclear reactions (*dotted line*) has been determined in separate runs. The insert shows the yield of ^{187}Os ions following insertion of stripper gas in the storage ring. Shown are the results of 10 individual measurements; the full curve gives the sum of Gaussians for the separate measurements. Reprinted with permission from Bosch et al. (1996). Copyright 1996 by the American Physical Society.

the last electron, changing the charge and allowing the now fully stripped ^{187}Os to be magnetically diverted into a detector (see fig. 3.5). The event rate of this detector, after the attrition due to scattering in the storage ring, gives the rate at which bound-state β-decay occurs (Bosch et al. 1996; Klepper 1997). Another study made simultaneous measurements of β-decay to both bound and continuous electron states (Ohtsubo 2005).

3.6 Cross Section for $p + p \to d + e^+ + \nu_e$

Application of the β-decay formalism, with a slight extension, allows calculation of the cross section for a reaction that at present is impossible to measure but that is critically important for our understanding of stellar processes. This is the cross section for $p + p \to d + e^+ + \nu_e$, the first reaction in the so-called pp-chain burning in the Sun. For calculating a cross section, time-dependent perturbation theory can again be used, but now with a factor of velocity, to give

3.6.1 $$d\sigma = (2\pi/\hbar)v_i^{-1}|\langle f|H|i\rangle|^2 dN/dE,$$

where v_i is the center-of-mass velocity between the two protons and H is the β-decay operator, which can be written, as in the preceding β-decay discussion, as the product $g^2\rho(Z, r, W)M^{space}M^{spin}$. The phase space factor dN/dE is the same as for β-decay, since it is that for the positron and the neutrino. The primary difference between the calculation of this cross section and that used for β-decay is that the initial wave function in the former problem is a two-proton scattering state, whereas in β-decay it is that of the initial nucleus. The Pauli principle requires that the two-proton scattering state be a spin singlet state at the low energies at which this reaction occurs in stars. Thus, because the final state is the ground state of the deuteron, a spin triplet state, this must be a Gamow-Teller transition. Thus M_{fi}^{spin} can be evaluated by simple considerations. It is just the sum over the final spin states and the average over the initial states, or $(2J_f + 1)/(2J_i + 1) = 3$. However, there is an additional factor of 1/2 (see sec. 3.9) because the two particles in the initial state are identical, so the multiplier becomes 3/2.

Evaluation of M_{fi}^{space}, the overlap integral of the initial and final state wave functions, involves a numerical integration. The value of the integral is boosted considerably by the fact that the deuteron is so loosely bound that its wave function extends well beyond the distance usually thought to characterize the matter distribution for that system. Nonetheless, the cross section for this reaction at 1 MeV is found to be around 10^{-47} cm^2, well below the level where it might be measured with present technology. Thus the cross section for this reaction is necessarily determined only theoretically. The details of this calculation are provided in a study by Schiavilla et al. (1998). Several types of wave functions were tried in the numerical integrations. Their consistency suggests that the uncertainty in the cross section is at the 1 %–2 % level.

3.7 Neutrinos and Neutrino Oscillations

Neutrinos, the chargeless and nearly massless particles that are emitted in the β-decays discussed above, are produced in a variety of situations in nuclear physics and astrophysics and detected as signatures of astrophysical processes in others, so they deserve some general discussion. We will encounter neutrinos in a number of contexts in subsequent chapters, such as in their signature of the processes going on in the Sun, of their effects in stellar collapse, and of the information they will provide when detected from a supernova.

There are three known flavors of leptons, the electron, e, the muon, μ, and the tau, τ, and each has its corresponding antiparticle. The antiparticle of each lepton has the opposite charge of that of the particle. Corresponding to each lepton flavor is a neutrino, the electron neutrino, ν_e, the mu neutrino, ν_μ, and the tau neutrino, ν_τ, and each also has its corresponding antiparticle.

However, since neutrinos are not charged, the possibility exists that they are their own antiparticles, that is, that they are "Majorana" particles. (If their particles and antiparticles are distinct states they are referred to as "Dirac" particles.) This is an important issue in neutrino physics and would permit "neutrinoless double β-decay," a nuclear process, although its consequences for astrophysics are not so clear. However, the possibility of sterile neutrinos (i.e., an additional neutrino that interacts even more weakly with matter than the conventional ones) might have very important astrophysical consequences. For example, a sterile neutrino might make the r-process of nucleosynthesis, which makes half of the nuclides heavier than iron, possible in some sites (we return to this in chap. 7).

One curious feature of neutrinos is that they are able to change flavors, a result of the fact that the neutrino flavor eigenstates are not the same as the neutrino mass eigenstates. The possibility of this effect was first noted by Gribov and Pontecorvo (1969), but has assumed major astrophysical significance much more recently in several situations, most notably in the detection of solar neutrinos, to which we will refer in subsequent chapters. Because of its importance, it is reasonable to devote some discussion to the general subject of neutrino oscillations, or neutrino mixing.

In order to simplify the mathematics, we will assume initially that the neutrinos oscillate only between two flavors (which can be a good approximation, as discussed below), so that the neutrino flavor eigenstates are represented by the vectors ν_e and ν_μ, and the mass eigenstates by the eigenvectors ν_1 and ν_2. These vectors are orthonormal and so are related by a unitary transformation as

3.7.1 $$\nu_e = +\cos\theta\,\nu_1 + \sin\theta\,\nu_2$$

3.7.2 $$\nu_\mu = -\sin\theta\,\nu_1 + \cos\theta\,\nu_2,$$

where θ is the mixing angle. If an electron neutrino is emitted from a β-decay or a reaction, it is emitted as a flavor eigenstate but evolves in time as a mass eigenstate, as determined by the Schrodinger equation (assuming for the moment that $\hbar$ and c are unity), as

3.7.3 $$\nu_e(t) = \cos\theta\,\nu_1(0)\exp(-iE_1t) + \sin\theta\,\nu_2(0)\exp(-iE_2t)$$

3.7.4 $$\nu_\mu(t) = -\sin\theta\,\nu_1(0)\exp(-iE_1t) + \cos\theta\,\nu_2(0)\exp(-iE_2t).$$

The probability that the emitted e-neutrino will have oscillated into a μ-neutrino after time t is given by

3.7.5 $$\text{Oscillation probability} = |\langle\nu_e(0)|\nu_\mu(t)\rangle|^2,$$

which can readily be shown to be

3.7.6 $$\text{Oscillation probability} = \sin^2 2\theta \sin^2[(E_2 - E_1)t/2].$$

This can be cast in its usual form by recognizing that the neutrino masses are usually sufficiently small compared with the neutrino energies that the neutrinos are extremely relativistic. Then this expression becomes

3.7.7 $$\text{Oscillation probability} = \sin^2 2\theta \sin^2(\Delta m^2 x/4E),$$

where $\Delta m^2 = m_2^2 - m_1^2$ and x is the distance through which the neutrinos have traveled since they were created. These are referred to as vacuum oscillations, as they do not depend on the presence of matter in the region through which the neutrinos have passed. Of particular note is that at least one of the neutrinos must have a nonzero mass if, as is the case, oscillations are observed. In addition, the mixing angle θ must be relatively large, or the distance through which the neutrinos travel x must be large, for vacuum oscillations to occur. The oscillation length is defined as the distance over which the oscillation completes half a cycle, so that

3.7.8 $$L = 4\pi\hbar E/\Delta m^2 c^3 = 2.48(m)\, E(\text{MeV})\Delta m^2(\text{eV}^2),$$

where $\hbar$ and c have now been put back in.

However, there is another type of oscillation that can occur; these are called matter-enhanced oscillations, or Mikheyev-Smirnov-Wolfenstein (MSW) oscillations after the three physicists who first developed their oscillation formalism (Mikheyev and Smirnov 1985; Wolfenstein 1978). In this instance, it is useful to develop the oscillation scenario (see, e.g., Kayser 2005) in a matrix notation by writing the Schrodinger equation as

3.7.9
$$i(d/dt)\begin{pmatrix} \nu_e(t) \\ \nu_u(t) \end{pmatrix} = \begin{pmatrix} \cos\theta\, \nu_1 E_1 + \sin\theta\, \nu_2 E_2 \\ -\sin\theta\, \nu_1 E_1 + \cos\theta\, \nu_2 E_2 \end{pmatrix} = H_V \begin{pmatrix} \cos\theta\, \nu_1 + \sin\theta\, \nu_2 \\ -\sin\theta\, \nu_1 + \cos\theta\, \nu_2 \end{pmatrix}$$

Performing the matrix multiplication, and taking advantage of the fact that ν_1 and ν_2 are linearly independent, allows determination of the Hamiltonion H_V as

3.7.10 $$H_V = p + (1/2p)\begin{pmatrix} \cos^2\theta E_1 + \sin^2\theta E_2 & \cos\theta\, \sin\theta\, (E_2 - E_1) \\ \cos\theta\, \sin\theta (E_2 - E_1) & \sin^2\theta E_1 + \cos^2\theta E_2 \end{pmatrix}$$

Performing the relativistic approximations for E_1 and E_2 permits this matrix to be written

3.7.11
$$H_V = \left[p + \left(m_1^2 + m_2^2\right)/4p\right]\begin{pmatrix}1 & 0\\ 0 & 1\end{pmatrix} + \Delta m^2/4p\begin{pmatrix}-\cos 2\theta & \sin 2\theta\\ \sin 2\theta & \cos 2\theta\end{pmatrix}$$

As with the above nonmatrix analysis, this analysis was done for vacuum oscillations; hence the subscript V on H_V. However, when the neutrinos are passing through matter, they will interact with the matter, albeit weakly. For μ- (or τ-) neutrinos at energies too low to form μ- (or τ-) leptons, the interaction will be only through neutral-current interactions, but e-neutrinos and antineutrinos can interact through both neutral-current and charged-current interactions. This can introduce an effect in the matrix for these neutrinos that will not exist for the μ- (or τ-) neutrinos; it can be taken into account by introducing potential terms in the Hamiltonion matrix as

3.7.12
$$H_M = H_V + V_Z\begin{pmatrix}1 & 0\\ 0 & 1\end{pmatrix} + V_W\begin{pmatrix}1 & 0\\ 0 & 0\end{pmatrix}$$

where V_Z represents the neutral-current interactions, which exist for all neutrino flavors, through exchange of a Z-boson, and V_W represents the additional interactions of the e-neutrinos and antineutrinos with matter through exchange of a W-boson. V_W is equal to $2^{1/2}\ G_F\ n_e = 7.6 \times 10^{-14}\ (Z/A)\ \rho(\text{g/cm}^3)$ eV (Kayser 2005), where G_F = Fermi constant = 1.436×10^{-49} ergs cm^3. H_M now becomes

3.7.13
$$H_M = [p + \left(m_1^2 + m_2^2\right)/4p + V_Z\begin{pmatrix}1 & 0\\ 0 & 1\end{pmatrix} + \begin{pmatrix}-(\Delta m^2/4p)\cos 2\theta + 2^{1/2}G_F n_e & (\Delta m^2/4p)\sin 2\theta\\ (\Delta m^2/4p)\sin 2\theta & (\Delta m^2/4p)\cos 2\theta\end{pmatrix}$$

where the extra term in the H_{11} matrix element, which was not there for the vacuum oscillations, accounts for the interactions of the e-neutrinos or antineutrinos with the matter.

The terms in equation 3.7.13 that multiply the identity matrix introduce a phase in the time dependence of the wave functions for the different flavors that will be the same for all flavors, so they are of no physical consequence. However, the second matrix will produce the MSW mixing. One can work through the algebra (see Kayser 2005) to show that there exists a resonant oscillation length for the MSW oscillation scenario that differs from that in the case of vacuum oscillations:

3.7.14 $$L_M = L_V \Delta m^2 / \{(\Delta m^2 \cos 2\theta - 2^{3/2} G_F n_e E)^2 + \Delta m^4 \sin^2 2\theta\}^{1/2},$$

which produces the resonance condition:

3.7.15 $$\Delta m^2 \cos 2\theta = 2^{3/2} G_F n_e E.$$

If the electron number density n_e in the star exceeds that required for this resonance condition at the site of the neutrino production, the neutrinos will eventually pass through the region of density at which MSW oscillations will occur as they exit the star, guaranteeing that MSW oscillations will occur (if the density does not change too rapidly; see below). The expression that describes the oscillation length at resonance is

3.7.16 $$L_M = L_V / \sin 2\theta = 4\pi E / \Delta m^2 \sin 2\theta.$$

Note that this resonance condition can give maximal mixing even when the mixing angle is small; the oscillations *will* occur if the resonance condition is met, which is not necessarily the situation for vacuum oscillations.

Since this transition must be adiabatic, a condition exists on the rapidity of variation of the electron density n_e that produces the resonance, that is,

3.7.17 $$(1/\rho)(d\rho/dx) L_V / \sin^2 2\theta << 1,$$

where ρ is the matter density. Thus the density must vary sufficiently slowly that there are many oscillations within the resonance length. This can have important consequences for the oscillations that occur in stars (Bahcall 1989).

That solar neutrinos do undergo flavor oscillations has now been established unequivocally in several experiments; we return to this in chapter 5. However, it is worth noting at this point that a recent result, that from the *Kamioka Liquid Scintillator AntiNeutrino Detector, KamLAND* (Araki et al. 2004), has definitely established the existence of neutrino oscillations from a very different source. This experiment utilized a neutrino detector, located in the Kamioka mine in Japan, that had a basic design that looked a lot like that of the *Sudbury Neutrino Observatory* (see chap. 2) but that had inside its 13-m diameter inner-sphere liquid scintillator, and between its 18-m outer sphere and the inner sphere, a buffer oil to minimize backgrounds in the inner sphere. Observing the scintillation from the interactions in the inner sphere were 1900 photomultiplier tubes. The signals observed in this detector resulted from interactions of neutrinos, technically the electron antineutrinos, $\bar{\nu}_e$s, from the many Japanese reactors (nuclear power plants are a major source of electricity in Japan), with the protons in the liquid scintillator. Their result

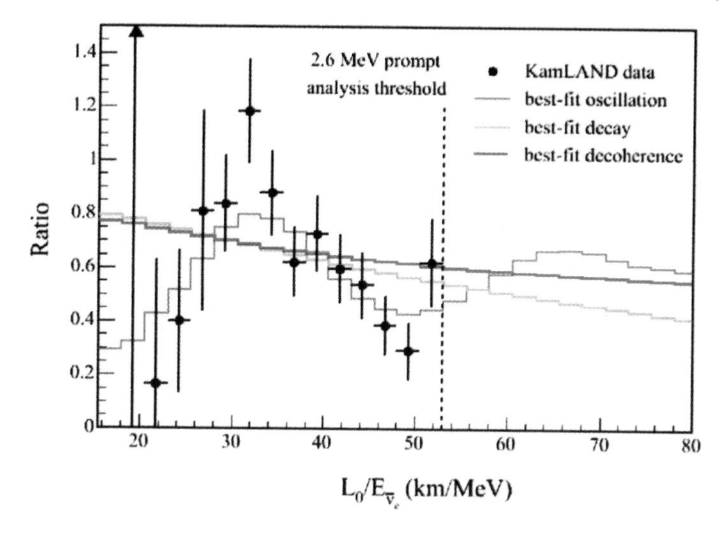

Fig. 3.6. Ratio of the observed ν_e spectrum to that expected were there no oscillations versus L_o/E. The curve shows the expectation for the best fit assuming oscillations, the best fit assuming the neutrino decay model (Stuart et al. 1999), and the best fit in a decoherence model (Barger et al. 1999), taking into account the individual time-dependent flux variations of all reactors and detector effects. The data points and models are plotted with $L_o = 180$ km, as if all antineutrinos detected in KamLAND were due to a single reactor at this distance. Reprinted with permission from Araki et al. (2005). Copyright 2004 by the American Physical Society.

is shown in figure 3.6. The oscillatory nature of the data is a result of dividing both the data and the best-fit spectrum by the expected no-oscillation result. This is complicated by the fact that different reactors located at different distances from the detector are running at different levels at different times, and these factors must all be included in the expected yield. However, the magnitude of the deviation from the nonoscillation flux, and from two other possible neutrino disappearance scenarios, is unmistakable. The range of the data in $L_o/E_{\bar{\nu}}$, which is obtained because the electron antineutrinos are detected from the reaction $\bar{\nu}_e + p \rightarrow e^+ + n$, so that the positron energy gives a good estimate of the energy of the incident $\bar{\nu}_e$, is used to produce the distribution seen, since the reactors and the detector are fixed. The total number of neutrinos observed above the threshold, 258, was well below that expected, 365.2 ± 23.7, so some sort of disappearance mechanism had to have been operating. Thus, given this result and the solar neutrino results discussed in chapter 5, there can be no doubt that a large fraction of the reactor neutrinos have oscillated away from $\bar{\nu}_e$s to some other—undetectable—type of neutrino and that, because of the oscillatory nature of the data, the explanation is neutrino oscillations and not one of the other possibilities.

Although the above discussion has been predicated on two-neutrino mixing, that is not the most general case. The general form of the mixing matrix, where equations 3.7.1 and 3.7.2 are written in matrix form and extended to three neutrino species by $\nu_\alpha = \Sigma_i U^*_{\alpha i} \nu_i$ (the ν_α are the flavor eigenstates and the ν_i are the mass eigenstates), and including basic features of the experimental results, is given by

$$U = \begin{pmatrix} 1 & 0 & 0 \\ 0 & c_{23} & s_{13} \\ 0 & -s_{23} & c_{23} \end{pmatrix} \begin{pmatrix} c_{13} & 0 & s_{13}e^{-i\delta} \\ 0 & 1 & 0 \\ -s_{13}e^{i\delta} & 0 & c_{13} \end{pmatrix} \begin{pmatrix} c_{12} & s_{12} & 0 \\ -s_{12} & c_{12} & 0 \\ 0 & 0 & 1 \end{pmatrix} \tag{3.7.18}$$

where, for example, c_{12} is cos θ_{12} and s_{12} is sin θ_{12}, and θ_{12} characterizes the mixing of the $i = 1$ and $i = 2$ mass eigenstates (see eqs. 3.7.1 and 3.7.2), and δ is the possible charge-parity (CP) nonconserving part of neutrino mixing. At present θ_{13} is known to be small, that is, it has not been observed to be nonzero, and CP violation, which appears to explain the prevalence of matter over antimatter in the universe, has not been observed conclusively in the neutrino sector as of the date of this book. Note from the form of the matrices that the first matrix, which represents atmospheric neutrino mixing, mixes the mass and flavor eigenstates in a different way than solar neutrino mixing does, which is represented by the third matrix. The solar mixing angle θ_s is approximately the same as θ_{12}. Thus, the solar mixing problem is well described by a two component matrix, as was suggested above (Kayser 2005). From the experimental results, discussed in more detail in section 5.5, it can be concluded that ν_e is essentially just a mixture of ν_1 and ν_2. Furthermore, these mass eigenstates will remain unchanged as the neutrinos that occupy them propagate from Sun to Earth. The cross-mixing matrix, the middle one, allows for the possibility of mixing between mass eigenstates other than those described in the first and third matrices. CP violation requires that all the mixing angles be nonzero (Kayser 2005), so they are usually cast as being associated with the mixing angle θ_{13}, which is certainly small and may turn out to be zero. Not included in this composite matrix is yet a fourth matrix that describes the possibility that the neutrinos are their own antineutrinos, that is, that neutrinoless double β-decay can occur.

We will return to the observations for these mixings in chapter 5. For the moment, we note that the mixing is large in the neutrino sector, very much unlike that in the quark sector.

3.8 Scattering Theory: Direct and Resonant Reactions

This section is intended to provide a sketch of the general features from which the equations of scattering theory are derived. Students who have had a course in advanced quantum mechanics will be able to browse through this section quickly, while those who have not may either find it necessary to just accept the results for now or to spend considerable time and effort looking up the references if they wish to have a better understanding about how the equations are derived. In any event, this section is not intended to replace a course in scattering theory but rather to provide the student of nuclear astrophysics who has not had a course in scattering theory with at least a feeling for the origin of the various expressions of the cross sections that are encountered in nuclear astrophysics.

Two extremes of scattering are those of direct reactions and resonant reactions. Resonant reactions occur as a result of scattering at energies below the peak of a barrier that can produce quasi-bound states of the nucleus. Indeed, the resonances observed are associated with the quasi-bound levels, as is discussed below. Resonant scattering in nuclear physics is a relatively low-energy phenomenon. Since the energies associated with the reactions that occur in stars fall into the low-energy regime, it is an especially relevant subject for our considerations. Direct reactions, by contrast, can occur at both low and high energies, but in either case they involve nonresonant scattering. Thus at the energies at which astrophysical reactions occur, both types of scattering can occur, and both often contribute importantly to the observed cross sections. In addition, in some cases they can interfere, although it is usually assumed that they do not do so.

Figure 3.7 indicates the general situation. The example it represents is proton radiative capture (a very important type of reaction for nuclear astrophysics) on nucleus (A, Z) to form nucleus $(A + 1, Z + 1)$ and a gamma ray. The energy levels indicated above zero energy in the $A + \mathrm{p}$ system are quasi-bound, or "compound nuclear," levels in the nucleus $(A + 1, Z + 1)$. If the energy associated with the interaction (i.e., the center-of-mass energy) is not close to one of the compound nuclear states of this system, then the scattering will be direct. If, on the other hand, the energy is close to one of the compound nuclear states, the result may be resonant scattering. That will not necessarily be the case, as the reaction channel might not be connected to the resonance state at that energy; this possibility will become more transparent below. The quantity Q is the reaction Q-value. The cross section for scattering in this system might look as indicated on the right. The gamma ray emitted in the radiative capture might, in principle, be produced from deexcitation from the

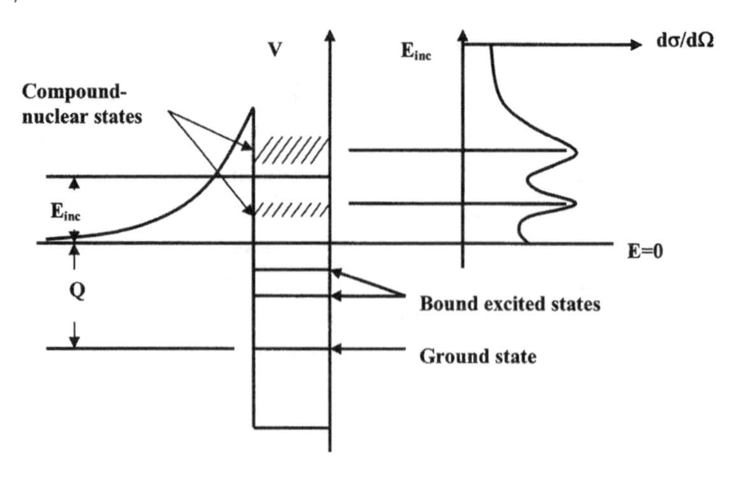

Fig. 3.7. Schematic diagram of scattering that could be either direct or resonant. The states in the (assumed for this figure to be a hard sphere) nuclear + Coulomb potential that can lead to scattering resonances are indicated with cross-hatching. The cross section at some scattering angle as a function of energy E_{inc} is shown to the right. In the usual case in astrophysical reactions involving two charged particles the kinetic energy of the reaction is below the Coulomb barrier so that quantum mechanical tunneling is required for interactions between the two nuclei involved to proceed.

incident energy to the ground state or to any of the excited states of the $(A+1,$ $Z+1)$ system.

If the scattering is direct, then the reaction cross section will be determined by a matrix element that connects the initial state of the proton and the nucleus (A, Z), the entrance channel, directly with the final state of the nucleus $(A+1,$ $Z+1)$ and the outgoing photon, the exit channel, via some transition operator T in a single-step process. Thus the cross section can be represented as

3.8.1 $$\sigma \propto |\langle A\, p|T|A+1\, \gamma\rangle|^2.$$

For the moment we have assumed that the spins of all particles are zero for simplicity; we will include the effects of nonzero spin later. Elucidation of the matrix elements of equation 3.8.1 requires formal reaction theory; this is the subject of advanced quantum mechanics and will not be dealt with in detail here.

However, if the reaction is resonant, the cross section is determined by a product of two matrix elements, the first connecting the initial state to the compound nuclear state through transition operator T_1 and the second connecting the compound nuclear state to the final state through transition operator T_2, as

3.8.2 $$\sigma \propto |\langle A\, p|T_1|A+1^*\rangle|^2 |\langle A+1^*|T_2|A+1\, \gamma\rangle|^2,$$

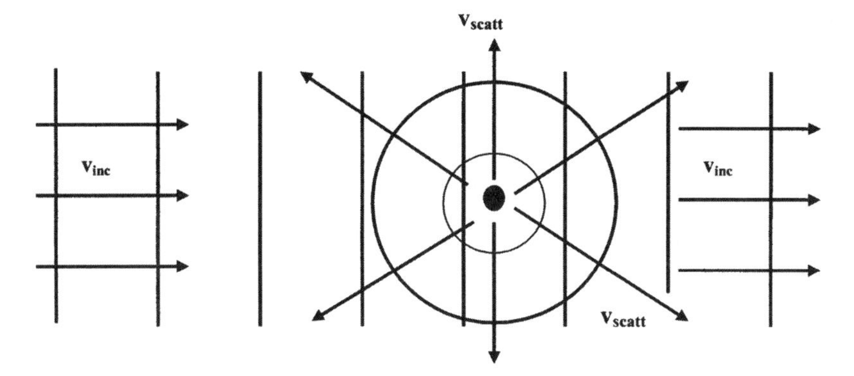

Fig. 3.8. Wave description of scattering in which the incident wave is a plane wave and the scattered is an angle-modulated outgoing spherical wave.

where $\langle A + 1^{*}|$ denotes an excited state—the quasi bound, or compound nuclear, state—of the $A + 1$ nuclear system. Indeed, the half-lives of these quasi-bound states are usually such that they live a long time compared with the nuclear crossing time of the projectile, which is $\sim 10^{-21}$ seconds, and the width of the resonance ΔE is related to its lifetime via the Heisenberg Uncertainty Principle: lifetime $\approx \hbar/\Delta E$. Thus, when a compound nuclear state is formed, it loses all memory of the process by which it was formed. Therefore, the decay of that state will be independent of how it was formed and, once formed, will always proceed with the same decay probabilities into its different possible decay channels.

This section will first present some of the general considerations needed to describe scattering. It will say very little about direct reactions; these are usually well covered in courses on scattering theory and for present purposes we only need to know that they occur. However, it will give a considerably more extensive discussion of resonant reactions; these are less frequently discussed. For more detailed developments of both, the interested student can go to the general references given for the respective cases (Austern 1970 or Satchler 1983 for direct reactions; Lane and Thomas 1958 or Preston 1962 for resonant reactions).

Generally, the picture that describes a scattering problem has, as indicated in figure 3.8, an incident plane wave and a scattered wave that is an angle-modulated spherically outgoing wave. What we will actually be describing is the scattering of spinless chargeless particles: spinless neutrons. The reasons for this are that we can put in the effects of spin by hand at the end and that treating neutrons avoids the complications of dealing with Coulomb wave functions, the interpretation of which is somewhat less transparent than if the scattering describes neutrons. Of course, the plane wave solutions we discuss would have to be replaced by the regular and irregular Coulomb wave functions if we were dealing with protons (as we will be the case when we discuss Thomas-Ehrman energy shifts later in this section).

Formally this situation can be described by the equation

3.8.3 $$\psi = \psi_{\text{inc}} + \psi_{\text{scat}}$$

3.8.4 $$= e^{ikz} + (e^{ik'r}/r)f(\theta),$$

where k is the wave number of the incident plane wave in the center of mass and k' is that for the scattered wave. Since the differential cross section is determined by the yield of particles into some solid angle, it is given as the absolute square of the (scattering angle dependent) coefficient $f(\theta)$ of the spherically outgoing wave, that is,

3.8.5 $$d\sigma(\theta)/d\Omega = |f(\theta)|^2.$$

Asymptotically, the expression for a plane wave can be written as a partial wave expansion in angular momentum ℓ:

3.8.6 $$e^{ikz} = (\pi^{1/2}/kr)\sum_{\ell=0}^{\infty}(2\ell+1)^{1/2}i^{\ell+1}[e^{-i(kr-\ell\pi/2)} - e^{+i(kr-\ell\pi/2)}]Y_{\ell 0},$$

where $Y_{\ell 0}$ is the appropriate spherical harmonic. The first wave in the brackets is a radially incoming wave, and the second is a radially outgoing wave.

For elastic scattering, on which we will focus for the moment, the total wave function will also have both radially incoming and outgoing waves. Causality dictates that the former cannot be affected by a scattering potential at the origin. However, in the total scattering wave function, the radially outgoing wave will be affected by the scattering potential. The result of the scattering center—the nucleus—will therefore be a shift in the phase of the outgoing wave that would not have existed had the nucleus not been there. This phase shift, of course, must be measured outside the nucleus; it varies with radius inside the nucleus.

Thus we can rewrite the above expression to include the effects of the potential as (still assuming that the wave number for the incoming and outgoing waves are the same, i.e., that we are describing elastic scattering):

3.8.7 $$\psi = (\pi^{1/2}/kr)\sum_{\ell=0}^{\infty}(2\ell+1)^{1/2}i^{\ell+1}[e^{-i(kr-\ell\pi/2)} - \eta_\ell e^{+i(kr-\ell\pi/2)}]Y_{\ell 0}.$$

3.8.8 $$= \sum_{\ell=0}^{\infty}(I_\ell + \eta_\ell O_\ell),$$

where η_ℓ ($|\eta_\ell| \leq 1$), in general a complex coefficient, characterizes the effect of the nucleus on the outgoing wave, and I_ℓ and O_ℓ are, respectively, the asymptotic forms of the incoming and outgoing ℓth partial waves and are defined by this equation. Substituting equation 3.8.7 into equation 3.8.4 gives

3.8.9 $$f(\theta) = (\pi^{1/2}/k)\sum_{\ell=0}^{\infty}(2\ell+1)^{1/2}i^{\ell+1}e^{-i\ell\pi/2}(1-\eta_\ell)Y_{\ell 0}.$$

If the situation being described is solely elastic scattering, then there are no scattering losses (i.e., inelastic or reaction processes), so we can write $\eta_\ell = \exp(2i\delta_\ell)$, where δ_ℓ is the phase shift associated with the ℓth partial wave. (The existence of reaction and inelastic processes can be represented in the elastic channel by an imaginary, or absorptive, term in the phase shift. In this case, η_ℓ could be taken to be $\exp(-\alpha + 2i\delta_\ell)$, where α is real and positive, so that $|\eta_\ell| < 1$.) This phase shift would be zero if the scattering center were not there, that is, there would be no scattered wave. This allows writing the elastic scattering cross section for the lth partial wave as

3.8.10 $$d\sigma(\theta)/d\Omega_\ell = (4\pi/k^2)(2\ell+1)\sin^2\delta_\ell|Y_{\ell 0}|^2.$$

Of course, this expression only describes how the waves scatter in a formal way. It has nothing to do with the properties of the nucleus but only with the fact that it exists. The above expressions apply to the general case, either direct elastic scattering or resonant elastic scattering.

To include the nucleus explicitly in an elastic scattering calculation in which only direct reactions occur, one could solve the Schrodinger equation, including a potential of the form given in equation 3.2.1, to solve for the wave function and, hence, the phase shifts. To include resonant scattering, however, a formalism has been developed that allows explicit inclusion of the details of the nucleus; we will now develop that formalism for elastic scattering, following to some extent the derivation in Preston (1962).

Scattering through Resonances

GREEN'S THEOREM. We first derive a form of Green's theorem that will allow us to obtain the resonance form for the cross sections. To do so, consider the Schrodinger equation for scattering wave functions $u_1(=r\psi_1)$ and u_2 at their respective energies, E_1 and E_2. Multiply each equation by the other wave function, then integrate the resulting equations over the volume of the nucleus, assumed to be contained within $r \leq a$ (and where, e.g., u_1'' is the second derivative of u_1):

3.8.11 $$\int_0^a [u_1'' + (2M/\hbar^2)(E_1 - V)u_1]u_2 dr = 0,$$

3.8.12
$$\int_0^a [u_2'' + (2M/\hbar^2)(E_2 - V)u_2]u_1 dr = 0.$$

Since $u(r)$ must vanish at the origin in order to be a physical solution of the Schrodinger equation, subtracting one equation from the other and integrating by parts gives

3.8.13
$$[u_1'(a)u_2(a) - u_1(a)u_2'(a)] + (2M/\hbar^2)(E_1 - E_2)\int_0^a u_1 u_2 dr = 0.$$

Note that if both u_1 and u_2 are orthonormal eigenfunctions,

3.8.14
$$\int_0^a u_1 u_2 dr = \delta_{12}.$$

Expanding u_E, the wave function at arbitrary energy E, in terms of the complete set of eigenfunctions $\{u_\lambda\}$ with eigenvalues E_λ,

3.8.15
$$u_E(r) = \sum_\lambda A_\lambda u_\lambda(r) \text{ for } 0 < r < a,$$

where

3.8.16
$$A_\lambda = \int_0^a u_\lambda u_E dr.$$

These states represent the quasi-bound states of the nucleus referred to above, that is, the bound states that result from a constraining potential, for example, a Coulomb barrier for a bound proton, and for which the wave functions are maximized inside the nucleus and minimized outside the nucleus. This requires that the derivative of the wave function be close to zero at the nuclear radius, that is, that $u_\lambda'(a) \approx 0$.

To obtain an expression for A_λ, apply the above result from equation 3.9.13, identify $u_\lambda = u_1$ and $u_E = u_2$:

3.8.17
$$A_\lambda = (\hbar^2/2M)u_\lambda(a)(E_\lambda - E)^{-1}(du_E/dr)_{r=a},$$

from which

3.8.18
$$u_E(r) = (\hbar^2/2M)\sum_\lambda u_\lambda(r)u_\lambda(a)(E_\lambda - E)^{-1}(a\,du_E/dr)_{r=a}.$$

Returning now to equation 3.8.7, the logarithmic derivative of each partial wave must be continuous at $r = a$, now assumed to be at the nuclear surface, so that

3.8.19
$$(u_\ell/ru_\ell')_{r=a} = (I_\ell - \eta_\ell O_\ell)/[\rho(I_\ell' - \eta_\ell O_\ell')]_{r=a} = R_\ell,$$

where $\rho = kr$. Note that R_ℓ must be real, since $u_\ell(0) = 0$. Solving for η_ℓ,

3.8.20
$$\eta_\ell = \{(I_\ell/O_\ell)[1 - R_\ell(\rho I_\ell'/I_\ell)]/[1 - R_\ell(\rho O_\ell'/O_\ell)]\}_{r=a}.$$

Now define (Preston 1962) the real and imaginary components of the logarithmic derivatives of O_ℓ and I_ℓ:

3.8.21 $$\rho\, O'_\ell / O_\ell = L_\ell = S_\ell + iP_\ell, \text{ and } \rho\, I'_\ell / I_\ell = L^*_\ell = S_\ell - iP_\ell.$$

where S_ℓ is the shift function and P_ℓ is the penetrability, the names being chosen to represent the effects of the two parameters, as will be elucidated below. Since I_ℓ and O_ℓ are the asymptotic incoming and outgoing waves, the phase φ_ℓ can be written as

3.8.22 $$I_\ell / O_\ell = \exp(-2i\varphi_\ell).$$

Then, substituting equations 3.8.21 and 3.8.22 into equation 3.8.20,

3.8.23 $$\eta_\ell = \exp(-2i\varphi_\ell)\{[1 - R_\ell(S_\ell - iP_\ell)]/[1 - R_\ell(S_\ell + iP_\ell)]\}.$$

But this can also be written (problem 3.10) as

3.8.24 $$\eta_\ell = \exp(-2i\varphi_\ell)\exp\{2i\tan^{-1}[R_\ell P_\ell/(1 - R_\ell S_\ell)]\}.$$

If, as before, we consider only elastic scattering, then $\eta_\ell = \exp(i2\delta_\ell)$, which gives

3.8.25 $$\delta_\ell = \tan^{-1}[R_\ell P_\ell/(1 - R_\ell S_\ell)] - \varphi_\ell.$$

Returning now to equation 3.8.18, and assuming that E is sufficiently close to a resonance at E_λ that the sum collapses to only one term (that for the ℓth partial wave), then

3.8.26 $$u_E(r) = (\hbar^2/2M)u_\lambda(r)u_\lambda(a)(E_\lambda - E)^{-1}(a\,du_E/dr)_{r=a}.$$

Defining $\gamma_\lambda(a) = [(\hbar^2/2M]^{1/2}\, u_\lambda(a)$ and dividing both sides by $du_E(a)/dr$ gives

3.8.27 $$(u_E/ru'_E)_{r=a} = R_\ell = \gamma_\lambda(a)^2/(E_\lambda - E).$$

Finally, δ_ℓ can now be written as

3.8.28 $$\delta_\ell = -\varphi_\ell + \tan^{-1}[\gamma^2_\lambda P_\ell/(E_\lambda - E)]/[1 - \gamma^2 S_\ell/(E_\lambda - E)]$$

3.8.29 $$= -\varphi_\ell + \tan^{-1}[(\Gamma_{\lambda\ell}/2)/(E_{\lambda\ell} - E - \Delta_{\lambda\ell})],$$

where $\Gamma_{\lambda\ell} = 2\gamma^2 P_\ell$, and $\Delta_{\lambda\ell} = -\gamma^2 S_\ell$. Inserting this into equation 3.8.10 (assuming $\varphi_\ell = 0$, although that is not a necessary assumption) immediately gives the Breit-Wigner cross section resonance form for elastic scattering:

3.8.30 $$\sigma = (2\ell + 1)\pi(1/k)^2 g_J \Gamma_{\lambda\ell}\Gamma_{\lambda\ell}/[(E - E_{\lambda\ell} + \Delta_{\lambda\ell})^2 + (\Gamma_{\lambda\ell}/2)^2],$$

where $k = p/\hbar$ is the wave number in the center-of-mass system. It should be noted that the parameters of this equation, that is, the widths and shift, can depend sensitively on the choice of the radius a, from which these quantities are derived. Nonetheless, the significance of this result is clear: the derivation demonstrates the relationship between the observed resonances and the properties of the nuclear levels from which they arise. This expression clearly produces a cross section with a resonance at E_R, which is shifted from $E_{\lambda\ell}$, the energy at which the quasi-bound state exists, by $\Delta_{\lambda\ell}$ (which is proportional to the shift function S_ℓ, which thus earns its name). The width of the resonance is characterized by $\Gamma_{\lambda\ell}$. Note that, since $\Gamma_{\lambda\ell}$ describes the width of the nuclear level in energy, as observed in the resonance scattering, as noted qualitatively above, it is related to the half-life $t_{1/2}$ of the level via the Heisenberg Uncertainty Principle as

3.8.31 $$t_{1/2} = \ln(2)\hbar/\Gamma_{\lambda\ell}.$$

Note also that the Γs are energy dependent; they depend both on a nuclear structure factor and the penetrability, which, as discussed in greater detail below, depends on a quantum mechanical tunneling probability and so is highly energy dependent. There are two factors of $\Gamma_{\lambda\ell}$ in this expression; these represent the two matrix elements that connect the initial state to the compound nuclear state and the compound nuclear state to the final state, to which we referred in the introduction to this section. In the present discussion, the latter has represented a reconnection back to the entrance channel, but that need not necessarily be the case; we will return to this point shortly.

Equation 3.8.29 shows that a resonance occurs when the denominator of the expression for the phase shift δ_ℓ goes to zero, resulting in a sudden sign change, or δ_ℓ passes through $(n + 1/2)\pi$ at $E_{\lambda\ell} - \Delta_{\lambda\ell}$, corresponding to a continuum state at E_λ where n is an integer.

It will be shown below that $\Gamma_{\lambda\ell}$ is simply a factor associated with the nuclear structure of the level denoted by $(\lambda\ell)$, via the γ^2 (structure) factor and the penetrability P_ℓ. Furthermore, although we have written equation 3.8.30 only for elastic scattering, as noted above, it applies also to any reaction if one of the two $\Gamma_{\lambda\ell}$ factors in the numerator is changed to one that reflects the structure of the reaction channel. This can be seen readily by recognizing that the wave function of the compound nuclear state will determine the probabilities for decay to all possible (energetically allowed and allowed by the structure of the state in the reaction channel) reaction channels, so that each $\Gamma'_{\lambda\ell}$ factor characterizes the decay to the particular channel it represents. Thus equation 3.8.30 can be generalized to be

3.8.32 $$\sigma = (2\ell + 1)\pi(1/k)^2 g_J \Gamma_{\lambda\ell}\Gamma'_{\lambda\ell}/[(E - E_{\lambda\ell} + \Delta_{\lambda\ell})^2 + (\Gamma_\ell/2)^2],$$

where $\Gamma'_{\lambda\ell}$ is the partial width associated with the reaction channel to which it applies. Note also that $\Gamma_{\lambda\ell}$ in the denominator has been replaced by just Γ_λ and that $\Gamma_\lambda = \Gamma_{\lambda\lambda} + \Gamma'_{\lambda\lambda} +$ the sum of all other possible partial widths in the λth partial wave. The spin statistical factor, g_j, has been inserted into in equations 3.8.30 and 3.8.32; its form will be discussed in the next section.

Energy Shifts

Note the shift in the resonance energy $\Delta_{\lambda\ell}$ from the actual energy of the state that results from this analysis; this is something that must be taken into account when utilizing data from indirect measurements, for example, nuclear transfer reactions, to predict astrophysical reaction rates. It is also important in the use of spectroscopic information from isobaric analog states, mentioned in the isospin discussion in section 3.5. As discussed also, in order to calculate the reduced widths, one must know the structure of the states involved. This may be known for one member of an isobaric analog multiplet and can then be used to calculate the reduced widths, for the state of interest in the analog nucleus. However, one must also be able to calculate the shifts; this can be done using the formalism developed by Thomas (1952). These shifts are referred to as Thomas-Ehrman shifts; an application can be seen in detail in Langanke et al. (1986).

The expressions from which one can calculate the Thomas-Ehrman shift result naturally from the formalism presented in section 3.8 (Thomas 1952), but its results will just be quoted here. There are energy shifts Δ_b associated with bound states, at binding energy E_b, and given by the expression

3.8.33
$$\Delta_b = -(3\hbar^2/2\mu R^2)\Theta_b^2 \rho\, W'_\lambda / W_\lambda,$$

where $\rho = kr$, μ is the nucleon-nucleus reduced mass, and the W_λ and W'_λ are the Whittaker function and its derivative with respect to ρ. The quantity $\Theta_b (0 < \Theta_b < 1)$ represents the spectroscopic strength of the relevant component in the wave function of the compound nuclear state. The energy shift for an unbound state at an energy E_r relative to the nucleon threshold ($E_r > 0$) is

3.8.34
$$\Delta_r = -(3\hbar^2/2\mu R^2)\Theta_r^2 P_\lambda (F_\lambda F'_\lambda + G_\lambda G'_\lambda),$$

where F_λ and G_λ are the regular and irregular Coulomb wave functions (treating the scattering of neutrons won't do for this calculation!) and the primes denote their derivatives with respect to ρ. Langanke et al. (1986) considered the case of the $^{19}Ne(p, \gamma)^{20}Na$ reaction, applying spectroscopic information from ^{20}F, the analog nucleus of ^{20}Na, to perform as definitive a determination for the $^{19}Ne(p, \gamma)^{20}Na$ reaction rate as possible in the absence of reaction data that resulted in ^{20}Na as the final state nucleus. Since they could assume that Θ_p for the case of interest is the same as Θ_n for analog states in ^{20}F, they could then determine, from their calculations of Δ_b and Δ_r, the locations of the states of interest in ^{20}Na from the expression

3.8.35 $$E_b - E_r = \Delta_b - \Delta_r.$$

A great deal is known about this excitation energy region from spectroscopic studies previously performed that led to the analog nucleus ^{20}F. From these data, Langanke et al. (1986) were able to determine that what had been previously identified as a broad low-lying resonance (which is unlikely, as noted in sec. 3.9) was probably a triplet of states. Langanke et al. (1986) were able to determine the resonance energies and, hence, the partial widths for the entrance channel protons. With estimates of the partial widths for the γ-decays, they were then able to calculate the reaction rate for the ^{19}Ne(p, γ)^{20}Na reaction, of importance to the burning of hydrogen at high temperatures. We will return to this type of burning, the so-called rp-process, in chapter 8.

Partial Widths for Radiative Capture Reactions

Equally important to determining the locations of the resonances and the particle partial widths is determining the partial widths for γ-decays, since reactions involving γ-decays are extremely important to nuclear astrophysics. Unfortunately, this is a more difficult task even than determining the essential information for the charged particles. Determining the transition probabilities requires, clearly, knowledge of the spins and parities, as these will limit the possibilities of the decays. Beyond that, calculation of the γ-decay rates requires knowledge of the nuclear wave functions, obviously a near impossibility unless a measurement can be made. However, one can at least make an educated guess as to those values from the following systematics; for this see the table below (from Blatt and Weiskopf 1962, who performed estimates of the integrals given in sec. 3.4). In the table the partial widths are given in eV, E_γ is given in MeV, and A is given in atomic mass units. As observed in Segre (1977) and Rolfs and Rodney (1988), these were calculated assuming that only a single particle actually took place in the transition. Thus this assumes an extreme single-particle wave function, that is, two states with no configuration mixing. Since that represents a nuclear extremum, these Γs might be regarded as an *upper* limit. Complex wave functions, however, may have many small components that can produce large enhancements through the collectivity they can produce; this will greatly enhance these widths. Thus, these definitely are not upper limits on the partial widths; the actual widths may be considerably larger even than indicated here.

Transition type	Energy and mass dependence
Electric dipole $E1$	$\Gamma(E1) = 6.8 \times 10^{-2} A^{2/3} E_\gamma{}^3$
Magnetic dipole $M1$	$\Gamma(M1) = 2.1 \times 10^{-2} E_\gamma{}^3$
Electric quadrupole $E2$	$\Gamma(E2) = 4.9 \times 10^{-6} A^{4/3} E_\gamma{}^5$
Magnetic quadrupole $M2$	$\Gamma(M2) = 1.5 \times 10^{-8} A^{2/3} E_\gamma{}^5$

Note that the charged-particle partial widths are often the order of an MeV, but those for the gamma rays are the order of an eV. The transition energies, indicated in the above table as E_γ, do not indicate proximity of either the initial or final state with respect to the Coulomb barrier; indeed, that clearly has no bearing on the values of Γ_γ. However, the Coulomb barrier does have an enormous effect on particle partial widths; as discussed in section 3.10, they would be large near the top of the Coulomb barrier but rapidly decreasing as the bombarding energy decreases because of the changing barrier penetrability. At some point, the Coulomb barrier reduces the barrier penetrability of the particle partial widths to the point at which the Γ_γ-value exceeds that of the particle partial widths, as discussed above.

Rarely in nuclear astrophysics is it possible to measure cross sections at energies as low as they occur in stars (although technology is approaching this, as is discussed in chap. 5). Nonetheless, in many of the cases of interest the states associated with the contributing resonances can be observed, and their energies and total and partial widths can be measured. Specifically, the direct reaction studies referred to in section 3.2 have provided essential information for many astrophysically interesting cross sections, often for reactions for which those cross sections could not possibly be observed directly. With the spectroscopic information from these studies it has been possible to calculate the contributions of these resonances to the reaction rates (see next section) just as if they had been measured directly. In many cases the resonances dominate the reaction rate, so this approach has proved to be extremely valuable. However, it should be noted that uncertainties can have a large effect on the results; as noted above the reduced widths are very sensitive to the choice of the radius a, and the uncertainties on spectroscopic strengths are rarely better than 20 %. Furthermore, the energies of the resonances are crucial; a slight shift in the energy can translate into a huge uncertainty in the reaction rate (as discussed in the next section). Nonetheless, these estimated rates are often the only information that can be obtained; examples of such rates and the analytical forms they take will be discussed in section 3.12.

3.9 Thermonuclear Reaction Rates

Thermonuclear reaction rates are crucial to nuclear astrophysics. Through them, experimental cross sections are included in the equations that go into the network codes that are used to describe the processes of energy generation, nucleosynthesis, and evolution in stars. The laboratory cross sections must be determined as excitation functions, that is, cross sections over a range of energies. Of course, most stellar environments are characterized by a temperature, which describes the range of energies of the constituents of that environment. Assume that we have nuclei A_1, initially at rest, interacting with nuclei a_1, incident on the A_1 target nuclei with velocity v, which can produce nucleus A_2 from their interaction. We may characterize the double differential rate at which A_2 nuclei are produced as

3.9.1
$$d[dN(A_2)]/dt = \sigma(v)v\, dN(a_1)N(A_1),$$

where the Ns are the number densities of the constituents and $dN(a_1)$ is the number of nuclei a_1 with velocity v with respect to the A_1 nuclei, which are assumed for the moment to be fixed. The a_1 nuclei will have a velocity distribution

3.9.2
$$dN(a_1) = N_1\varphi(v)d^3v,$$

where

3.9.3
$$\int_0^\infty \varphi(v)d^3v = 1.$$

This gives the expression

3.9.4
$$dN(A_2)/dt = N(a_1)N(A_1)\int_0^\infty \sigma(E)v\varphi(v)d^3v$$

as the rate at which A_2 nuclei are produced per unit volume in this specialized situation. Of course, both A_1 and a_1 are in thermal equilibrium, so both are characterized by distributions. If these are characterized by Maxwell-Boltzmann distributions:

3.9.5
$$\varphi(v) = 4\pi v^2[m/(2\pi k_B T)]^{3/2}\exp[-mv^2/2k_B T],$$

where k_B = the Boltzmann constant = 1.38×10^{-16} ergs K^{-1}. The resulting expression for $dN(A_2)/dt$ then becomes a double integral over the velocities of A_1 and a_1. If we transform this to an integral over the relative velocities and the velocity of the center of mass, the integral over the latter coordinates is seen to just be the normalization integral for the distribution, and the integral over the relative velocities gives the result

3.9.6 $$dN(A_2)/dt = N(A_1)\,N(a_1)\langle \sigma v\rangle,$$

where

3.9.7 $$\langle \sigma v\rangle = (8/\pi\mu)^{1/2}(k_B T)^{-3/2}\int_0^{\infty} E\sigma(E)\exp[-E/k_B T]dE$$

and is known as the thermonuclear reaction rate for the process $a_1 + A_1 \rightarrow A_2$ + anything else. In the above equation, E is the center-of-mass energy, μ is the reduced mass of a, and A, and σ is the cross section for the reaction being considered.

Nonresonant Form of the Reaction Rate

The thermonuclear reaction rate is often desired for nonresonant interactions between two charged particles. In this case the form that $\langle \sigma v\rangle$ takes will depend on whether or not resonances exist in the low-energy cross section; they often do. However, since the actual rate may depend both on resonances and the thermonuclear reaction rate in the absence of resonances, both forms are required. For interactions between two charged particles, the Coulomb barrier dominates the low-energy cross section, which drops sharply as the center-of-mass energy decreases. On the other hand, the Maxwell-Boltzmann distribution drops sharply as the center-of-mass energy increases. The penetrability of the Coulomb barrier by charged particles of charges $z_1 e$ (for nuclei a_1) and $Z_1 e$ (for nuclei A_1) moving with relative velocity v can be shown to be approximately (see sec. 3.10)

3.9.8 $$\text{penetrability} = \exp[-2\pi z_1 Z_1 e^2/\hbar v] = \exp(-b E^{-1/2}).$$

In equation 3.9.8, $b = (2\mu)^{1/2}\pi e^2 z_1 Z_1/\hbar$. It is useful to parameterize the cross section between the two charged particles in terms of this penetrability and the astrophysical S-factor as

3.9.9 $$\sigma(E) = [S(E)/E]\exp(-b E^{-1/2}]).$$

Since cross sections are generally proportional to $1/E$ and the exponential term effectively accounts for the Coulomb barrier, $S(E)$ will be relatively independent of energy if the cross section is nonresonant. The basic problem that has existed throughout the history of nuclear astrophysics is that the measurements of nuclear cross sections that are performed in the laboratory rarely can be extended to the low energies that are relevant to stars. Thus, use of the astrophysical S-factor helps greatly in extrapolating the laboratory cross sections to the energies at which they are relevant to stellar processes.

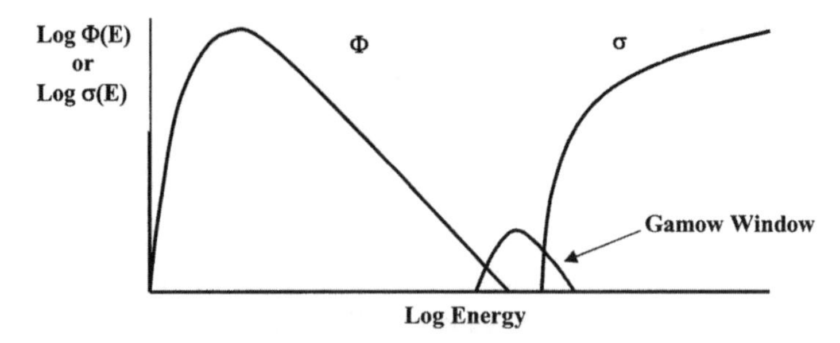

Fig. 3.9. The contributors to the nonresonant thermonuclear reaction rate. Shown are the cross section between two charged particles σ and the Maxwell-Boltzmann distribution, Φ. Their product results in the Gamow window.

The reaction rate now takes the form

3.9.10 $$\langle \sigma v \rangle = (8/\pi\mu)^{1/2}(k_B T)^{-3/2}\int_0^{\infty} S(E)\exp(-E/k_B T - bE^{-1/2})dE.$$

If the cross section is nonresonant, the term $\exp(-E/k_B T - bE^{-1/2})$ defines the energy region over which the entire contribution to the integral arises. This is known as the Gamow window. The two factors that give rise to it, the Boltzmann distribution and the cross section for two charged particles, are indicated in figure 3.9. The region in which their product peaks is usually in the region where both have become small. The Gamow window can be approximated by a Gaussian. The energy at which the Gaussian peaks, E_o, and its full width at $1/e$, Δ, are given as

3.9.11 $$E_o = (bk_B T/2)^{2/3} = 1.22\left(z_1^2 Z_1^2 \mu T_6^2\right)^{1/3} \text{keV},$$

3.9.12 $$\Delta = (4/3^{1/2})(E_o k_B T)^{1/2} = 0.749\left(z_1^2 Z_1^2 \mu T_6^5\right)^{1/6} \text{keV},$$

where T_6 is the temperature expressed in millions K; for example, if the temperature is 10^7 K, $T_6 = 10$ and μ is the reduced mass in atomic mass units, u. If $S(E)$ is fairly constant over the range of that Gaussian and does not vary rapidly at other energies, it can be taken outside the integral. The resulting integral over a Gaussian gives the analytic expression for the nonresonant reaction rate

3.9.13 $$\langle \sigma v \rangle = (2/\mu)^{1/2}\Delta(k_B T)^{-3/2} S(E_o)\exp(-3E_o/k_B T).$$

3.9.14 $$= (2/\mu)^{1/2}(k_B T)(4/9)(3)^{-1/2} E_o^{-3/2} S(E_o)\tau^2 \exp(-\tau),$$

where

3.9.15 $$\tau = 3E_o/k_B T = 42.46\left[z_1^2 Z_1^2 \mu / T_6\right]^{1/3}.$$

This can be cast into numerical form to give

$$\langle \sigma v \rangle = 7.20 \times 10^{-19} \, [\mu z_1 Z_1]^{-1} S(E_o) \tau^2 \exp[-\tau] \, \mathrm{cm}^3/\mathrm{s}. \tag{3.9.16}$$

In this expression, $S(E_o)$ is in keV·b, μ is in atomic mass units, and z_1 and Z_1 are the (integer) proton numbers for their respective nuclei.

Of course, this is the zeroth-order approximation for the reaction rate, but it is often very close to the actual rate. Refinements to account for slow energy variations of the S-factor and for the differences between the Gaussian and the term it approximates can be introduced; these are discussed below. They will produce the additional terms that one finds in the tabulated nonresonant reaction rates in section 3.12.

This expression also allows us to infer the basic dependence of the reaction rate on temperature and nuclear charge. Since the Gamow window arises from the exponentially decreasing high-energy tail of the energy distribution of the interacting nuclei and the exponentially increasing low-energy tail of the charged-particle cross section, anything that affects either of those distributions can greatly impact the reaction rates. A temperature increase will both increase the number of nuclei at the higher energies (exponentially) and allow the Gamow window to shift to a somewhat higher energy, thus sampling a larger cross section. Conversely, an increase in the charge of one of the interacting nuclei will both decrease the cross section at low energy because of the increased Coulomb barrier and force the Gamow window to shift to higher energy and hence to energies with fewer nuclei. Understanding these qualitative features of the nonresonant reaction rate will allow us to infer some important general results later on.

Resonant Form of the Reaction Rate

By contrast, if the cross section is resonant, most of the contribution to the integral will arise from the energy region near the resonance, provided it is within the Gamow window. Resonances, as described earlier, are related to nuclear states that are above zero energy but that can be excited in the compound nucleus if the two projectiles are near the incident energy that corresponds to such a state. In the previous section, we obtained the resonance form of the cross section; this is the Breit-Wigner cross section for an interaction through the ℓth partial wave:

$$\sigma_{\mathrm{BW},\lambda} = (2\ell + 1)\pi(1/k)^2 \Gamma_1 \Gamma_2 [(E - E_R)^2 + (\Gamma/2)^2]^{-1}, \tag{3.9.17}$$

where, as discussed above, Γ is the total width of the resonance, E_R is the resonance energy, k is the wave number in the center of mass ($=p/\hbar$), and Γ_1

and Γ_2 are the partial widths of the entrance and exit channels of the reaction. As noted above, the partial width for each reaction channel is a measure of the probability that channel will be populated in the decay of the resonance, so it is related to the nuclear structure of the associated level. The probability that the resonance will be formed is proportional to the partial width for the entrance channel. The partial widths also contain the barrier penetrability factor where that is appropriate. For example, in a (p, γ) reaction, Γ_1 will be related to the component of the wave function in the compound state formed in the entrance channel that looks like a proton coupled to the ground state of the target nucleus; Γ_1 does contain a factor of the barrier penetrability for charged-particle induced reactions. There may be several $\{\Gamma_{2i}\}$ for γ-decay of the compound state thus formed to lower lying states of the composite nucleus, each with its respective probability of being populated, which is related to its Γ_{2i}.

If the Breit-Wigner expression for the resonance cross section is inserted into the general expression for the thermonuclear reaction rate given in equation 3.9.7, and the resonance is sufficiently narrow that little variation in the other terms in the integrand will result over its width (as is often the case, especially for resonances at low center-of-mass energies), then the other terms can be set to their value at E_R and taken outside the integral, leaving only an integral over the resonance cross section. This results in the standard expression for the narrow resonance reaction rate of

3.9.18 $$\langle \sigma v \rangle = \hbar^2 (2\pi/\mu k_B T)^{3/2} g_J (\Gamma_1 \Gamma_2/\Gamma) \exp(-E_R/k_B T).$$

Note that the spin statistical factor, g_J, has been included in equations 3.8.32 and 3.9.18 to include the effects of spins. Reaction theories usually assume that all transitions between any two spin substates are equally probable. Then g_J simply accounts for the fact that, for target nucleus of spin J_i, incident projectile of spin j_i, and resonance of spin J, there are $(2J_i + 1)(2j_i + 1)$ possible spin projections in the entrance channel; reaction theories average over these. Then, when a compound nuclear state is formed, there will be $2J + 1$ possible ways the state associated with the resonance can be formed; reaction theories sum over these possibilities. Thus, for resonance reactions, g_J is given by

3.9.19 $$g_J = (2J + 1)/[(2J_i + 1)(2j_i + 1)].$$

Equation 3.8.18 is usually recast, using the definition

3.9.20 $$\omega\gamma = \omega\Gamma_1\Gamma_2/\Gamma,$$

with $\omega = g_J$, as

3.9.21 $$\langle \sigma v \rangle = \hbar^2 [2\pi/\mu k_B T]^{3/2} (\omega\gamma) \exp[-E_R/k_B T] f.$$

3.9.22 $$= 8.09 \times 10^{-12} (\omega\gamma)_3 \mu^{-3/2} T_6^{-3/2} \exp(-11.605\, E_3/T_6) f \text{ cm}^3 s^{-1}.$$

The f added to equation 3.9.22 is a factor that multiplies the cross section to account for shielding by the electrons; it is essentially equal to 1.0 except at low energy (see sec. 4.2). However, such energies often include those in the region of the Gamow window. The factor $\omega\gamma$ depends critically on the Coulomb barrier. Near the top of, or above, the Coulomb barrier it will be expected that $\Gamma_\gamma \ll \Gamma_{\text{particle}} \approx \Gamma$, so $\omega\gamma \approx \omega\Gamma_\gamma$. At lower energy, $\Gamma_\gamma \approx \Gamma \gg \Gamma_{\text{particle}}$, so $\omega\gamma \approx \omega\Gamma_{\text{particle}}$. Also, $\omega\gamma_{\text{higher}E}$ is usually $\omega\gamma_{\text{lower}E}$ because of the Coulomb barrier and the fact that Γ_γ is driven by the electromagnetic interaction, which is intrinsically considerably weaker than the strong interaction, which drives Γ_p. This suggests that resonances are likely to be more difficult to observe at energies well below the top of the Coulomb barrier; this is borne out by experiment. Since those resonances of greatest interest to astrophysics are usually well below the top of the Coulomb barrier, however, this presents a continual challenge to the experimentalists in this field.

Note also that in general both direct and resonant reactions can occur and in principle they can interfere. This interference is usually difficult to observe since the different components of the reaction strength are themselves difficult to observe, so it is often just assumed that no interference between the different components occurs and the two components are added incoherently. However, this may not be the case; it is definitely important in some cases that we will discuss in subsequent chapters.

Note that the spin statistics factor for the direct reaction case is given by the sum over the final states and an average over the initial states:

3.9.23 $$g_J = [(2j_f + 1)(2J_f + 1)] / [(2j_i + 1)(2J_i + 1)],$$

The spin of the compound nuclear state does not appear, as it obviously must not.

Once the thermonuclear reaction rate has been computed, the expression for the rate at which the final nucleus is formed is, as given above,

3.9.24 $$\text{Rate}_{aA} = N(a_1) N(A_1) \langle \sigma v \rangle / (1 + \delta_{aA}),$$

where the number densities of a_1 and A_1 are now included. Note that if the two interacting particles are identical, then only one new particle is produced for each two incident particles of that type. This is taken into account by the term that contains a Kronecker delta in equation 3.9.24.

In some situations the resonances will be broad. Then the approximations made for narrow resonances cannot be used, so full numerical integration may be required to obtain the reaction rates.

Species Lifetimes under Stellar Burning

A concept often used to discuss the processes involving nuclei is that of the lifetime τ of an individual species that is subject to a particular destruction process; it is obviously given by the inverse of its reaction rate for that process, that is,

3.9.25 $$\tau = \text{Rate}_{aA}^{-1}.$$

Inverse Reactions

Inverse reactions may be important for some reactions and/or under some stellar conditions. Generally, the reactions that dominate nuclear astrophysics are required to have positive Q-values since the incident (thermal) energies are usually low. However, if the Q-value is small and/or the temperature is high, a reaction may proceed in both directions. In these cases, the net reaction rate is what is important for nucleosynthesis and energy generation. In most cases, a tiny inverse reaction rate will occur simply because only the particles on the high-energy tail of the Maxwell-Boltzmann distribution can create the inverse reactions. Fortunately, the inverse reaction rate is calculable, usually to a good approximation, from the forward rate, so it can be included in the expression for the net rate with some additional spin and correction factors.

Consider the reaction $1 + 2 \to 3 + 4$, where the numbers denote the nuclei involved. We showed that we can write the cross section for this process, assuming it proceeds through a compound nuclear state, as

3.9.26 $$\sigma_{12\to34} = \pi(k_{12})^{-2}\{(2J+1)/[(2J_1+1)(2J_2+1)]\}|\langle 12|H_{\rm I}|C\rangle|^2 \\ |\langle C|H_{\rm II}|34\rangle|^2,$$

where in this case J is the spin of the compound nuclear state. If the process goes by a direct reaction, the spin statistical factor changes somewhat, as indicated in equation 3.9.23, but the basic argument to be presented does not change. The matrix elements involve transition operators that take the two particles in the entrance channel to the compound nuclear state and then to the two projectiles in the exit channel. The quantity k_{12} is the wave number = $p_{12}/\hbar$ for the entrance channel.

Similarly we can write

3.9.27 $$\sigma_{34\to12} = \pi(k_{34})^{-2}\{[(2J+1)/[(2J_3+1)(2J_4+1)]\}|\langle 34|H_{\rm II}|C\rangle|^2 \\ |\langle C|H_{\rm I}|12\rangle|^2,$$

where the various factors are defined similarly to those for equation 3.9.26 and the matrix elements are the same. This is the principle of time-reversal invariance.

Thus since, for example, $\mu_{12} = m_1 m_2/(m_1 + m_2)$, we can write

3.9.28
$$\sigma_{12}/\sigma_{34} = [(m_3 m_4 E_{34})/(m_1 m_2 E_{12})][(2J_3 + 1)(2J_4 + 1)]/[2J_1 + 1)(2J_2 + 1)].$$

Then

3.9.29
$$\langle \sigma v \rangle_{12} = (8/\pi \mu_{12})^{1/2} (k_B T)^{-3/2} \int_0^\infty \sigma_{12} E_{12} \exp(-E_{12}/k_B T) dE_{12},$$

and similarly for $\langle \sigma v \rangle_{34}$. If we now define $E_{34} = E_{12} + Q$, where Q is the Q-value in the nucleus 1 + nucleus 2 channel, this produces the result

3.9.30
$$\langle \sigma v \rangle_{34}/\langle \sigma v \rangle_{12} = [(2J_1 + 1)(2J_2 + 1)]/[(2J_3 + 1)(2J_4 + 1)] \times (\mu_{12}/\mu_{34})^{3/2} \exp(-Q/k_B T).$$

The inverse rates are generally given along with the forward-going rates in the reaction rate tabulations (see sec. 3.12).

The total rate r at which a reaction proceeds is clearly the difference between the forward going rate and the inverse rate, that is,

3.9.31
$$r = r_{12 \to 34} - r_{34 \to 12} = [N_1 N_2/(1 + \delta_{12})] \langle \sigma v \rangle_{12} - [N_3 N_4/(1 + \delta_{34})] \langle \sigma v \rangle_{34},$$

This can be rewritten as

3.9.32
$$r = [\langle \sigma v \rangle_{12}/(1 + \delta_{12})]\{N_1 N_2 - N_3 N_4[(2J_1 + 1)(2J_2 + 1)]/[(2J_3 + 1) \times (2J_4 + 1)](\mu_{12}/\mu_{34})^{3/2} \times [(1 + \delta_{12})/(1 + \delta_{34})] \exp(-Q/k_B T)\},$$

where the N_i are the number densities of the various particle species in the stellar environment. At equilibrium, the N_i adjust so that $r = 0$, and (in the absence of other processes)

3.9.33
$$[(1 + \delta_{12})/(1 + \delta_{34})](N_3 N_4/N_1 N_2) = [(2J_3 + 1)(2J_4 + 1)]/[(2J_1 + 1)(2J_2 + 1)](\mu_{12}/\mu_{34})^{-3/2} \times \exp(Q/k_B T).$$

Note, however, that this might never occur. For example, consider successive reactions involving multiple light particles, as will be seen in chapter 5 to arise in hydrogen burning:

3.9.34
$$X + {}^1\mathrm{H} \to Y + \gamma,$$

3.9.35
$$Y + {}^3\mathrm{He} \to Z + \gamma,$$

and so on.

The first reaction might not ever come to equilibrium because Y is being both synthesized and consumed. If one of these reactions equilibrates slowly, or if the abundances of both components in one reaction are changing (as might be the case in eq. 3.9.35), equilibrium of the set of reactions may never occur. Thus the equilibrium conditions discussed in the context of equation 3.9.33 apply to a restricted case. We will see more complicated examples of such equilibrium processes in subsequent chapters.

It is important to note that the expression in equation 3.9.28 technically only applies to forward and inverse reactions between two specific states in the entrance and exit channels. Since reactions can also occur between, for example, the ground state in the entrance channel and excited states in the exit channel and vice versa, equations 3.9.30 and 3.9.32 are only an approximation to what actually occurs. Furthermore, at high temperature, excited states of the entrance channel may be populated, and reactions including those states in the entrance channel can occur. This further complicates achieving an accurate description of the total reaction rate.

Inclusion of the inverse reaction rates in the total reaction rate can be especially important to radiative capture reactions, for example, $^{13}\mathrm{N}(\gamma, \mathrm{p})^{12}\mathrm{C}$ as the inverse reaction for $^{12}\mathrm{C}(\mathrm{p}, \gamma)^{13}\mathrm{N}$. For the latter reaction, $Q = 1.95$ MeV, so at very high temperatures the inverse reaction can contribute appreciably to the reaction processes. Unfortunately, the structure of the two energy-averaged cross sections involving two particles or one particle and a photon are not the same, so the handy relationship that was derived in equation 3.9.30 for two particles in each channel is not so easily obtained for the case in which one of the "particles" is a photon. One can estimate the importance of the inverse reactions by simply calculating the result. But first one must calculate the rate of photodisintegration.

To indicate how this comes about, denote (following Clayton 1983) the rate at which composite nucleus 3 is photodisintegrated back into nuclei 1 and 2 by $r_\gamma = N_3\lambda_\gamma$, where λ_γ is the rate at which that process occurs and N_3 is the density of composite nuclei N_3. We wish to determine an expression for λ_γ, the photon-induced counterpart to the $\langle\sigma v\rangle$ values we have determined above for thermally averaged cross sections involving two particles. In thermodynamic equilibrium, the values of N_1, N_2, and N_3 would adjust so that the rates at which nuclei 3 are created and destroyed would be equal, that is,

3.9.36 $$\lambda_\gamma(N_3)_e = (N_1)_e(N_2)_e\langle\sigma v\rangle_{12}.$$

In this situation the Saha equation would apply:

3.9.37 $$(N_1)_e(N_2)_e/(N_3)_e = [(2\pi\mu k_\mathrm{B}T)^{3/2}/h^3][G_1G_2/G_3]\exp(-Q/k_\mathrm{B}T),$$

where Q is the reaction Q-value for the $1 + 2 \rightarrow 3 + \gamma$ reaction, μ is the reduced mass in the $1 + 2$ system, and G_1, G_2, and G_3 are the nuclear partition functions given, for example, for the ith state in nucleus 1, as

3.9.38 $$G_{1i} = (2J_{1i} + 1)\exp(-E_{1i}/k_B T)/\sum_j (2J_{1j} + 1)\exp(-E_{1j}/k_B T),$$

where E_{1j} is the excitation energy of the $(j-1)$th excited state of nucleus 1 ($j = 1$ is the ground state) and J_{1j} is its nuclear spin. However, the cross-section relationship must apply even if thermal equilibrium has not been achieved, so we can write

3.9.39 $$\lambda_\gamma = [(2\pi\mu k_B T)^{3/2}/h^3][G_1 G_2/G_3]\exp(-Q/k_B T)\langle\sigma v\rangle_{12}.$$

Thus the net rate at which particles 1 and 2 combine to form particle 3 in this situation will be given by

3.9.40 $$r = N_1 N_2 \langle\sigma v\rangle_{12} - N_3 \lambda_\gamma.$$

One can compare the rate λ_γ to that for the forward-going reactions given in equation 3.9.26. As noted above, one can determine the importance of the photodisintegration rate at any given temperature by simply calculating the magnitude of the effect. At $T_9 > 0.75$, the photodisintegration of ^{13}N is faster than its β-decay (Clayton 1983). This could impact the s-process (see chap. 7), as it relies on ^{13}N as a source of neutrons, although the temperatures would not be expected to be as high as $T_9 = 0.75$ during s-processing.

Energy Production in Thermonuclear Reactions

Clearly, the rate at which energy is produced in reactions such as $1 + 2 \rightarrow 3 + 4$, ϵ_{12} is (the rate at which the reaction proceeds) × (the energy produced per reaction), that is,

3.9.41 $$\varepsilon_{12} = Qr_{12}(\text{MeV cm}^{-3}\text{s}^{-1})$$

3.9.42 $$= QN_1 N_2 \langle\sigma v\rangle_{12}.$$

The net rate at which energy is produced will be given by the difference between this rate and its inverse reaction rate, times Q, that is,

3.9.43 $$\text{Net energy production rate} = \varepsilon_{12} - \varepsilon_{34}.$$

Temperature Dependence of Reaction Rates

It is often useful to compare the relative dependence of reaction rates on temperature. Knowledge of this dependence often makes it possible to obtain at least a qualitative understanding of the way in which the various reactions of stellar burning depend on temperature and assume different relative importance as the temperature varies. If it is assumed that the form for the nonresonant reaction rate is proportional to $\tau^2 e^{-\tau}$ (from eq. 3.9.14) and that rate has the power law proportionality T^{α}, it can be shown that (see problem 12)

3.9.44 $$\langle \sigma v \rangle \propto T^{(\tau-2)/3}.$$

We will return to this expression in subsequent chapters.

3.10 Penetrability through a Barrier and Spectroscopic Coefficients

The "penetrability" refers to the ability of particles trapped inside the potential barrier that results from the combination of the Coulomb potential and angular momentum terms in the Hamiltonion to tunnel through the barrier and escape from the nucleus. We will see that treating the penetrability separately will allow us to relate the partial widths $\Gamma_{\lambda\ell}$ directly to the nuclear structure of the states of interest. If we write

3.10.1 $$\psi_{\ell m}(r, \theta, \varphi) = [u_\ell(r)/r] Y_{\ell m}(\theta, \varphi),$$

for partial wave ℓ, then the radial component of the solution to the Schrodinger equation for $u_\ell(r)$ becomes

3.10.2 $$-(\hbar^2/2\mu) d^2 u_\ell(r)/dr^2 + [\ell(\ell+1)\hbar^2/2\mu r^2 + V(r) - E] u_\ell(r) = 0.$$

The potential $V(r)$ has the form

3.10.3 $$V(r) = \begin{cases} V_{\text{Coulomb}} = z_1 Z_1 e^2/r, & r > R_n \\ V_C + V_{\text{nuc}}, & r < R_n, \end{cases}$$

where μ is the reduced mass of the $a_1 + A_1$ system and R_n is the (usually assumed hard-sphere) radius of the nuclear part of the interaction potential between a_1 and A_1. This situation is illustrated in figure 3.7, which shows schematically the sum of the two potentials.

Although the wave function can be determined from the above equation, to obtain a solution in terms of some nuclear model, for example, the shell model, it is necessary to do considerably more work. We will not pursue that approach for the present but rather will just note that the actual wave function may in general involve many configurations regardless of which nuclear model is used, for example,

3.10.4 $$u_\ell = \sum_j \Theta_{\ell j} u_{\ell j},$$

where the $\{u_{\ell j}\}$ might be, for example, states in the particle-core coupling model (see, e.g., eq. 3.3.1). In equation 3.10.4, the coefficients $\Theta_{\ell j}$ ($0 < \Theta_{\ell j} < 1$) indicate the amplitudes of each of the configurations in the wave function. Since $\Theta_{\ell j}^2$ gives the probability of the configuration $u_{\ell j}$, it also gives the probability of decay of each compound nuclear state to each state in the residual nucleus; they represent the spectroscopic strength of their specific wave function configuration. For example, in the particle-core coupling model, this might represent the decay to a continuum proton plus the core state indicated in $u_{\ell j}$. In some cases several of the $u_{\ell j}$ might have the same core state. (For example, a $d_{3/2}$, $d_{5/2}$, and $s_{1/2}$ neutron coupled to a 2^+ core state could all form $J^\pi = 3/2^+$ or $5/2^+$ states.) Then all such states could decay to that core state. Of course, some of those decays might not be energetically allowed, even though the configurations might be possible in the wave function. Nonetheless, this approach does allow us to develop a relationship between the width of the state $\Gamma_{\ell j}$ and the coefficients in the wave function, the $\Theta_{\ell j}$.

This spectroscopic strength is related to the spectroscopic factor introduced in section 3.2 by

3.10.5 $$\Gamma_p = C^2 \mathcal{S} \Gamma_{\mathrm{sp}},$$

where Γ_{sp} is the calculated proton width for a pure single-particle state, that is, it would be the value for Γ_{p} if the value of Θ in equation 3.10.4 for a single particle coupled to the ground state of the target nucleus were 1.0 and all other $\Theta_{\ell j}$ were 0.0. Note that this S is not the same as the S used for the astrophysical S-factor. Although the standard notation uses the same letter for both, we will use the script S for the spectroscopic factor to distinguish between them.

Since $\Gamma_{\ell j}/\hbar$ is the inverse of the lifetime of the compound nuclear state, it is just the rate at which the probability leaks out of the nucleus when it is in that state. This can be approximated (semiclassically) by

3.10.6 $$\Gamma_{\ell j}/\hbar = v|u_{\ell j}(\infty)|^2 \approx v\, P_\ell(v)|u_{\ell j}(R_n)|^2,$$

where v is the speed of the particle at infinity, $P_\ell(v)$ is the penetrability of the barrier at the energy (or velocity at infinity) of the compound nuclear state, and $|u_{\ell j}(R_{\mathrm{n}})|^2$ is the probability of finding the particle within dr of the nuclear surface in the $u_{\ell j}$ state. This is just the ratio of the surface to the volume, which is $3/R_{\mathrm{n}}$ so, using that approximation

3.10.7 $$\Gamma_{\ell j} = (3\hbar v/R_n)\, P_\ell(v)\Theta_{\ell j}^2.$$

A more sophisticated calculation would take into account the actual wave function of the state, but the present estimate will suffice for our purposes.

Since the total width Γ_ℓ will be the sum of the partial widths, that is, it represents the decay of the compound nuclear state into all exit channels,

3.10.8 $$\Gamma_\ell = \sum_j \Gamma_{\ell j} = (3\hbar v/R_n)\,P_\ell(v)\sum_j \Theta^2_{\ell j}.$$

Note also that, since $\Gamma_{\lambda\ell} \propto 2\gamma^2 P_\ell$ (see eq. 3.8.27) in the Breit-Wigner resonance cross section, γ^2 is just proportional to the spectroscopic strength $\Theta^2_{\lambda\ell}$, giving the spectroscopic strengths the same indices for this comparison to the $\Gamma_{\ell\lambda}$ of section 3.8.

One can also solve directly for the penetrability. Although the differential equation can be solved numerically it can also be solved by the WKB approximation (Clayton 1983); it is useful to do so to observe the astrophysical consequences of the various terms. The general form of the solution in the WKB approximation is

3.10.9 $$P_\ell(E) = (E_c/E)^{1/2}\exp\left\{-2^{3/2}\mu^{1/2}/\hbar\int_R^{R_o}[z_1 Z_1 e^2/r + \ell(\ell+1)\hbar^2/(2\mu r^2) - E]^{1/2}dr\right\}$$

which can be integrated to give the result

3.10.10 $$P_\ell \approx (E_c/E)^{1/2}\exp(-2\pi z_1 Z_1 e^2/\hbar v + 4\{E_C/[\hbar^2/(2\mu R_n^2)]\}^{1/2} - 2\ell(\ell+1)[\hbar^2/(2\mu R_n^2 E_C)]^{1/2}),$$

where $E_C = z_1 Z_1 e^2/R_n$ and z_1 $(Z)_1$ is the charge number of nucleus 1 (2). Note that the first term inside the brackets, which is usually the dominant term, is that which is factored out of the cross section to give the much less energy-dependent astrophysical *S*-factor (see eqs. 3.9.9 and 3.9.10). Note, however, that the other terms can be significant and that they do occur in the exponent. Since the ability of a charged particle to enter a nucleus at low energy in, for example, a proton capture reaction, depends on its ability to penetrate the Coulomb barrier, the above expression indicates the extent to which $\ell = 0$ waves are favored over higher angular momentum waves. Indeed, it is useful to rewrite the above expression as

3.10.11 $$P_\ell(E)/P_0(E) \approx \exp\{-2\ell(\ell+1)[\hbar^2/(2\mu z_1 Z_1 e^2 R_n)]^{1/2}\},$$

3.10.12 $$= \exp[-7.62\ell(\ell+1)/(\mu z_1 Z_1 R_n)^{1/2}],$$

where μ is in amu and R_n in femtometers in equation 3.10.12. For example, for the reaction $^{12}C(\alpha,\gamma)^{16}O$, a very important reaction in helium burning, to which we will return in chapter 6, $\ell = 0$ is favored over $\ell = 1$ by a factor of about 5. But $\ell = 0$ is favored over $\ell = 2$ by 2 orders of magnitude.

For many decades the information needed to calculate the resonance contributions to reaction rates has been obtained through direct reactions, as discussed in section 3.2. For example, a ${}^{A}Z({}^{3}\mathrm{He}, \mathrm{d})^{A+1}(Z + 1)$ reaction will transfer a proton to nucleus (A, Z). If the transfers can be studied to states in $(A + 1, Z + 1)$ that are important to astrophysical processes in (p, γ) reactions on (A, Z), then the strengths at which those states are populated in, for example, a $({}^{3}\mathrm{He}, \mathrm{d})$ reaction will correspond to their (p, γ) reaction strengths. Furthermore, one might be able to measure directly the widths Γ of the states. In many cases, this is sufficient to calculate the contributions of those states to the resonance reaction rate.

3.11 Statistical Model Nuclear Cross Sections

It is often useful to be able to calculate nuclear cross sections from their general properties. This has led to the development of a statistical model of cross sections, often called Hauser-Feshbach cross sections after their originators (Hauser and Feshbach 1952). These take into account the fact that the nuclear level densities for heavier nuclei will be sufficiently high at the typical energies of incident projectiles that there will be many overlapping resonances and possibly several reaction channels. In this case, an average over the properties of the resonances can be performed. Since we are talking about resonance scattering, we can write the scattering, as was done in equation 3.8.2, as a product of the cross section that forms the compound nuclear state followed by a decay probability, that is,

3.11.1 $$\sigma \propto |\langle A\, p | T_1 | A + 1^* \rangle|^2 |\langle A + 1^* | T_2 | A + 1\, \gamma \rangle|^2.$$

With constants added, and ignoring spins for the moment, the first half of the right-hand side is just the cross section for formation of the compound nucleus. The second half is just proportional to the probability that the compound nucleus will decay into a particular channel, in this case, the nucleus $A + 1$ and a gamma ray. If we assume the Breit-Wigner cross section, given in equation 3.9.17, and write the maximum value for it, which assumes that the energy is at the resonance energy, it becomes

3.11.2 $$\sigma_{\mathrm{BW}}^{\max} = (2\ell + 1)(4\pi / k^2)\Gamma_1 \Gamma_2 / \Gamma_2$$

where, as noted in section 3.9, $\Gamma = \Sigma_i \Gamma_i$. This clearly defines Γ_2/Γ to be the probability of the decay of the compound nuclear state into channel 2.

In equation 3.10.7, the partial widths were written in such a way as to contain spectroscopic information. However, it is generally assumed in the Hauser-Feshbach formalism that there are so many levels being averaged over in any realistic situation, that is, the level density is so high, that all the strength of a particular type will be contained within the energy sample. This is rarely true, especially for light nuclei, but that assumption, together with the assumption of the maximum value for the Breit-Wigner cross section, sets an upper limit for the cross section we will obtain. If it assumed that the spectroscopic strengths in all the partial widths are unity, then the partial widths can be written as factors times the penetrability, which is usually written as a "transmission coefficient" T_ℓ. Then the Hauser-Feshbach cross section can be written, in its simplest form, as

3.11.3

$$\sigma_{\mathrm{HF}} = (\pi/k^2) T_1(\ell_1) T_2(\ell_2) / \sum_i T_i(\ell_i),$$

where now the dependences of the different transmission factors on their respective orbital angular momenta are indicated explicitly. Of course we have assumed so far that the spins are all zero. We can easily insert the spins as was done above, by assuming an average over the initial states and sum over the final state, which in this case is the compound nuclear state. Indeed, this confirms the statement made above that the spin statistical factor to be used when a compound nuclear state is involved is just $(2J+1)/[(2j_{\mathrm{p}}+1)(2J_{\mathrm{t}}+1)]$, where J is the spin of the compound nuclear state and j_{p} and J_{t} are the spins of the incident projectile and the target nucleus.

Of course we must sum over the orbital angular momenta, or, equivalently, the total angular momenta and parities, of all possible resonances. Including also the spin statistical factors then gives

3.11.4

$$\sigma_{\mathrm{HF}} = (\pi/k^2)[2j_p+1)(2J_t+1)]^{-1} \eta \sum_{J,\pi} (2J+1) T_1(J,\pi) T_2(J,\pi) / \left[\sum_i T_i(J,\pi) \right],$$

where the transmission coefficients can be calculated with an expression such as 3.10.9 or with a typical nuclear + Coulomb potential. The summing indices i include all relevant entrance channel and exit channel spins and angular momenta. The coefficient η is added to account for the fact that the statistical model cross sections generally overestimate the actual cross sections somewhat, so that η is generally of order 0.1 to 1.0 (see, e.g., Somorjai et al. 1998; Sauter and Kappeler 1998; Chloupek et al. 1999; Ozkan et al. 2002).

3.12 Standardized Reaction Rate Formulae

Over the past several decades the basic forms of the different terms in the reaction rates have been well established, and (often rather complicated) analytical forms were given. Two groups of modern nuclear astrophysicists have now compiled tabulated data sets of reactions that are relevant to nuclear astrophysics. Of course, the tabulations greatly simplify the effort needed to do computations involving astrophysical reaction rates. However, it is worthwhile understanding how these rates arise, since one does sometimes have to use the analytical forms, and in any event it is useful to understand the origins of the terms in the analytical expression.

Higher order terms in the analytical expressions for the reaction rates than those indicated in equations 3.9.16 and 3.9.22 arise both from the variation of the astrophysical S-factor with energy and from the imperfection of the Gaussian approximation of the integrand in equation 3.9.10. The time-honored approach (see, e.g., Rolfs and Rodney 1988) is to define the correction factor to the approximation to the integrand, after $S(E)$ has been moved to the left of the integral, as

3.12.1
$$\int_0^\infty \exp(-E/k_\mathrm{B}T - bE^{-1/2})dE = F(\tau)\pi^{1/2}(\Delta/2)\mathrm{e}^{-\tau},$$

where $F(\tau)$ is the correction. One can then expand $\exp(-E/k_\mathrm{B}T - bE^{-1/2})$ in a Maclaurin series in $1/\tau$ (since $\tau \gg 1$) and perform the integral to get $F(\tau)$ (see Clayton 1983). It is (Rolfs and Rodney 1988)

3.12.2
$$F(\tau) = 1 + (5/12)(1/\tau) - (35/288)(1/\tau)^2 + \cdots$$

To obtain some indication of the importance of this correction, for the $\mathrm{p} + \mathrm{p}$ reaction at $T_6 = 15$, roughly the temperature at the core of the Sun, $F(\tau)$ is 1.03, a testimonial to the accuracy of the Gaussian approximation.

Similarly the energy variation of the astrophysical S-factor can be represented by an expansion, that is,

3.12.3
$$S_\mathrm{eff}(E_\mathrm{o}) = S(0) + S'(0)E + (1/2)S''(0)E^2 + \cdots,$$

where, for example, $S''(0)$ is the second derivative of $S(E)$ with respect to energy at $E = 0$. The expansion is generally done around the origin for convenience. Although an expansion around E_o would be technically more appropriate, E_o is energy dependent, so this could unnecessarily complicate the expansion.

Combining these two corrections into an effective S-factor now gives

3.12.4
$$S_\mathrm{eff}(E_\mathrm{o}) = S(0)\{1 + (5/12\tau)[S'(0)/S(0)][E_\mathrm{o} + (35/36)k_\mathrm{B}T] + (1/2)[S''(0)/S(0)][(E_\mathrm{o}^2 + (89/36)E_\mathrm{o}k_\mathrm{B}T]\}.$$

Because of the dependence of E_o on T, this expansion ends up depending on powers of $T^{1/3}$, so that, finally, $\langle \sigma v \rangle$ can be expanded as

$$\langle \sigma v \rangle = (2/\mu)^{1/2} \Delta (k_B T)^{-3/2} S_{eff}(E_o) \exp(-3 E_o / k_B T), \tag{3.12.5}$$

$$= A T^{-2/3} \exp(-B T^{-1/3}) \sum_{n=0}^{5} \alpha_n T^{n/3}, \tag{3.12.6}$$

where the coefficients A and B depend on the reaction, as do the coefficients α_n in the expansion.

The higher order terms usually are small, but that is not always the case. In such cases, different analytical expressions have sometimes had to be assumed. Perhaps the most obvious example is for $^{12}C + {}^{12}C$, where the energy dependence of the low-energy cross section imposes a very unusual form for the S-factor:

$$\underline{S} = S(E) \exp(-gE), \tag{3.12.7}$$

where g is usually taken to be around 0.46 MeV^{-1} and S is defined in the usual way (see eq. 3.9.9). Other low-energy heavy-ion fusion reactions, such as $^{16}O + {}^{16}O$ also appear to exhibit unusual behavior, but $^{12}C + {}^{12}C$ seems to provide the most extreme example. A great deal of attention has been given this reaction by nuclear physicists because of its obvious importance to the carbon-burning phase of stellar evolution (see chap. 6). Of course this cross section translates into an unusual thermonuclear reaction rate, since it must contain all the odd behavior of the cross section. The reaction rate for the $^{12}C + {}^{12}C$ reaction, as well as for the other heavy-ion reactions, is given at (http://www.phy.ornl.gov/astrophysics/data/cf88/analyt_rates.html). Specifically the rate for $^{12}C + {}^{12}C$ is

$$r = 4.27 \times 10^{26} T9A^{5/6} T9^{3/2} \exp(-84.165/T9A^{1/3} - 2.12 \times 10^{-3} T9^3), \tag{3.12.8}$$

where $T9 = T_9$, the temperature in GK, and $T9A = T9/[1.0 + 0.0396^* T9]$. The interested student can pursue this cross section and the other heavy-ion cross sections further by looking into the references given in the section on carbon burning in Rolfs and Rodney (1988).

One set of compiled reaction rates is to be found on the Nuclear Physics Compilation of REaction (NACRE) rates Web site, http://pntpm.ulb.ac.be/nacre.htm. In these compilations, the data are presented in tabular form, in which the rates are calculated at a variety of temperatures from an analytical expression.

For examples, the analytical expression for the $^{14}N(p, \gamma)^{15}O$ reaction rate to the ^{15}O ground state and for its inverse reaction are given in the NACRE data set (in the NACRE notation) as

$$\begin{aligned} \mathrm{N14pg}_{gs} &= 4.83 \times 10^7/T^{2/3}\exp(-15.231/T^{1/3} - (T/0.8)^2)(1 - 2.00T \\ &\quad + 3.41T^2 - 2.43T^3) + 2.36 \times 10^3/T^{3/2}\exp(-3.010/T) \\ &\quad + 6.72 \times 10_3 T^{0.380}\exp(-9.530/T), \end{aligned} \tag{3.12.9}$$

where T is in GK. The rev. ratio = $2.699 \times 10^{10}\ T^{3/2}\exp(-84.677/T)$.

Those for the ^{17}O(p,α)^{14}N reaction to the ^{14}N ground state are given as

$$\begin{aligned} \mathrm{O17pa}_{gs} &= 9.20 \times 10^8/T^{2/3}\exp(-16.715/T^{1/3} - (T/0.06)^2) \\ &\quad (1 - 80.31T + 2211T^2) + 9.13 \times 10^{-4}/T^{3/2}\exp(-0.7667/T) \\ &\quad + 9.68/T^{3/2}\exp(-2.083/T) + 8.13 \times 10^6/T^{3/2}\exp(-5.685/T) \\ &\quad + 1.85 \times 10^6 T^{1.591}\exp(-4.848/T), \end{aligned} \tag{3.12.10}$$

for $T_9 \leq 6$. The rev. ratio = $0.6759\exp(-13.829/T)$. Note that the first rate is relatively uncomplicated, while the second, because it can depend on several resonances, contains a larger number of terms.

Another thermonuclear reaction rate data set is that compiled by astrophysicists at Oak Ridge National Laboratory. This Web site can be found at http://www.phy.ornl.gov/astrophysics/data/data.html. One important aspect of this Web site is that it includes one of the more extensive nuclear astrophysics data sets ever compiled, that of Caughlan and Fowler (1988). Other older data compilations also exist (in some cases, just as the analytical expressions for the reaction rates): Caughlan et al. (1985); Fowler, Caughlan, and Zimmerman (1967, 1975).

3.13 Neutron-Induced Reaction Cross Sections

All the preceding sections on reactions have dealt with those occurring between a charged particle and a (charged) nucleus. However, neutron-induced reactions, especially (n, γ) reactions, also play an important role in nuclear astrophysics, so their systematics also need to be developed. Obviously the major difference between reactions with charged particles and those with neutrons is the absence of the Coulomb barrier for the latter. Indeed, although reactions between charged particles decrease sharply at low energies because of the Coulomb barrier, those between neutrons and nuclei increase as the energy goes to zero.

The standard prescription for a cross section (see eq. 3.8.10) has a $1/E$, that is, a $1/v^2$, factor multiplying the rest of the terms. While this would be the dominant term for neutron-induced reaction cross sections if only s-wave scattering occurred, this is usually moderated by low-energy resonances and the presence of some p-wave processes, both of which conspire to produce an overall $1/\nu$ dependence for most of the neutron-induced reactions, especially if the data are energy averaged.

Of course, energy averaging is relevant for (n, γ) cross sections for astrophysics, as the stellar environment automatically produces this effect. What is needed in nuclear astrophysics is the thermonuclear reaction rate, and this requires determining $\langle \sigma \mathrm{v} \rangle$. However, for neutrons, $\sigma \propto 1/\nu$ so, usually to a good approximation,

3.13.1 $$\langle \sigma \mathrm{v} \rangle \approx \text{constant} \approx \langle \sigma \rangle v_T,$$

where ν_T is usually assumed, for experimental convenience, to be $\nu_T = (2k_{\mathrm{B}}T/m)^{1/2}$ for $k_{\mathrm{B}}T \approx 30$ keV. Thus, for any specific reaction, the neutron capture cross section can be measured at one energy, thereby determining, to a good approximation, that cross section for the entire energy spectrum relevant to astrophysics. This has been applied in the measurements required for the s- and r-processes, to be discussed in chapter 7.

In some cases it has turned out to be important to measure the cross sections over a wide energy range, as there are significant variations from the $1/\nu$ dependence. Since it is not known in advance the specific cases to which this will apply, it is important to measure as many of the (n, γ) cross sections as possible in order to provide an accurate characterization of the s- and r-processes. This has been done for many of the reactions of importance to the former process, but it is virtually impossible to do so for the r-process reactions, simply because they involve very short-lived nuclei.

Note that what has been discussed is the energy dependence for "typical" neutron capture reactions. However, the actual cross section can vary widely with energy; there can be many low-energy resonances that somehow often average out to produce the approximate $1/\nu$ dependence assumed. Furthermore, the magnitude of these cross sections varies widely over the periodic table and in ways that can be extremely important to the processes of nucleosynthesis. This is particularly important to the r-process.

As noted above, what is important in astrophysics is actually energy-averaged cross sections, with temperatures assumed to be those of the environment in which the neutron captures are taking place. In order to obtain cross sections of the accuracy required for detailed simulations of the nucleosynthesis that occurs in the different processes of nucleosynthesis, especially the s-process, two approaches have been used. One is to measure the cross sections over a wide energy range in order to measure them as a function of energy and then perform the required energy averaging after the cross sections have been determined. This approach is used at the CERN (Rubbia

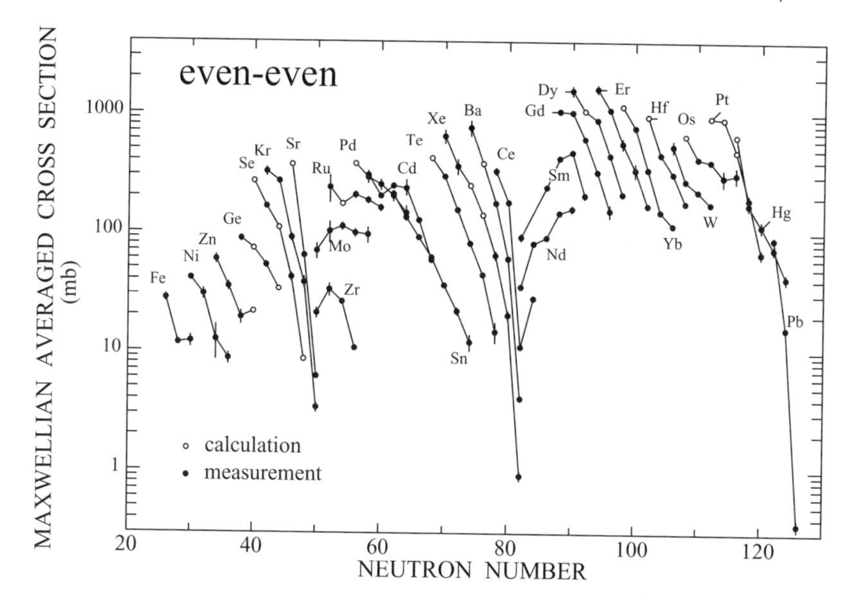

Fig. 3.10. Maxwellian averaged stellar (n, γ) cross sections at 30 keV throughout the periodic table. This figure is intended to show the mass dependence of these cross sections, along with the sharp changes, in some cases spanning orders of magnitude, that occur at the neutron shell closures. Odd-A nuclei tend to have considerably larger cross sections. From http://pntpm3.ulb.ac.be/Trento/talks/pdf/fkaeppeler.pdf. Reprinted with permission from F. Kaeppeler.

et al. 1998; Tagliente and n_TOF Collaboration 2004), and Oak Ridge National Laboratory (Koehler et al. 1996) facilities. The other is to produce a neutron beam that has an energy distribution similar to that of a thermal environment. This latter approach has been used extensively in many measurements at the facility at Karlsruhe (Beer, Kaeppeler, Reffo, and Venturini 1983).

The Oak Ridge Electron Linac Accelerator (ORELA) was used to bombard a Ta or Be neutron production target with 180-MeV electrons to produce a white neutron source, that is, one with a flux that is fairly uniform over a wide range of energy. (The CERN neutron source utilizes a 20-GeV proton beam on a lead target to produce the neutrons.) The ORELA electron beam is pulsed at a rate of around 500 Hz, and the time of flight of the neutrons, and hence their energy, was measured over a flight path of 9–200 m (longer flight paths are used to produce higher energy resolution) by determining the time difference between the event time and the beam burst on the neutron production target. In this way, the yield as a function of neutron energy can be measured. The energy averaging over a Maxwellian distribution can then be performed. Note that this approach can measure the details of the energy-dependent cross section, most notably deviations from the $1/v$ dependence and the presence of resonances.

The other approach, that used at Karlsruhe, utilizes a proton beam to bombard a ^{7}Li target, and neutrons are produced by the (very prolific) ^{7}Li(p, n) reaction. All the neutrons are emitted in a forward cone of 120° and, because of reaction kinematics, the energy distribution of those neutrons presents a good approximation to a Maxwellian distribution. With this facility, the reaction yields that are measured are already energy averaged. Of course, details of the cross sections would be difficult to determine with this approach; its beauty is in the ease at which it can produce the cross sections of interest. Indeed, an enormous amount of work has resulted in an extraordinary body of knowledge of neutron capture cross sections as a function of mass and isotope. These are shown in figure 3.10. Especially noteworthy in this figure is the sharp drop in neutron capture cross section at the neutron closed shells at $N = 50$, 82, and 126.

CHAPTER 3 PROBLEMS

1a. Consider an e^{-} decay $(Z, N) \rightarrow (Z+1, N-1)$, where Z is even and $N = Z + 1$. Derive a general expression, using the atomic mass formula, for the energy available to the decay products.

1b. Assuming that nuclei have a uniform charge distribution with hard-sphere radius given by $R = 1.2A^{1/3}$ fm, determine all the constants in your expression from part 1a.

1c. Now consider an e^{+} decay $(Z, N) \rightarrow (Z-1, N+1)$. Derive, as before, a general expression for the available energy for this process and determine the energy available for $^{13}\text{N} \rightarrow {}^{13}\text{C}$. How does your value compare to reality?

2a. Consider the ordering of the shell-model orbitals. What would you guess would be the spin and parity of the first excited state of ^{40}Ca (there are several possible answers), assuming the ground state (as it is for all even-even nuclei) spin and parity are 0^{+} and the first excited state is made by a single-particle hole excitation?

2b. What would you expect the ground state spin and parity of ^{41}Sc to be? Why?

2c. What spins and parities for ^{41}Sc could you make by coupling the maximum spin value you got for part 2a to the value you got from part 2b? All of these states have been found to exist!

2d. Consider the shell-model structure of the state of maximum spin from part 2a and the state of maximum spin from part 2c. Are there any possible configurations of the state of maximum spin and parity that are precluded by the Pauli principle?

3. How many states does ^{9}Be have that are bound to proton or neutron emission? You will need to determine its (γ, n) and (γ, p) thresholds.

4. Nucleons in nuclei have their spin and orbital angular momentum strongly coupled through the spin orbit force, so a "jj-coupling" scheme is appropriate for describing the angular momentum states in nuclei. If that were not the case, however, an "LS-coupling" scheme would be appropriate. Verify that the number of nucleons permitted to occupy the states in each scheme is the same when both total angular momenta in the jj-coupling scheme for each value of orbital angular momentum are included and given the fact that two spin states can occur for each orbital angular momentum value in the LS-coupling scheme.
5. Explain the statement made in section 3.2 that the negative parity states in ^{17}O that have two neutrons in $d_{5/2}$ orbits and a neutron hole in the $1p_{3/2}$ or $1p_{1/2}$ orbits would be strongly populated in a $^{18}O(d, ^3H)^{17}O$ reaction but would be expected to be weakly populated in a $^{16}O(d, p)^{17}O$ reaction.
6. Determine the Q-value, using atomic mass tables or the masses from the Internet, for the $^{12}C(d, p)^{13}C$ reaction.
7. Work through the matrix algebra to derive equations 3.7.10 and 3.7.11.
8. Derive equation 3.7.7, the expression for the two-state neutrino oscillation probability.
9. Derive the expression for the resonant oscillation length for MSW oscillations, as in equation 3.7.14.
10. Show that, with η_ℓ defined as in equation 3.8.24, it is possible also to write δ_ℓ as in equation 3.8.25. Hint: write $1 - R_\ell S_\ell - iR_\ell P_\ell = e^{i\alpha}$. Then $\tan \alpha = R_\ell P_\ell/(1 - R_\ell S_\ell)$.
11. Show that the ratio of the forward and inverse cross sections, equation 3.9.28, follows from equation 3.9.27. Then derive equation 3.9.30, the relationship between the forward and inverse thermonuclear reaction rates.
12. Assume the nonresonant form of the reaction rate, given in equation 3.9.14, and assume that it is proportional to T^x. By differentiating both sides of the proportionality, derive equation 3.9.44.
13. Show that equation 3.8.29 can be derived from equation 3.8.28, that is,

$$\begin{aligned}\delta_\ell &= -\varphi_\ell + \tan^{-1}[\gamma_\lambda^2 P_\ell/(E_\lambda - E)]/[1 - \gamma^2 S_\ell/(E_\lambda - E)] \\ &= -\varphi_\ell + \tan^{-1}[(\Gamma_{\lambda\ell}/2)/(E_{\lambda\ell} - E - \Delta_{\lambda\ell})]\end{aligned}$$

14a. Assuming $\exp(-E/k_B T - b/E^{1/2}) = I_{max} \exp[-(E - E_o)^2/(\Delta/2)^2]$, derive equation 3.9.11, the expression for E_o.

14b. Matching the second derivatives in the expression in problem 4, derive equation 3.9.12, the expression for Δ.

14c. Then perform the integral for $\langle \sigma v \rangle$ to obtain equation 3.9.13.

15. The $^{12}C(p, \gamma)^{13}N$ reaction cross section has a $J^{\pi} = 3/2^{-}$ resonance at 424 keV in the center of mass and a peak cross section at that energy of 1.2×10^{-4} barns. The resonance has a width, essentially the proton width, since the only other possible channel, γ-decay, will have a width much smaller than that, of $\Gamma = 40$ keV. What is the value for the dimensionless reduced width θ_p^2 for that state?

16. Show that if some nuclear species X in a star can be destroyed by several reactions (or possibly β-decay), its total lifetime is given by $\tau^{-1} = \Sigma_i \tau_i{}^{-1}$.

17. Compute the energy of the Gamow peak for the $^{12}C(p, \gamma)^{13}N$ reaction for a temperature at the center of the Sun, $T_6 = 15$. Then check the NACRE S-factors to determine the S-factor and dS/dE for the $^{12}C(p, \gamma)^{13}N$ reaction at that temperature. Having determined those values, compute the lifetime of ^{12}C against this reaction in a stellar environment having 50 % hydrogen by weight, a density of 80 g cm^{-3}, and a temperature of 15×10^6 K. Ignore electron screening.

18. Combine the factors that go into expanding the differences among the actual Gamow window and a Gaussian and the S-factor, as discussed in section 3.12, to derive equation 3.12.6.

19. Explain the comment in section 3.1 that the Sun would have burned up long ago if there had been a stable-mass 5 u nucleus.

20. Show that the slope of the electron energy spectrum for allowed β-decays is zero near its endpoint if the neutrino has zero rest mass but becomes infinite if it has a finite rest mass. To do so you will have to go through the derivation at the beginning of section 3.5, but without the assumption that the neutrino rest mass is zero. This endpoint behavior has been used to attempt to measure the neutrino rest mass.

21. Two types of experiments can be used to determine if neutrinos oscillate: appearance experiments, in which a neutrino of a type not in the original beam appears, or disappearance experiments, in which the flux of the original beam falls off faster than would be expected (as in the *KamLAND* experiment). Using the mixing matrix given in equation 3.7.18, show that the disappearance probability that would be measured in a reactor experiment (in which $\bar{\nu}_e$s are emitted) is given by

$$P[\bar{\nu}_e \rightarrow \mathrm{Not}\bar{\nu}_e] \approx \sin^2 2\theta_{13} \sin^2 \Delta_{31} + \cos^4 \theta_{13} \sin^2 2\theta_{12} \sin^2 \Delta_{21}$$

where, for example,

$$\Delta_{21} = 1.27 \Delta m_{21}^2 (\mathrm{eV})^2 L(\mathrm{km})/E\ (\mathrm{GeV}), \text{ and } \Delta m_{21}^2 = m_{\bar{\nu}_2}^2 - m_{\bar{\nu}_1}^2.$$

4

STELLAR BASICS OF NUCLEAR ASTROPHYSICS

4.1 Photon-Electron Scattering Processes

The scattering of photons from electrons will generally dominate the interactions of the photons as they pass through matter. These will fall into four basic categories:

- *Bound-bound absorption.* This is the process in which a photon is absorbed by a bound atomic or ionic electron, producing an excited state of the atom or ion. This is a true absorption process, as there is no scattered outgoing photon. The inverse of this process is deexcitation by emission of a photon.
- *Bound-free absorption.* In this process, a photon is absorbed by a bound electron, producing either an ion or a more highly ionized state plus an unbound (continuum state) electron. This is also a true absorption process, as there is no photon in the final state. The inverse of this process is radiative recombination of the electron into an atom or ion.
- *Free-free absorption.* This is a process in which a photon interacts with an unbound (continuum) electron and promotes it to a more highly excited continuum state. This is also a true absorption process, as there is no photon in the final state. Its inverse is bremsstrahlung.
- *Scattering from electrons.* This process is not an absorption process, as the photon exchanges energy with the electron from which it scatters and can either gain or lose energy in the scattering process. There is a photon in the final state. This process is called Compton scattering when the process occurs between a photon and an electron.

The cross sections for all these processes have been well studied and are generally calculable through quantum electrodynamics. They will not be dealt

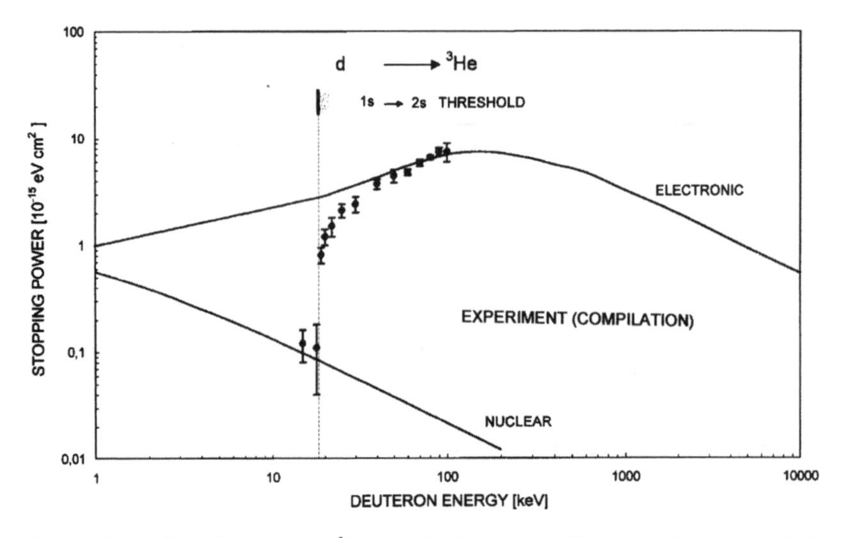

Fig. 4.1. Energy loss of deuterons in ^{3}He gas. The dramatic cutoff occurs at the point at which the incident deuterons can no longer ionize the ^{3}He gas atoms. From A. Formicola et al. (2000).

with in this text; the interested reader is referred to Clayton (1983), in which the calculations of these different processes are discussed. Empirical formulae are also given in Pagel (1997).

4.2 Debye Shielding

The thermonuclear reaction rates presented in chapter 3 assume interactions between bare nuclei, that is, it is assumed that the Coulomb interaction between the charged particles exists to infinity. Of course, the targets in which the cross sections are determined experimentally are atoms, so they are shielded to a small extent by their electrons. This is usually a minor effect, although at very low energies that is not the case. Indeed, in some classic experiments by Formicola et al. (2000), it was shown that the energy loss of ions as they moved through matter suddenly dropped precipitously as the ion energies decreased in energy. This dramatic effect is shown in figure 4.1; it results from the fact that the ions are losing their energy to atomically bound electrons until the ion's energy becomes too low to separate the electrons from the atoms in the medium. While this does not usually affect astrophysical considerations, the differences between the atoms on which reactions are measured and the fully stripped ones that exist in stars can sometimes be extremely important and must be taken into account in some situations.

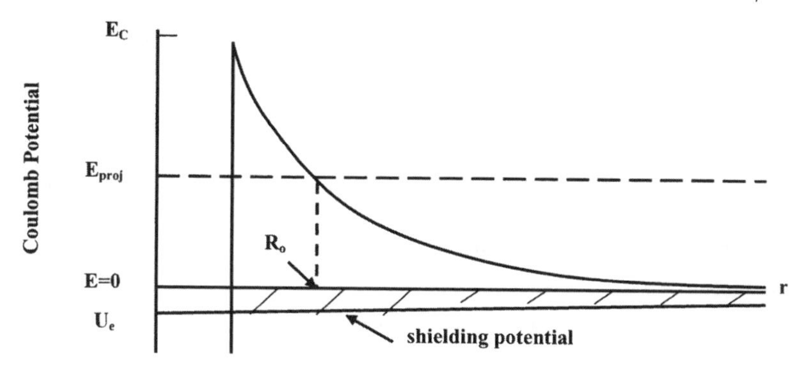

Fig. 4.2. Schematic of the effect of the electron cloud on the Coulomb potential in a dense plasma. The potential is reduced at all distances, but most importantly at the nuclear radius and goes essentially to zero beyond the Debye-Huckel radius. This increases the penetrability through the barrier and hence the cross section. The energy of the incident projectile is E_{proj}.

Although nuclei in stars are often fully ionized, the continuum electrons tend to clump around the nuclei, so they also provide some screening. However, the effect of the atomic shielding is not the same as that provided by the clumping of the continuum electrons, so one needs to distinguish between the screening effects that exist in stars and those that exist in the experimental conditions under which the nuclear reaction cross sections are measured.

We first consider the effects of electron screening on stellar reaction rates. Technically, this is the shielding that occurs at low density, that is, the shielding that occurs at the densities that occur in normal stellar burning. The shielding that occurs, for example, in neutron star crusts can be quite different. Figure 4.2 indicates the situation that results from the screening produced by the electron clumping, that is, there will be a shift in the Coulomb potential between the unshielded Coulomb potential and the shielded potential. Because the reactions that occur in stars depend so sensitively on quantum mechanical tunneling, a small shift in the Coulomb barrier can cause large corrections in the thermonuclear reaction rate determined by correcting measured cross sections that have shielding by the atomic electrons, to the bare nuclear case, to the case that actually exists in stars, which includes plasma electron shielding.

It was shown by Clayton (1983) that the shielding introduced an additional potential term into the Schrodinger equation that accounted for the screening. For the moment we will proceed through the analysis of the solution with that assumption and then return later to the actual form of the screening potential.

This will affect the penetrability of the charged projectiles, having charges Z_1e and Z_2e, as they enter or exit through the barrier. The expression that needs to be solved, using the WKB approximation (and following Clayton 1983), and denoting the added potential term as U_e, is

4.2.1 $$P_\ell(E) = (E_c/E)^{1/2} \exp[-2^{3/2}\mu^{1/2}/\hbar] \int_R^{R_0} [(Z_1 Z_2 e^2/r + U_e - E)^{1/2} dr],$$

where R_0 is the radius at which the projectile energy intersects the potential curve (see fig. 4.2).

Since U_e is negative, it will act in the opposite direction of the (repulsive) Coulomb potential resulting from the two positively charged particles. We will show below that it usually will vary slowly over the region where it is important, which is equivalent to saying that the incident energy is usually considerably larger than U_e. This is illustrated in figure 4.2.

Following Clayton (1983), in the absence of the electron shielding the reaction rate integral can be written as

4.2.2 $$r_{12} = N_1 N_2 \int_0^\infty \Phi(E) v\sigma(E) dE,$$

where $\Phi(E)$ is the probability that the center of mass kinetic energy at large separation is E. Because of the screening potential, however, the penetrability will not be that for the actual center of mass energy but rather will be that for that energy $-U_e$. This will make the penetrability at any energy the equivalent of the penetrability at a slightly higher energy than would be the case without U_e, thereby increasing the penetrability. A similar shift will occur inside the nucleus, so the penetrability of particles escaping the nucleus will be shifted by roughly the same amount (since U_e varies slowly with radius). If the cross section is rewritten as

4.2.3 $$\sigma(E) = P(E) F_{\text{nuc}},$$

where $P(E)$ is the penetrability and F_{nuc} is a purely nuclear factor (see sec. 3.10), equation 4.2.2 can be rewritten, now with the screening correction, as

4.2.4 $$r_{12} = N_1 N_2 \int_0^\infty \Phi(E) v P(E - U_e) F_{\text{nuc}}(E - U_e) dE.$$

The integral can be recast by defining $E' = E - U_e$, by noting that $v \propto E^{1/2}$ and by replacing $\Phi(E)$ by the Maxwell-Boltzmann distribution expression at the energies around the Gamow window, $\exp(-E/k_B T)$, which gives

4.2.5 $$\int_{-Ue}^\infty (E' + U_e)^{1/2} \exp[-(E' + U_e)/k_B T] P(E') F_{\text{nuc}}(E') dE'.$$

However, U_e is generally small, so the lower limit of the integral can be set to 0, and the U_e term can be pulled out of the only term in which it matters, the exponential, to give

4.2.6 $$R_{12} = \exp(-U_e/k_B T) N_1 N_2 \int_0^\infty \Phi(E) v\sigma(E) dE.$$

Thus the shielding factor is just the exponential in front of the integral, and the effect of the shielding can be written as

4.2.7 $$\langle\sigma v\rangle_{\text{shielded}} = f\langle\sigma v\rangle_{\text{bare}},$$

where the "electron shielding factor" f is

4.2.8 $$f = \exp(-U_e/k_B T).$$

Because U_e is negative, $f \geq 1$, that is, it will enhance the stellar reaction rates from their unshielded values.

We will now determine the form of U_e. It has been shown that the electrons in the plasma will cluster into spherical shells around a nucleus at the Debye-Huckel radius R_D (Salpeter 1954; Bahcall et al. 2002), given by

4.2.9 $$R_D = \left[k_B T/(4\pi e^2 \rho N_A \xi)\right]^{1/2},$$

where N_A is Avogadro's number. The quantity ξ is given as

4.2.10 $$\xi = \sum_i (Z_i^2 + Z_i) X_i / A_i,$$

where the sum is over all positive ions and X_i is the mass fraction of nuclei of type i. The Debye-Huckel radius is often the order of the atomic size, as it is in the Sun, but it can be much smaller at high densities, since the electrons are in continuum states.

The effect of the screening is (Salpeter 1954) to introduce an additional exponential attenuation to the normal Coulomb potential for an ion of charge Z_1:

4.2.11 $$\phi(r) = Z_1 e/r \exp(-r/R_D).$$

If $r \ll R_D$ for all radii of interest, that is, if the energy of the incident ions intersects the potential curve inside of R_D (see fig. 4.2), this becomes

4.2.12 $$\phi(r) = Z_1 e/r - Z_1 e/R_D.$$

Then the effect of the shielding is to produce a decrease in the usual Coulomb potential in the form of a simple shift of the height of the barrier, as indicated in figure 4.2.

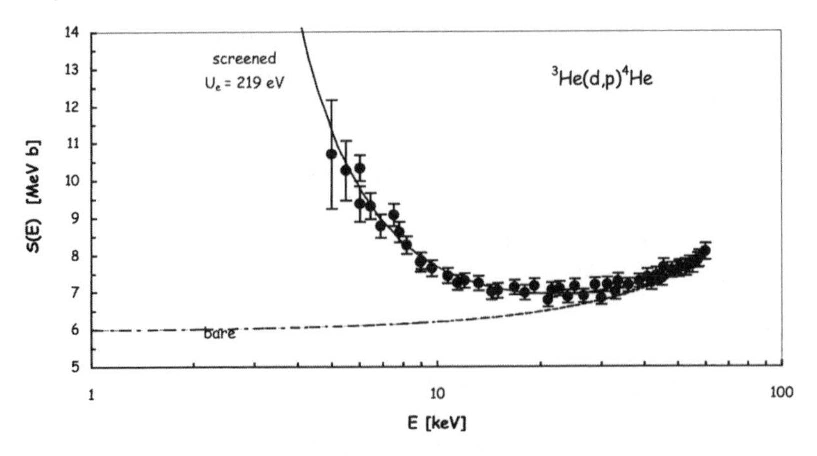

Fig. 4.3. Astrophysical *S*-factor and data for the ^{3}He(d,p)^{4}He reaction, done with the LUNA facility. The errors shown represent only statistical and accidental uncertainties, which were used in the fits. The dashed curve represents the *S*-factor for bare nuclei and the solid curve that for screened nuclei with $U_e = 219$ eV. Reprinted Aliotta et al. [2001], with permission from Elsevier.

Salpeter (1954) determined that the height of the barrier was simply related to the concentration of electrons in the neighborhood of Z_1 by the Boltzmann factor, which is proportional to $\exp(-Z_1 e\phi / k_B T)$, so that the magnitude of the electron shielding factor *f* is, for nucleus of charge Z_2

4.2.13 $$f = \exp\left(Z_1 Z_2 e^2 / k_B T R_D\right).$$

The quantity *f* generally varies (Salpeter 1954) between 1 and 2 for typical stellar densities and compositions. It is usually close to 1.0 except at low energies and for large Z_1. However, this factor can become huge in high densities. Indeed, when the density becomes sufficiently high that degeneracy effects become important, this analysis breaks down (Brown and Sawyer 1997), as would be expected, since the electrons were assumed to be in nondegenerate continuum states.

Some examples of the magnitude of the electron shielding factor on the cross sections for the $p + p \rightarrow d + e^+ + \nu_e$ reaction are as follows (from Rolfs and Rodney 1988): $E = 100$ keV, $f = 1.000027$; $E = 20$ keV, $f = 1.0021$; $E = 5$ keV, $f = 1.023$; $E = 1$ keV, $f = 1.33$.

Recent work, using the underground accelerator facility *LUNA* at the Laboratori Nazionali del Gran Sasso, has shown that the effects can sometimes be very dramatic. Specifically, the *LUNA* group has measured the ^{3}He(d,p)^{4}He reaction cross section down to extremely low energies. What

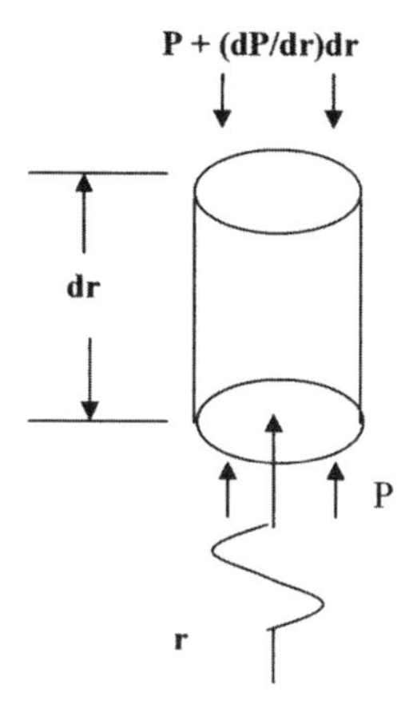

Fig. 4.4. Sketch of the different forces acting on a volume element of gas in a star.

is represented by the unshielded thermonuclear reaction rate is that which would result from a deuteron interaction with a bare ^{3}He nucleus. What is actually measured in the laboratory, however, is the reaction rate that results from a deuteron interacting with a ^{3}He atom, that is, a nucleus shielded by atomic electrons. Generally (Clayton 1983) $r_D \ll r_A$, so the incident d will be unaware of the Debye-shielded ^{3}He nucleus to a much smaller radius than would be the case in a laboratory scattering experiment, in which the ^{3}He nucleus would be shielded by atomic electrons.

In any event, these effects have all been taken into account in figure 4.3. There it is seen that the effects of the electronic shielding are only significant at low energies, at least at solar densities, but that they will have a large effect on the low-energy cross sections.

4.3 Hydrodynamic Equilibrium

It is instructive to consider the conditions that prevail, and the equations that result, from hydrodynamic equilibrium. These conditions actually apply to most of the phases of a star's life; hydrostatic equilibrium is only violated in the final phase of stellar collapse (and, of course, where turbulence exists!). This is not intended to be a complete description of the situation but to give the student of nuclear astrophysics some sense of the origins of the equations and concepts that govern a star's existence.

For equilibrium to exist with a given pressure P and stellar mass, hence gravity (see fig. 4.4),

4.3.1 $$\{P - [P + (dP/dr)dr]\}ds = \rho \; dr \; ds \; GM_r/r^2,$$

where

4.3.2 $$M_r = \int_0^r \rho \, 4\pi r^2 dr,$$

assuming spherical symmetry. The quantities in the expressions are: ρ = density, ds = an area element, P is the pressure, and the other quantities are as defined in the figure. Note that the distribution of mass satisfies the equation

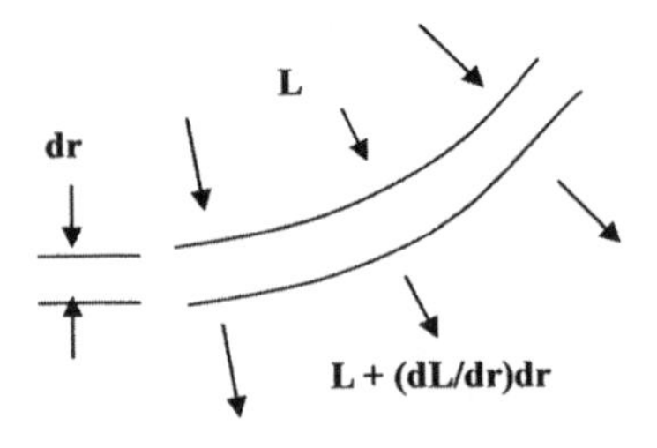

Fig. 4.5. Sketch of the different aspects of energy transfer in the volume of a star.

4.3.3 $$dM_r/dr = 4\pi r^2\rho.$$

Thus

4.3.4 $$dP/dr = -\rho GM_r/r^2.$$

This is one of the basic results that govern stellar structure and evolution.

4.4 Luminosity and Energy Balance

The second equation relates the luminosity into and out of a region of a star to its sources of energy: nuclear, radiative, and that resulting from PdV work. For 1 g of an ideal gas in the shell from r (to the center of the star) to $r + dr$,

4.4.1 $$d(\text{energy})/dt = d(3k_B T/2m)/dt = \varepsilon - \varepsilon' - PdV/dt,$$

where ε represents the nuclear energy generated in the shell, m is the molecular mass, and ε' is the net radiative energy brought in. The expression for ε' is obtained from considering figure 4.5, which shows that

4.4.2 $$[-L + (L + (dL/dr)dr)] = 4\pi r^2\varepsilon'\rho\Delta r, \text{ or}$$

4.4.3

$$\varepsilon' = [4\pi r^2\rho]^{-1}dL/dr$$
= net energy brought into the shell per gram per second via the luminosity.

Assuming that we are dealing with 1 g of matter, $V = M/\rho = 1/\rho$. Thus

4.4.4 $$dV/dt = -\rho^{-2}dr/dt,$$

so that

4.4.5 $$d(3k_B T/2m) = \varepsilon - [4\pi r^2\rho]^{-1}dL/dr + [P/\rho^2]d\rho/dt = d(3P/2\rho)dt,$$

from the ideal gas law. Combining the last two terms gives

4.4.6 $$dL/dt = 4\pi r^2\rho[\varepsilon - (3/2)\rho^{2/3}d(P\rho^{-5/3})dt].$$

This equation shows the relationship between the luminosity and the nuclear energy that is responsible for it.

4.5 Convection versus Radiation?

Considerations of stellar burning and evolution depend strongly on whether or not convection can occur; if it does, it will be the dominant mode of energy transfer. The following considerations show how the conditions within a star determine whether convection can occur (refer to fig. 4.6).

Consider a bubble at location 1. Pressure $P_1 = P_1^*$, and density $\rho_1 = \rho_1^*$. Assume the bubble is displaced a distance dr. If after displacement the bubble continues to move in the same direction, then the result will be convection. If it reverses direction, then convection will not occur, and the mode of energy transport will be radiation.

Assume the bubble is displaced to location 2, $P_2 = P_2^*$. But ρ_2^* can be either greater or less than ρ_2. If $\rho_2^* < \rho_2$, the bubble will continue to displace in the same direction, and there will be convection.

Since this is an adiabatic process, $PV^\gamma =$ constant, so $\rho/P^{1/\gamma} =$ constant′. Thus

4.5.1 $$\rho_2^* = \rho_1^*(P_2^*/P_1^*)^{1/\gamma} = \rho_1^*(P_2/P_1)^{1/\gamma}.$$

We can also write

4.5.2 $$P_2 = P_1 - (dP/dr)dr \text{ and } \rho_2 = \rho_1 - (d\rho/dr)dr.$$

The condition for no convection will be for $\rho_2^* > \rho_2$, that is,

4.5.3 $$\rho_2^* = \rho_1(P_2/P_1)^{1/\gamma} = \rho_1[1 - (1/P_1)(dP/dr)dr]^{1/\gamma}$$
$$\approx \rho_1[1 - (\gamma P)^{-1}(dP/dr)dr] > \rho_2 = \rho_1 - (dP/dr)dr,$$

or

4.5.4 $$(1/\rho)(d\rho/dr) > (\gamma P)^{-1}(dP/dr) \text{ (condition for no convection).}$$

4.6 Chemical Composition and Mean Molecular Weight

The density ρ of any medium can be defined as

4.6.1 $$\rho = N\mu\, M_\mathrm{u},$$

where N is the number of free particles per unit volume, μ is the mean molecular weight in atomic mass units u, and M_u is the mass of 1 atomic mass unit in grams. Thus

4.6.2 $$\mathrm{N} = \rho/(\mu\, M_\mathrm{u}) = N_\mathrm{A}\,\rho/\mu,$$

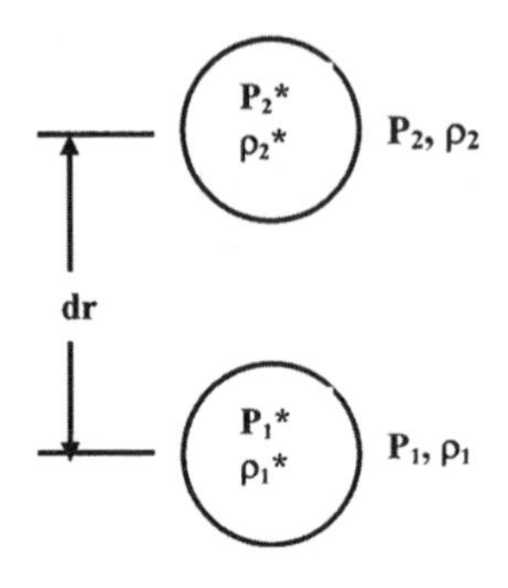

Fig. 4.6. Relevant parameters affecting the motion of gas in a star.

since $N_A = 1/M_u$ = Avogadro's number. The pressure of a gas may be written

4.6.3 $$P_{gas} = Nk_B T = (N_A k_B/\mu)\rho T.$$

Generally, gases will be comprised of a variety of different constituents, and the mean molecular weight can be written in terms of these constituents. Define the fraction by weight of element Z by the X_Z, so that

4.6.4 $$\sum_Z X_Z = 1.$$

Each atom of Z contributes n_Z free particles per cubic centimeter to the gas (this will be equal to $Z+1$ for complete ionization). Denote by N_Z the number density of atoms of Z in the gas and by ρ_Z the partial density due to Z. Then

4.6.5 $$N_Z = (\rho_Z/A_Z)\, N_A = (\rho X_Z/A_Z)\, N_A.$$

Since the number density of particles in the gas due to Z is $N_Z n_Z$, the total number density of particles in the gas can be written as

4.6.6 $$N = \sum_Z N_Z n_Z = \rho N_A \sum_Z X_Z n_Z/A_Z = \rho N_A/\mu.$$

Thus

4.6.7 $$\mu^{-1} = \sum_Z X_Z n_Z/A_Z.$$

If X = weight fraction of hydrogen, Y = the weight fraction of helium, and Z = the weight fraction of all other atoms, we can write the mean molecular weight in its usual form:

4.6.8 $$\mu = [X n_H/1.008 + Y n_{He}/4.004 + (1 - X - Y)\langle n_Z/A_Z\rangle]^{-1},$$

where $\langle n_Z/A_Z\rangle$ denotes the mean value of n_Z/A_Z for all constituents of the gas except hydrogen and helium.

In the case of complete ionization, it can be shown that, to a good approximation,

4.6.9 $$\mu = 2/(1 + 3X + Y/2).$$

The mean molecular weight will be used in some of our subsequent deliberations about stars.

4.7 Degeneracy and White Dwarfs

An example of an astrophysical site that utilizes many of the concepts that we have developed in the last two chapters is a calculation of the conditions that exist within a white dwarf. As noted above, a white dwarf is like a huge atom, which is supported by electron degeneracy pressure and for which the Pauli principle therefore needs to be taken into account. The density of electron states dN_e can be written, using the normalization in a box of side L, as discussed in section 3.5,

4.7.1 $$dN_e = 4\pi n(ndn),$$

where $n_x = p_x L/h$ = the quantum number along the x-axis, p_x is the momentum along that axis, and $n = [n_x^2 + n_y^2 + n_z^2]^{1/2}$. If nonrelativistic particles are assumed, then this can be shown to be

4.7.2 $$dN_e = 4\pi m_e^{3/2}(2E)^{1/2}(L/h)^3 dE.$$

The lowest energy states will be successively filled with electrons until all electrons occupy a state; this defines the Fermi energy, E_F. Taking into account that $L^3 = V$, the volume of the white dwarf, and including an extra factor of 2 for the two spin states of each electron,

4.7.3 $$N_e/V = 2\left(4\pi\, m_e^{3/2}/h^3\right)\int_0^{E_F}(2E)^{1/2}dE,$$

4.7.4 $$= \left(16\pi m_e^{3/2}2^{1/2}/3h^3\right)E_F^{3/2}.$$

We can also use the chemical composition of the star to get N_e, with X = the mass fraction of hydrogen, Y = that of helium, and Z = that of everything else (and assuming that the nuclear charge Q_{nuc} divided by the mass A is approximately 1/2):

4.7.5 $$N_e/V = (\rho/m_H)[X(1/1) + Y(2/4) + Z(Q_{nuc}/A)]$$

4.7.6 $$= (\rho/2m_H)(1 + X).$$

This then gives a solution for E_F:

4.7.7 $$E_F = \left[3\rho(1 + X)h^3/\left(32\pi\, m_H\, m_e^{3/2}2^{1/2}\right)\right]^{2/3}.$$

Note that this is not thermal energy; it cannot be radiated away because there are no unfilled lower energy states to which the higher energy electrons can go. Indeed, at high densities, $E_F \gg k_B T$, so this is very nonthermal.

We can also calculate the pressure due to the electrons. In that case, dP_e, the differential momentum transfer per unit area, is given by

$$dP_e = (2pv)dN_e/6V. \tag{4.7.8}$$

The velocity v is the length of a column "one second" of particles with velocity v, $2p$ is the momentum transfer each particle imparts to the side it strikes, and dN_e is the number of particles in that particular energy (or momentum) bin. Note that, as written, this expression takes into account the fact that the particles do not strike the walls at normal incidence. This expression can be integrated, including an extra factor of 2 to account for the electron spin, as

$$P_e^{nr} = (16\pi/3)(m_e^{3/2}\, 2^{1/2}/h^3) \int_0^{E_F} E^{3/2} dE, \tag{4.7.9}$$

$$= (32\pi 2^{1/2}/15h^3) m_e^{3/2}\, E_F^{5/2}, \tag{4.7.10}$$

$$= (1/40)(3/2\pi)^{2/3}[(\rho(1+X)/m_H)]^{5/3} h^2/m_e. \tag{4.7.11}$$

The superscript "nr" here is to indicate that nonrelativistic expressions were used in deriving the phase space factors from the particle in the box. Similarly, assuming extreme relativistic expressions for the particle in the box yields the expression

$$P_e^r = (1/8)(3/16\pi)^{1/3}[(1+X)\rho/m_H]^{4/3} hc. \tag{4.7.12}$$

4.8 Galactic Chemical Evolution

While most of the considerations of this textbook will involve the synthesis of the nuclides of the periodic table and the related conditions that are thought to exist in stars in order to produce that nucleosynthesis, another effect that must be considered is how the nuclides synthesized in stars are expelled from the star, thereby enriching their abundances in the interstellar medium. Obviously, they must get out of the star in order to be counted as part of the Galactic chemical abundance; those nuclei that remain within the white dwarf or neutron star at the end of a star's evolution cannot be observed. The modern formulation of Galactic chemical evolution (GCE) was developed by Tinsley (1975). However, an excellent textbook on GCE exists, by Pagel (1997), and a good write-up exists on the Web (Holtzman 2005), to which the interested student should turn. Much of what follows below is excerpted from Pagel's book. This discussion is intended only to give the student of nuclear astrophysics an introduction to the basic concepts of GCE.

This evolution of the elemental abundances, and the isotopes therein, as a galaxy ages is denoted by the name Galactic chemical evolution. In describing GCE, one must take into account

- the rate of star formation as a function of time (it may not have been the same in the early Galaxy as it is now),
- the distribution of stellar masses that are formed,
- the lifetimes of stars as a function of their masses,
- how much of their mass they expel into the interstellar medium (ISM),
- the nuclear constituents of that mass, and
- the flow into and out of the Galactic gas.

To set the stage for some of the things that we will have to consider, we will show data in chapter 7 that indicate that the earliest stars that we can observe have much larger abundances of some heavy nuclides compared with iron and oxygen than does our Sun. This effect is well understood, at least qualitatively, in the sense that massive stars go through all their stages of stellar evolution and end their lives (in a core-collapse supernova) much more rapidly than do less massive stars. And the very massive stars tend to produce relatively large abundances of heavy nuclides and less of the more common elements, oxygen and iron, than do less massive stars. Assuming that the present-day Galactic abundances are the result of the well mixed products of many stars of a wide range of masses, the early abundances would therefore be expected to be depleted in oxygen and iron compared with the present-day abundances because the less massive stars are relatively more responsible for their synthesis than are the massive stars and the less massive stars would not have had time to evolve enough to add their synthesized nuclei to the ISM. However, very small stars, even those made at the beginning of the Galaxy, may not even have had time to evolve to their final states. However, such stars are very useful in that, if they are relatively isolated from mixing with the Galactic gas, their composition may therefore reflect the composition of the interstellar gas that existed very early on in the universe. It is obvious that such stars have abundances that are very different from the 70 % hydrogen, 28 % helium, and 2 % metals of our solar system, but the way in which their abundances differ from those of our solar system gives important clues as to how GCE proceeded in the early universe and how stars evolve.

Iron is made in both medium-mass and heavier stars, so that the buildup of iron in the Galaxy can be used to measure the time since the beginning of the Galaxy. However, most of it is produced in the medium-mass stars, so its abundance did not increase as rapidly as those of some of the other elements, for example, oxygen, neon, and the other α-elements, in the early universe. Abundances are generally represented as logarithmic ratios. For example, for iron,

$$[\mathrm{Fe/H}] = \log[(\mathrm{Fe/Fe_{\odot}})/(\mathrm{H/H_{\odot}})], \tag{4.8.1}$$

where, for example, $\mathrm{Fe}_{\odot}$ is the iron abundance in the solar system. The relationship between [Fe/H] and time, or for that matter between any other element and time, is certainly not linear, but the abundances of the elements used as "clocks", for example, Fe and O, are at least monotonic. The relationships between [O/H] and time and between [Fe/H] and time, between Fe and O and between Mg and O, are shown in figure 4.7. There it can be seen that, although there is a lot of scatter of the data around their mean, there is indeed a steady growth of both Fe and O with time. The correlations between Fe and O, and between Mg and O, also shown, show considerably less scatter.

The determination of the time independent of the elemental abundance of the stellar atmosphere is nontrivial; it involves a vast amount of empirical evidence and theoretical information in extracting an age from the properties observed for each star. The details constitute a major treatise in their own right, so they will not be discussed further in this book. However, they are well described in the article by Edvardsson et al. (1993).

In devising a GCE model, one must include

- the initial conditions presumably (but not in all models) as imposed by the big bang,
- the nucleosynthesis products of stars of different mass, initial metallicity, and mass loss,
- the initial mass function (IMF), which describes the birthrates of stars of different masses,
- a description of the total star formation rate with time, which will depend on many parameters, such as gas mass and density, and
- descriptions of the other parameters affecting GCE, including galactic gas inflow and outflow.

These will be discussed below.

Nucleosynthesis Products of Stars

As described in subsequent chapters, stars of sufficient mass will pass through a series of evolutionary phases in which they first burn their core hydrogen to helium, then the helium to carbon and oxygen, then the carbon to neon and more oxygen, then the oxygen to magnesium and silicon, then those nuclides, through a complex set of nuclear reactions, to iron and nickel,

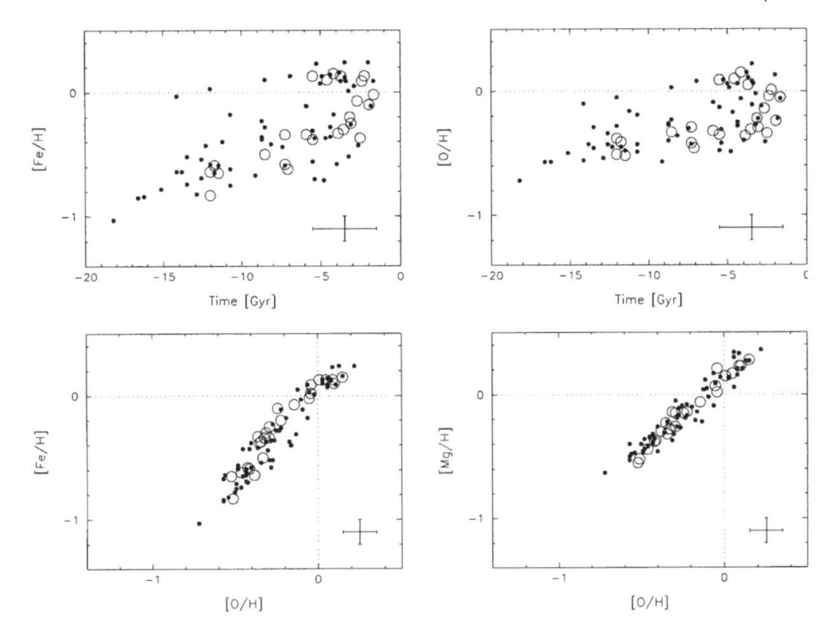

Fig. 4.7. Relationships between [O/H] and [Fe/H] and time and between O and Fe and Mg and Fe in main sequence stars (data from Edvardsson et al. 1993). Typical errors are indicated in the right-hand corner. Further details are given in van den Hoek and deJong (2004).

finally producing a core-collapse supernova and achieving a final state of either a neutron star or a black hole. Stars of lesser mass will not complete all these phases of evolution, as their mass will not be sufficient to generate high enough core temperatures, through conversion of gravitational potential energy to thermal energy, to burn all these nuclides. However, most stars will expel only a fraction of these nuclides into the interstellar medium to enrich the abundances therein. To complicate things further, roughly 50 % of the stars that one observes are actually binary systems, and these stars can evolve to another type of star that will enrich the galactic abundances in quite a different way. Finally, stars of different mass evolve on very different timescales. The results of many theoretical studies were summarized by Pagel (1997) (much of which is from Maeder 1992) and will be summarized here. Times quoted are for the completion of the phase in which the star's core hydrogen is converted to helium. Since that is generally roughly 90 % of the lifetime of the star, that number gives some idea how long the star lives. $M_\odot$ = the mass of the Sun.

- $M = 0.9\ M_\odot$. These stars, having masses slightly less than 1 $M_\odot$ spend 15.5 Gyr simply burning their core hydrogen, longer than the age of the universe, so they contribute nothing to GCE except to tie up Galactic mass. This applies to all less massive stars as well.

- $M = 1.0\ M_\odot$. Stars the mass of the Sun burn their core hydrogen for about 10 Gyr, so they also contribute nothing to GCE. Even if they did evolve on a shorter timescale, they would not evolve far enough in evolutionary phase to contribute much to GCE. Indeed, these statements apply to stars of mass up to 3 $M_\odot$. A 3 $M_\odot$ star will spend 350 Myr burning its core hydrogen. Stars with lower metallicity have slightly shorter lives.
- $M = 5\ M_\odot$. A 5 $M_\odot$ star will spend a little less than 100 Myr burning its hydrogen, more or less independent of its metallicity, so it evolves considerably more rapidly than a less massive star. It will produce about half a solar mass of newly synthesized helium and will leave a 1 $M_\odot$ C-O core at the end of helium burning. The final state of this star will be a white dwarf.
- $M = 10\ M_\odot$. A star of this mass will spend about 20 Myr burning its hydrogen, essentially independent of its metallicity. Stars of this mass and heavier can have significant stellar winds during their lifetimes, which will result in large mass loss prior to the final explosion of the star; this will influence the element production of such stars. To give an order of magnitude of some element production, a 10 $M_\odot$ star will produce roughly 1.4 $M_\odot$ of newly synthesized helium, 0.08 $M_\odot$ of carbon, 0.10 $M_\odot$ of oxygen, and 0.6 $M_\odot$ of heavier elements, which will be expelled into the ISM at the end of the star's life. Stars of this initial mass will ultimately produce a core-collapse supernova, with a neutron star as their final state.
- $M = 20\ M_\odot$. Stars of this mass will spend about 10 Myr burning their hydrogen, more or less independent of their metallicity. Such stars may also have significant stellar winds, which will also influence their nucleosynthesis. They would be expected to expel into the ISM a bit less than 2 $M_\odot$ of newly synthesized helium, 0.25 $M_\odot$ of carbon, 1.3 $M_\odot$ of oxygen, and 2.8 $M_\odot$ of heavier nuclides, the last two numbers being essentially independent of the mass loss. The end state of this star could be either a neutron star or a black hole.
- $M = 60\ M_\odot$ 60 $M_\odot$. stars are thought to spend about 4.0 Myr burning their hydrogen, more or less independent of their metallicity. The amount of newly synthesized material that is produced by such stars and ultimately expelled into the ISM varies hugely depending on the amount of mass loss by stellar winds. It can be as high as 13.5 $M_\odot$ of helium and 7.0 $M_\odot$ of carbon, if there is a great deal of mass loss, and 14 $M_\odot$

of oxygen and 17 $M_\odot$ of heavier nuclides if there is small mass loss. Stars as massive as these would be expected to end their lives as black holes.

One interesting observation on the above numbers is that virtually all the nucleosynthesis that is done by stars is done by stars of 5 $M_\odot$ or heavier, even though their abundance is lower than that of the less massive stars, and all of these stars live for 100 Myr or less. This is a tiny fraction of the age of the Galaxy, which is thought to be more than 10 Gyr, which has led to the assumption of "instantaneous recycling" of stars, that is, it is often assumed for convenience in GCE models that the stars that contribute to nucleosynthesis die as soon as they are born.

As noted above, many stars exist in binary systems. In these situations, if one of the stars evolves to a white dwarf, when its companion evolves to its helium-burning phase and the periphery of the star expands, the white dwarf may draw material from the surface of its companion. This may cause the dwarf to exceed the maximum mass it can sustain, following which it will undergo thermonuclear runaway in a deflagration, causing the entire star to explode as a type Ia supernova. This will expel slightly less than 1 $M_\odot$ of Fe and Ni into the ISM (Nomoto, Thielemann, and Yokoi 1984; Thielemann, Nomoto, and Yokoi 1986).

The Initial Mass Function

The IMF describes the relative birthrates of stars as a function of their mass. Again, following Pagel (1997), the IMF is represented in two ways, either as $\phi(m) \propto dN/dm$ or $\xi \propto dN/d(\log m)$, where the stellar masses are represented as $mM_\odot$. Since stars of different mass live different amounts of time, the present-day mass function (PDMF) is not the same as the IMF. Indeed, since stars having masses less than 0.9 $M_\odot$ have lifetimes longer than the age of the Galaxy, every star of that mass ever born in the Galaxy is still present. For such stars we can write

4.8.2 $$\text{IMF} = \text{PDMF}/[T\langle\psi(t)\rangle],$$

where T is the age of the Galaxy and $\langle\psi(t)\rangle$ is the mean star formation rate. For massive stars that live for $\tau_{ms}(m)$ on the main sequence of the Hertzsprung-Russell diagram (see sec. 1.6), that is, that burn their core hydrogen for that length of time:

4.8.3 $$\text{IMF} = \text{PDMF}/[\tau_{ms}(m)\psi(T)],$$

where $\psi(T)$ is the present-day star formation rate. For stars having masses in between these two limits,

4.8.4 $$\text{IMF} = \text{PDMF} \Big/ \int_{T-\tau_{ms}(m)}^{T} \psi(t)dt.$$

Through a number of studies and arguments it has been found that

4.8.5 $$0.5 < \psi(T)/\langle\psi(t)\rangle < 1.5,$$

that is, the star formation rate has been relatively constant over the history of the Galaxy. Note, however, that this would allow an exponentially decaying star formation rate if it had a decay time longer than about 10 Gyr.

Several forms exist for the IMF, but, since this is intended only as an initial exposure of the student to GCE, only the simplest will be given. This is from Salpeter (1954) and is

4.8.6 $$\varphi(m) \propto m^{-0.35}.$$

Other IMF formulations are considerably more complex than this one but still exhibit the same basic feature of the distribution that falls off with high mass.

Basic Galactic Chemical Evolution Equations

The total mass of the Galaxy, M, is comprised of two entities, that which exists as gas, g, and that which is locked up in stars, s. Thus,

4.8.7 $$M = g + s.$$

However, M is not constant but rather is influenced by the rate of the mass of gas accreted into the Galaxy, F, and the rate of ejected mass of gas, E. Thus,

4.8.8 $$dM/dt = F - E.$$

The mass of the gas is determined by

4.8.9 $$dg/dt = F - E + e - \psi,$$

where e is the mass of the gas ejected by stars and ψ is the star formation rate by mass. The last two terms are just minus the net rate at which mass goes into stars, that is,

4.8.10 $$ds/dt = \psi - e.$$

Finally, the abundance of any element, represented by its mass fraction Z, can be written as

4.8.11 $$d(gZ/dt) = e_Z - Z\psi + Z_F F - Z_E E,$$

where the first term on the right-hand side of the equation represents the amount of the element ejected from stars with whatever its mean enrichment is, the second represents the amount lost to the ISM by star formation, the third represents the amount inflowing into the Galaxy with the mass fraction of the intergalactic medium, and the last represents the amount flowing out in a Galactic wind. The last term may be heavily enriched in heavy elements, as core-collapse supernovae are thought to contribute in a major way to the Galactic wind.

There are many more details associated with GCE; some involve studies of GCE in other galaxies as well as details of GCE such as abundance gradients within galaxies (Chiappini, Romano, and Matteucci 2003). In addition, although GCE generally assumes that the abundances within the galaxy are well mixed; this is not necessarily the case and is certainly not the case for the early Galaxy. An application of GCE, that associated with nucleocosmochronology, will be developed further in chapter 7. For much more detail, however, the interested reader is referred to the book by Pagel (1997) and to the articles referenced therein.

CHAPTER 4 PROBLEMS

1a. Determine the mean molecular weight at the center of the Sun. Assume $X = 0.5$, $Y = 0.48$, $Z = 0.02$, and complete ionization.

1b. Estimate (very roughly) P at the center of the Sun. Use $T_6 = 15$, density $= 120$ g/cm^3.

2. Use the expressions leading up to equation 4.3.4 to obtain a crude estimate of the temperature at the core of the Sun. Take $M_r = M_\odot/2$, $\Delta r = r = R_\odot/2$ to obtain an average density in that region. This in turn gives, from the equation for dP/dr, a pressure at the core of the Sun. Then, from the ideal gas law, $P_{core} = (k_B/m)\rho T_c$, where m is the mean molecular mass in the fully ionized core of the Sun. Plugging these numbers into the ideal gas law gives a rough estimate of the temperature at the center of the Sun. The actual value is about 15.5 million K.

3. Obtain expression 4.6.9, the approximate expression for the mean molecular weight, and state the assumptions that go into obtaining that formula.

4. Perform the substitutions to derive equation 4.4.6, the relationship between luminosity and nuclear energy generated in a star.

5. Derive the expression for the density of states dN/dE assuming the electrons to be relativistic, that is, assume $E = cp$ instead of $E = mv^2/2$. Then use your expression to derive the result given in equation 4.7.12.

6. One can obtain some estimate of when the transition from the nonrelativistic to the relativistic situation occurs in white dwarfs by equating the expressions for nonrelativistic and relativistic pressures and solving for the critical density. Do so, and estimate the critical density for the transition.
7. Using the ideal gas law, $P = (k_B/\mu)\rho T$, rewrite equation 4.5.4 for the condition for no convection in terms of ρ, r, and T.
8. Assume the temperature and mass distributions of a star are such that the energy released from nuclear reactions per cubic centimeter per second, $\varepsilon\rho$, is given by

$$\varepsilon\rho = (\varepsilon\rho)_\circ \exp(-r/\beta),$$

where $\beta = 0.01\ R$ and R is the radius of the star. Assume the luminosity and radius of the Sun for this problem. What fraction of this star's energy generation occurs within $0.02\ R$?

5

HYDROGEN/BURNING

5.1 Introduction

Stars spend most of their lives burning the hydrogen in their cores. In this situation, indeed in each of the phases of stellar burning, the star exists in hydrodynamic equilibrium, with the gravitational force just balancing the pressure created by the thermal energy being produced in the core of the star by the nuclear reactions going on therein, as discussed in chapter 4. Stars that exist on the main sequence in the Hertzsprung-Russell plot, seen in figures 1.7 and 9.6, are in their H-burning phase; the long period of time spent in that phase explains why most of the stars observed are on the main sequence.

There are two sets of reactions that constitute hydrogen burning; these are known as the pp-chains and the CNO cycles. The rates at which these two processes proceed depend on a high power of the temperature; for the slowest reaction in the pp-chains (the weak interaction $^{1}H + {}^{1}H \rightarrow {}^{2}H + e^{+} + \nu_{e}$) it is about as T^4, and for the slowest reaction in the primary hydrogen burning reaction in the CNO cycles (the electromagnetic interaction $^{1}H + {}^{14}N \rightarrow {}^{15}O + \gamma$) it is as T^{20}, at the temperature at the core of the Sun. The CNO cycles operate on a much shorter timescale than do the pp-chains; this is basically because the rate at which the pp-chains can proceed is limited by a weak interaction. On the other hand, the higher Coulomb barriers of the CNO cycle reactions require that the latter processes operate at higher temperatures. Stars with the mass of our Sun, with a core temperature of about 15.5 million K, burn their hydrogen predominantly via the pp-chains. Stars that are significantly more massive than the Sun, however, have a higher core temperature, so they will operate primarily via the CNO cycles.

5.2 pp-Chains

The dominant reactions that govern the pp-chains and their respective Q-values are as given below. Note that, in the case of positron emission, the first reaction in the pp-chain, the Q-value given is actually that for electron capture; the endpoint energy of the positron, or of the neutrino, is actually 1.02 MeV less than the value indicated.

5.2.1 $$^1H + ^1H \rightarrow ^2H + e^+ + \nu_e,\ Q = 1.44\,\text{MeV};$$

5.2.2 $$^2H + ^1H \rightarrow ^3He + \gamma,\ Q = 5.49\,\text{MeV};$$

5.2.3 $$^3He + ^3He \rightarrow ^4He + 2p,\ Q = 12.86\,\text{MeV}.$$

Reactions 5.2.1 to 5.2.3 constitute the pp-I reaction chain. Each nucleus in those reactions has an associated differential equation that gives its rate of change of abundance. For example, that for ^2H is (where $[^2H]$ is, e.g., the density of ^2H)

5.2.4 $$d[^2H]/dt = (1/2)[^1H]^2\langle\sigma v\rangle_{1,1} - [^1H][^2H]\langle\sigma v\rangle_{1,2}.$$

Note the factor of 1/2 in front of the first term on the right-hand side of the equation to take into account that the reaction occurs between identical particles. While the reactions indicated in equations 5.2.1–5.2.3 are not the only ones that involve ^2H, they are the dominant ones. For ^2H, a star reaches quasistatic equilibrium, which means that for most of its life the ^2H abundance changes very slowly, so that $d[^2H]/dt$ is very close to zero. In this case one finds

5.2.5 $$[^2H]/[^1H] = \langle\sigma v\rangle_{1,1}/2\langle\sigma v\rangle>_{1,2}.$$

Nuclear astrophysicists have measured cross sections for virtually all of the reactions associated with hydrogen burning. However, that for $^1H + ^1H \rightarrow ^2H + e^+ + \nu_e$ is far below measurable limits, so it has been calculated in the way described in section 3.5.

The next set of reactions in the pp-chains is

5.2.6 $$^3He + ^4He \rightarrow ^7Be + \gamma,\ Q = 1.59\,\text{MeV};$$

5.2.7 $$^7Be + e^- \rightarrow ^7Li + \nu_e,\ Q = 0.861\,\text{MeV};$$

5.2.8 $$^7Li + p \rightarrow 2\,^4He,\ Q = 17.35\,\text{MeV}.$$

Reactions 5.2.1, 5.2.2, and 5.2.6–5.2.8 constitute the pp-II chain. Note that electron capture on atomic ^7Be occurs via capture of an inner-shell electron. However, ^7Be is almost always fully stripped at the temperatures at which the

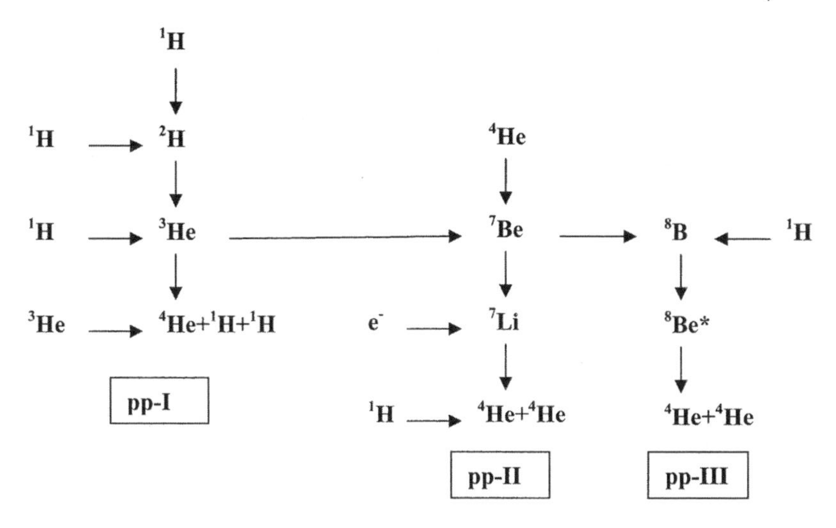

Fig. 5.1. The pp-chains of hydrogen burning. The three chains are as noted on the figure, although ^{3}He must be produced by the first two reactions in pp-I to produce the ^{7}Be that initiates the pp-II and pp-III chains.

cores of stars exist, so the electron capture in such an environment involves continuum electrons. ^{7}Be therefore has a considerably different half-life in the Sun, 125 days (which depends on density) (Bahcall 1962, 1994) than would be measured in a laboratory (53.3 days; Tuli 2000).

A third set of reactions within the pp-chain also occurs:

5.2.9 $$^7\text{Be} + \text{p} \rightarrow {}^8\text{B} + \gamma, Q = 0.137\,\text{MeV};$$

5.2.10 $$^8\text{B} \rightarrow {}^8\text{Be} + \text{e}^+ + \nu_\text{e}, Q = 17.98\,\text{MeV};$$

5.2.11 $$^8\text{Be} \rightarrow 2{}^4\text{He}, Q = 2.94\,\text{MeV}.$$

Reactions 5.2.1, 5.2.2, 5.2.6, and 5.2.9–5.2.11 constitute the pp-III chain. Of special interest here is the high endpoint energy produced in ^{8}B decay; we will return to this below. Technically, ^{8}B decays to the broad first excited state of ^{8}Be at 3.04 MeV, so the actual endpoint for that β-decay is reduced accordingly.

Each chain produces 26.73 MeV by converting four protons into one net ^{4}He nucleus (the pp-II and pp-III chains input a ^{4}He to create ^{7}Be, so the two ^{4}He nuclei in their bottom lines actually represent only one net ^{4}He nucleus), two positrons, and two electron neutrinos. Figure 5.1 shows pictorially the three pp chains.

Extraordinary effort has gone into measuring the cross sections associated with the reactions of the pp-chains. While it is not possible to discuss all these

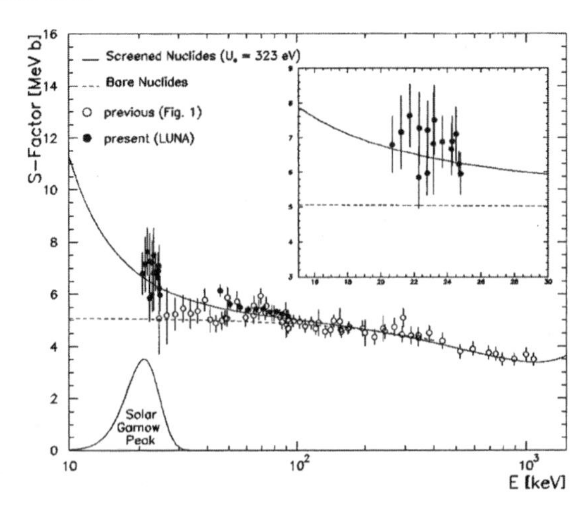

Fig. 5.2. Data and calculated astrophysical *S*-factor for the ^{3}He(^{3}He,2p)^{4}He reaction. The dashed and solid curves represent $S_{bare}(E)$ and $S_{screened}(E)$, respectively (see sec. 4.2). The solar Gamow peak is shown in arbitrary units. The upper right corner inset shows an expanded version of the underground *LUNA* (see sec. 2.12.6) data. Reprinted with permission from Junker et al. (1998). Copyright 2000 by the American Physical Society.

reactions, it is worth discussing at least one of them; ^{3}He(^{3}He,^{4}He)2 ^{1}H. This has a Gamow peak in the Sun of 24 keV, an energy that is well below the capability of virtually every accelerator-detector combination ever used to try to measure its rate, primarily because of the background radiation produced by cosmic rays and other sources of ambient radiation. However, the *LUNA* collaboration (see sec. 2.12.6) has built an accelerator underground in the Laboratori Nazionali del Gran Sasso in Italy to measure some of the cross sections of nuclear astrophysics in a vastly reduced background environment. Some of these data are shown in figure 5.2 (Arpesella et al. 1996); there it is seen that the data actually do achieve an energy at the Gamow window. Note the importance, shown in the two curves, of the electron shielding at low energies (see sec. 4.2).

Another reaction that deserves special attention is ^{7}Be(p, γ)^{8}B. The ^{8}B so produced is responsible for nearly all of the yield in one of the solar neutrino detectors discussed below, the chlorine detector. Since this was the world's only solar neutrino detector for many years, and it produced the famous discrepancy between the predicted solar neutrino flux and that observed (see below), this reaction has received a great deal of attention. Unfortunately, its cross section is difficult to measure, not only because it is small, but also because the target, ^{7}Be, has a half-life of 53.3 days and even because the cross section itself has several components. The cross section is dominated by a resonance at

0.63 MeV, although that has little effect on the cross section at the Gamow window. The situation is summarized in figure 5.3, in which the direct measurement of Junghans et al. (2003), together with the results of four other modern measurements, are exhibited.

This reaction has also been studied in detail by the Coulomb breakup approach (Kikuchi et al. 1998; Davids et al. 2001; Iwasa et al. 1999; Schumann et al. 2003). These measurements generally produced slightly, but significantly (~10 %), lower values for the *S*-factor at the Gamow window than did the extrapolated *S*-factors from the direct measurements. Although this apparent discrepancy has produced considerable controversy among nuclear astrophysicists, a recent theoretical paper (Esbensen, Bertsch, and Snover 2005) may have explained it. This paper attributes the problem to the lack of inclusion of higher order effects in the analysis of the Coulomb breakup measurements, which appear to lessen the discrepancy when they are taken into account. The asymptotic normalization coefficient (ANC) approach (see sec. 3.2) was also applied to this reaction by Trache et al. (2001).

5.3 Solar Neutrino Measurements

In the early 1960s, Raymond Davis, a chemist, interacting closely with John Bahcall, a theoretical nuclear astrophysicist, set out to test Bahcall's model of how the Sun works by measuring the flux of solar neutrinos that the model

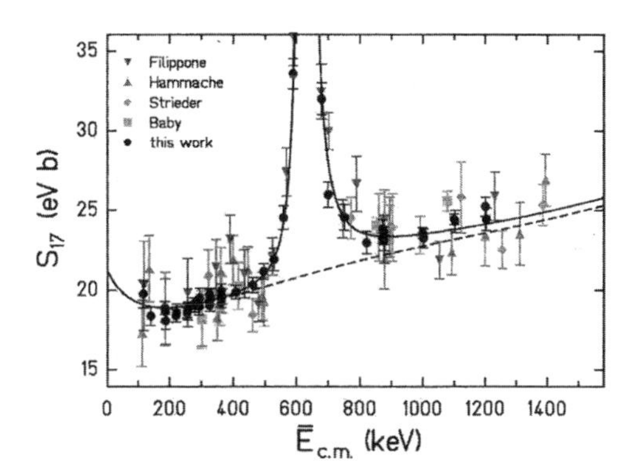

Fig. 5.3. Astrophysical *S*-factor for the $^7Be(p, \gamma)^8B$ reaction, from Junghans et al. (2003), (and the data indicated as "this work" are from that paper). The data are from the several modern measurements of this cross section, and the curve indicates the results of a calculation from Descouvemont and Baye (1994). The error bars include both statistical and systematic errors. References indicated on the graph are to earlier measurements of this reaction cross section; they are given in the references. Reprinted with permission from Junghans et al. (2003). Copyright 2003 by the American Physical Society.

predicted should be produced by the nuclear reactions that occur in the Sun. Davis devised an experiment (Davis, Harmer, and Hoffman 1968) that utilized 615 tons of perchloroethylene—cleaning fluid (C_2Cl_4)—to observe the neutrinos. The observations utilized the interactions of the neutrinos with the ^{37}Cl in the cleaning fluid. However, neutrinos interact so weakly with anything—their cross sections depend on energy and, in the case of interactions with nuclei, on the nucleus with which they are interacting, but they still are generally of magnitude 10^{-40} to 10^{-43} cm^2—that the measurements had to be performed in an environment in which the cosmic ray–induced backgrounds were minimized to as large an extent as possible. This meant siting the experiment far underground. The site selected for the experiment was the Homestake gold mine in Lead, South Dakota. The experiment was thus located under 4850 feet of granite. This experiment represented the birth of neutrino astronomy, a field that has grown enormously in significance since this first experiment. In recognition of Davis's contributions to this field, he was awarded the Nobel Prize in 2002.

The same thing that makes the neutrinos difficult to detect, their tiny cross sections, makes them invaluable as signatures of astrophysical processes. Detection of the solar neutrinos affords an opportunity to sample the conditions at the very center of the Sun, which cannot be done by direct detection of photons, which scatter many times before they reach the solar surface. Indeed, the photons emitted from the solar surface bear no relationship to those produced within the core of the Sun. The reactions by which these neutrinos are detected usually involve conversion of some nucleus to another via the charged-current weak interaction, producing detectable signatures in the process.

Davis's detector utilized ^{37}Cl as the detection medium, which provides signals via the following processes:

5.3.1 $$^{37}\mathrm{Cl} + \nu_e \rightarrow {}^{37}\mathrm{Ar} + e^-,\ Q = -0.816\,\mathrm{MeV};$$

5.3.2 $$^{37}\mathrm{Ar} + e^- \rightarrow {}^{37}\mathrm{Cl} + \nu_e,\ Q = 0.816\,\mathrm{MeV}.$$

The β-decay of ^{37}Ar (actually e^- capture) produces X-rays, which can be detected. The half-life of ^{37}Ar, 35 days, allowed the detector to build up its abundance of ^{37}Ar for a month, after which the Ar was filtered out of the detection fluid and observed for the desired decays. This represents an incredible chemical achievement, as there were but a handful of ^{37}Ar atoms produced per month—the cross sections of solar neutrinos are on the lower end of the range of cross sections noted above—and they had to be separated out from the 615 tons of C_2Cl_4. The results from this detector produced the famous solar neutrino problem, the factor of about 3 discrepancy between the observed detection rate in ^{37}Cl and that predicted from the standard solar model (see, e.g., Bahcall, Pinsonneault, and Basu 2001; Bahcall and Pinsonneault 1998).

The *Q*-values of these reactions are of great interest, as they determine the minimum energy of the neutrinos that can be detected. Figure 5.4 shows the predicted spectrum of neutrinos from the Sun. Unfortunately, the threshold energy for the Cl detector prevents it from observing the neutrinos produced in the $^{1}\text{H} + {}^{1}\text{H} \rightarrow {}^{2}\text{H} + e^{+} + \nu_e$ reaction. Indeed, most of the yield in Davis's detector came from the β-decay of ^{8}B, a result of the energy dependence of neutrino-induced charged-current weak interactions. Because this is a weak branch of the pp-chains (see sec. 5.7), it was long thought that some nuclear physics effects might be responsible for the solar neutrino problem. At the same time, this experiment demonstrated the power of neutrino astrophysics in actually measuring the details of the nuclear reactions that were going on in the core of the Sun; this would have been impossible if just the photons were detected. Note that it also confirmed that nuclear reactions do actually power the Sun!

There exist several other detectors of high-energy neutrinos; some of these are discussed in some detail in section 2.10. One is the *Kamiokande* detector. It was originally designed to observe baryon decay, but, since Supernova 1987a, its best-known scientific mission has become neutrino astronomy. It detects neutrinos from the relativistic electrons or positrons produced by their interactions with the nuclei or the electrons in the water; these subsequently produce Cherenkov light that is observed in the photomultiplier tubes. Although this detector was not designed to detect solar neutrinos, it is sensitive to the high-energy neutrinos from the Sun. It has produced a result in qualitative agreement with the Homestake result. One especially significant feature is that the events in which the solar neutrinos scatter elastically from electrons in the water are sufficiently directional that they point back to the Sun, a confirmation that these are indeed solar neutrinos. This facility has been upgraded (now *Super-Kamiokande*) and has become a mainstay of neutrino astronomy.

Several other detectors either recently in existence or being constructed are also capable of observing aolar neutrinos. Two of these, *GNO* (Altmann et al. 2000) and *SAGE* (Abdurashitov et al. 1999), use ^{71}Ga as the detecting nucleus. The incident neutrinos convert those nuclei to ^{71}Ge, which decays with a half-life of 11.43 days. The detectors are allowed to accumulate ^{71}Ge for several weeks, then the Ge is extracted and collected and the decays of the ^{71}Ge

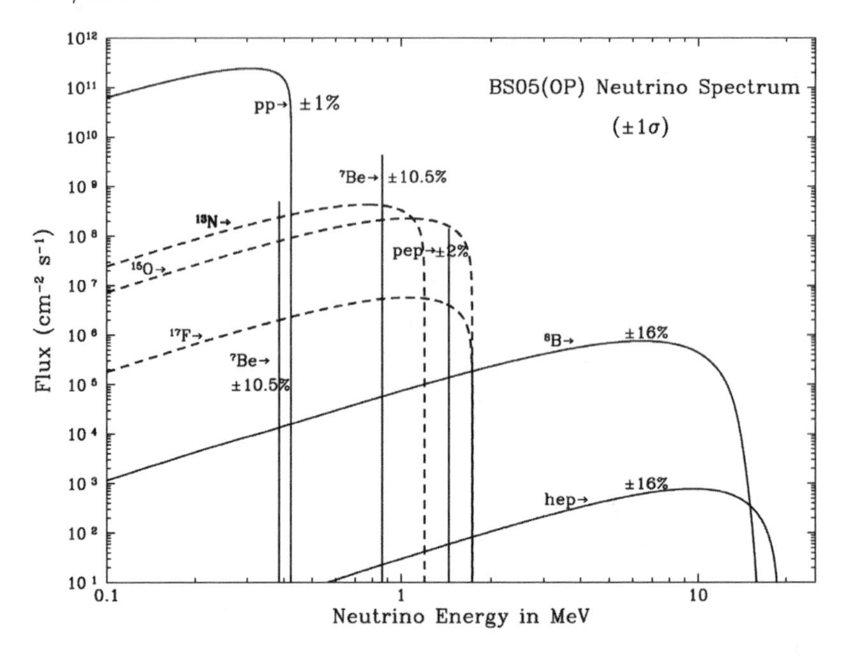

Fig. 5.4. Spectrum of neutrinos emitted from the Sun, according to the standard solar model (from Bahcall et al. 2005). The percentages indicate the uncertainties on each of the neutrino yields. Credit: *The Astrophysical Journal.*

observed. The advantage of the Ga detector is its energy threshold, 0.22 MeV, which makes it sensitive to the neutrinos from the $^1H + {}^1H \rightarrow {}^2H + e^+ + \nu_e$ reaction. These two detectors give results (Bahcall 1997) that agree with each other and confirm that there is an apparent solar neutrino deficit, although the magnitude of the deficit is less than that indicated from the higher energy neutrinos observed in the Homestake and *Super-Kamiokande* experiments (although that does not represent a conflict because of the energy differences). The standard solar model (Bahcall, Pinsonneault, and Basu 2001; Bahcall and Pinsonneault 1998) predicts that the Ga detectors should record about 135 SNU (1 SNU, or solar neutrino unit, is equal to 1×10^{-36} interactions per target nucleus per second.). The two Ga detectors observe about 71 ± 7 SNUs (Abdurashitov et al. 1999; Altmann et al. 2000), well below the predicted value. Indeed, this would be even below the value for just the neutrinos from the $^1H + {}^1H \rightarrow {}^2H + e^+ + \nu_e$ reaction if it were assumed that all the Sun's energy is produced from the pp-I chain (i.e., if the rate for the reaction $^3He + {}^4He \rightarrow {}^7Be + \gamma$ is set to zero), which is about 88 SNU. This strongly suggests that the solution to the solar neutrino problem cannot lie in the nuclear reactions. Because of this, the suggestion that the neutrinos may oscillate, either in departing from the dense core of the Sun (Mikaev-Schmirnov-Wolfenstein [MSW] oscillations; see sec. 3.7) or in making their way to our terrestrial detectors (vacuum oscillations), into a neutrino to which terrestrial detectors are not sensitive, has become the generally accepted explanation. The general details of neutrino oscillations were discussed in section 3.7, and solar neutrino oscillations will be discussed in section 5.6.

5.4 *Sudbury Neutrino Observatory*

Another solar neutrino detector, *Sudbury Neutrino Observatory* (*SNO*) (see sec. 2.10) was constructed at a depth of 2 km in the Creighton Mine in Sudbury, Ontario (see sec. 2.14), using heavy water as the detection medium. The heavy water is contained inside a 5-cm thick transparent acrylic vessel submerged in light water. The vessel itself is a 12-m diameter sphere. Because of its unique properties and sensitivities to solar neutrinos, a detailed discussion of this detector is warranted. The possibility that heavy water could be used to detect solar neutrinos was first suggested by Chen (1985). Deuterium has special properties in this context, because it is sensitive not only to charged-current interactions, as are the other detectors, but to neutral-current interactions as well. Furthermore, the signals produced by *SNO* determine the type of interaction that produced the event. Thus if the neutrinos produced in the Sun oscillated from one that would produce a charged-current interaction in the detector to one that would not, the latter neutrino would also be detectable in *SNO* as a neutral-current event, providing a unique test of the standard solar model (discussed in detail in sec. 5.7). The interactions that can be observed with *SNO* are as discussed below (for more detail, see http://www.sno.phy.queensu.ca/).

Charged-Current Reaction

5.4.1 $$\nu_e + {}^2\text{H} \rightarrow {}^1\text{H} + {}^1\text{H} + e^-.$$

As an electron neutrino interacts with the neutron in a deuterium nucleus, a *W* boson may be exchanged, converting the neutron into a proton and the ν_e to an electron (see fig. 5.5) via a charged-current interaction. The electron ends up with most of the neutrino's energy, from the kinematics of the reaction. The incident neutrino imparts sufficient energy to the electron that it will be ejected at nearly the speed of light, so it will produce Cherenkov light in moving through the water. The light will be detected by *SNO*'s photomultiplier

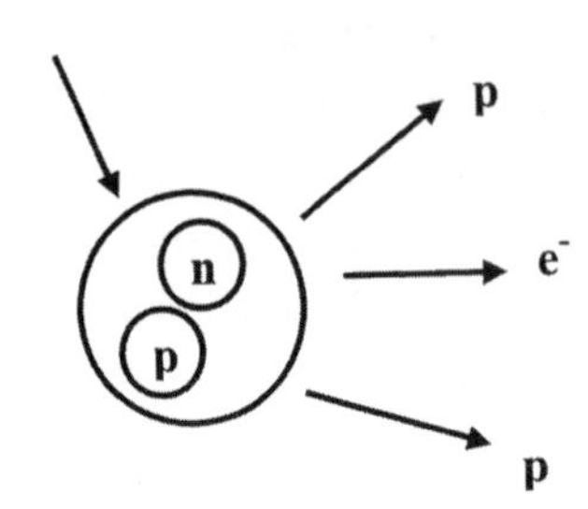

Fig. 5.5. A charged-current event, as would be observed in *SNO*.

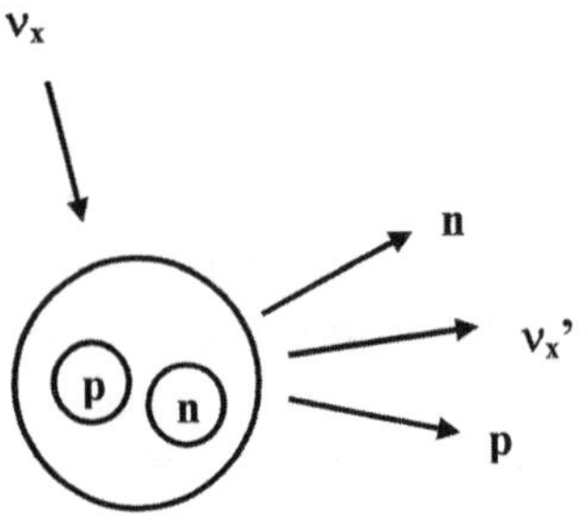

Fig. 5.6. A neutral-current event, as would be seen in *SNO*.

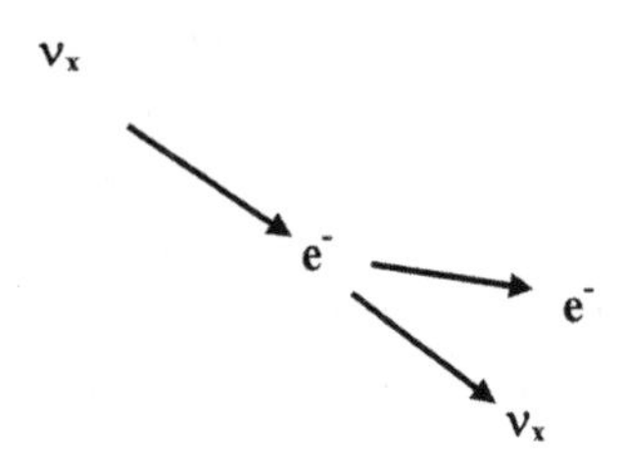

Fig. 5.7. An electron scattering event, as would be seen in *SNO*.

tubes; the amount of light produced is proportional to the energy of the incident ν_e.

The standard solar model predicts that *SNO* should have observed about 30 charged-current events per day.

Neutral-Current Reaction

5.4.2 $$\nu_x + {}^2\mathrm{H} \rightarrow {}^1\mathrm{H} + \mathrm{n} + \nu_x'.$$

In this reaction, a Z^o boson imparts energy to the deuteron via a neutral-current interaction, which, if the energy is sufficient, will break the deuteron apart. The liberated neutron is thermalized in the heavy water and then captured on another nucleus. The capture process yields a gamma ray, which will Compton scatter on an electron, which then produces detectable Cherenkov light.

The neutral-current reaction is equally sensitive to all three neutrino flavors; the detection efficiency depends on the neutron capture efficiency and the resulting gamma-ray cascade. Neutrons can be captured directly on deuterium, but this is not a very efficient process. Thus *SNO* used two other separate means to enhance the neutral-current reaction detection. The first utilized ^3He proportional-counter tubes hung in a grid within the D_2O. ^3He has a very large cross section for the capture of thermal neutrons, which produce an energetic proton-triton pair, which will produce an electrical pulse in the counter wire. The second detection scheme utilized chlorine salt in *SNO*'s heavy water. The chlorine (actually ^{35}Cl) has a high absorption cross section for thermal neutrons, resulting in a gamma-ray cascade with total energy peaked at around 8 MeV. The gamma rays produce detectable Cherenkov light, as described above.

The standard solar model predicts that *SNO* should have observed about 30 neutrons per day.

Electron Scattering

5.4.3

$$e^- + \nu_x \rightarrow e^- + \nu_x.$$

This reaction (see fig. 5.7) is not unique to heavy water, and it is the primary detection mechanism in light water detectors, for example, *Super-Kamiokande*. Although the reaction is sensitive to all neutrino flavors, the e-neutrinos dominate by a factor of 6. The final-state energy is shared between the electron and the neutrino, so very little spectral information can be obtained from this reaction. Good directional information however, is obtained; thus background events are identified as those that do not point back to the Sun.

The standard solar model predicts that *SNO* should have observed about three electron-scattering events per day.

5.5 The Solution to the Solar Neutrino Problem

The combined results from *Super-Kamiokande* and *SNO* have shown unequivocally that the solar neutrino problem is the result of neutrino oscillations, that is, that the flavor eigenstates of the neutrinos are not the same as the mass eigenstates. The way in which this is usually presented is as a plot of the quantity Δm^2 versus $\tan^2\theta$. The measurements of the solar neutrinos produce a well defined solution in that space, as is indicated in figure 5.8. This graph indicates the mixing angle solution for the solar neutrinos (the smaller mass solution), which are emitted as ν_es, a flavor eigenstate, but which travel

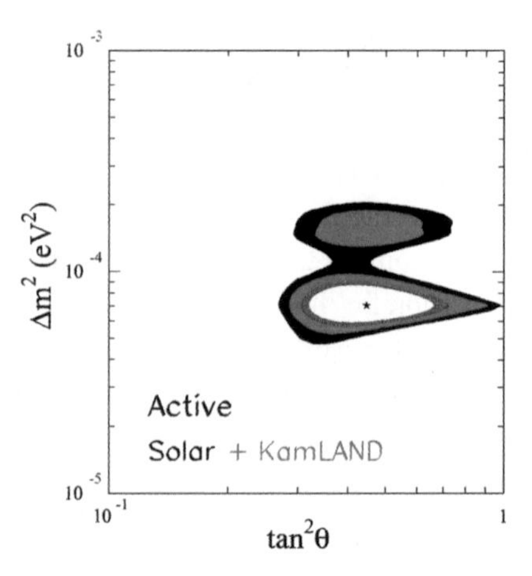

Fig. 5.8. The solutions to the neutrino mixing data. These assume that there are no sterile neutrinos. This solution includes the solar neutrino data from the Homestake, *GNO*, *SAGE*, *Super-Kamiokande*, and *SNO* experiments. The analysis also includes data from *KamLAND*, an experiment that measured the electron neutrino flux from Japanese reactors and that observed definitively a reduction in the expected neutrino flux that was due to oscillations. The different contours indicate the 90%, 95%, 99%, and 99.73% (innermost to outermost, respectively) confidence limits. Reprinted from Bahcall, Gonzales-Garcia, and Pena-Garay (2003). Copyright IOP Publishing Ltd.

as mass eigenstates. The graph also indicates the Δm^2-mixing angle solution (the larger mass solution) for ν_μs, as have been observed by several large underground detectors in the world but especially by *Super-Kamiokande*. These two solutions are sufficient to determine most of the parameters of the matrix that transforms the flavor eigenstates to the mass eigenstates; the probabilities of the different components of the resulting wave functions are shown schematically in figure 5.9.

These wave functions result from a combination of the data but especially from the *SNO* results. With these results and the formalism presented in section 3.7, we are in a position to understand how the wave functions shown in figure 5.9 were determined. This will follow closely a discussion in Kayser (2005). Solar neutrinos are born as ν_es, and since θ_{13} has been found to be extremely small, they are a mixture of ν_1 and ν_2. Thus they can be represented as a 2×2 matrix, as indicated in equation 3.7.18. The high-energy ν_es are influenced by the existence of the matter of the Sun and are subject to the MSW mixing, as discussed in chapter 3. The two mass states ν_1 and ν_2 make up two flavor states: ν_e and a state ν_x that is a coherent linear combination of ν_μ and ν_τ. The precise $\nu_\mu - \nu_\tau$ composition of ν_x depends on the $\nu_\mu - \nu_\tau$

mixing angle θ_{atm} measured in atmospheric neutrino oscillations; this appears to be very close to a 50:50 mixture. Thus ν_x can be taken to be $(\nu_\mu - \nu_\tau)/2^{1/2}$, and solar neutrino flavor change is the process $\nu_e \rightarrow \nu_x$.

The Hamiltonion for the solar neutrinos may be written as in equation 3.7.13, except that the diagonal terms introduce a common phase to all the eigenstates and can be ignored. Thus the Hamiltonion becomes

5.5.1 $$H = \{\Delta m^2/4p\}\begin{pmatrix} -\cos 2\theta & \sin 2\theta \\ \sin 2\theta & \cos 2\theta \end{pmatrix} + 2^{1/2} G_f n_e \begin{pmatrix} 1 & 0 \\ 0 & 0 \end{pmatrix}$$

The matrices in this equation are in the $\nu_e - \nu_x$ basis, and Δm^2 and θ are the mass difference squared and mixing angle relevant to solar neutrinos.

At the center of the Sun, where the solar neutrinos are created, the coefficient of the second, matter-interaction term in the Hamiltonion is, assuming a (typical) density at the center of the Sun of 100 g/cm^3,

5.5.2 $$2^{1/2} G_F n_e \approx 0.75 \times 10^{-5}\ \text{eV}^2/\text{MeV}.$$

By comparison, assuming a typical ^{8}B neutrino energy of roughly 8 MeV, the coefficient of the first vacuum term in this Hamiltonion is

5.5.3 $$\Delta m^2/4p \approx 0.25 \times 10^{-5} \text{eV}^2/\text{MeV}.$$

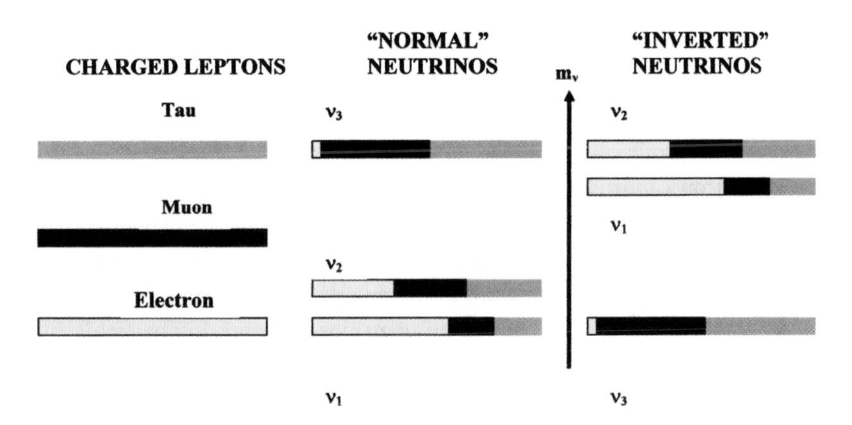

Fig. 5.9. The neutrino mass spectrum for the different flavor eigenstates in each of the three neutrino mass eigenstates. Yellow denotes an electron or an electron neutrino, black a muon or a muon neutrino, and gray a tau or a tau neutrino. The mass spacing between ν_1 and ν_2 is determined from solar neutrino oscillations, and the mass spacing between that doublet and ν_3 is determined by atmospheric neutrino oscillations. At the time of writing of this book, either of the two mass hierarchies shown, the normal hierarchy and the inverted hierarchy, is possible.

Thus we can approximate, very roughly, the situation at the core of the Sun by assuming that the matter-dependent interaction dominates. Within this approximation, the Hamiltonion is diagonal, so that the ν_e born at the core of the Sun is an eigenstate with eigenvalue $2^{1/2}G_F n_e$. Since the other eigenvalue is zero, the ν_e is born in the higher energy eigenstate of the Hamiltonion.

Under the conditions where the large mixing angle MSW effect occurs, the electron density n_e changes slowly enough with r that the propagation of high-energy neutrinos from the core of the Sun to the outer edge is adiabatic. It can be shown that the solutions of the Hamiltonion in equation 5.5.1 will not cross so that neutrinos will also be in the higher energy eigenstate upon emerging from the Sun. But this will just be the solution of the Hamiltonion in which the electron density is zero, that is, the first term in the Hamiltonion. This is the mass eigenstate denoted as ν_2; it is in this state that the neutrino remains for the duration of its journey to Earth. From this we can easily calculate the probability that the higher energy neutrinos produced in the core of the Sun will be detected in terrestrial detectors as ν_es; it is just the square of the ν_e component of ν_2, which has been found from the *SNO* results to be 1/3. MSW mixing removes 2/3 of the high-energy ν_e flux by the time the neutrinos have emerged from the Sun.

Using the same approach, one can calculate the probability that lower energy neutrinos emitted from the core of the Sun will be detected on Earth as ν_es, although the calculation is more complex, as the exact solutions to the Hamiltonion in equation 5.5.1 must be found and the distribution in neutrino energies must be taken into account (as was not done fully in the above analysis).

However, if one can detect the lower energy neutrinos, for example, those from electron capture on ^{7}Be, which produces monoenergetic neutrinos at 0.86 MeV, one should find that these neutrinos have also oscillated, but the amount by which their flux is reduced is less than that for the higher energy neutrinos. This depends on details of the mixing and the amount of vacuum oscillations and MSW oscillations the neutrinos undergo. In any event, the reduction in their flux due to the oscillations should be different from those of the ^{8}Be neutrinos (see Bahcall and Pena-Garay 2004 for more details). If this transition does occur, the reduction in flux of the ^{7}Be neutrinos should be 45 %, instead of the 2/3 factor from MSW oscillations noted above, for the large mixing angle solution. Both *Borexino* and *Kamland* are being put forth as potential detectors of the ^{7}Be neutrinos; their fluxes should provide an important test of our understanding of neutrino physics.

It should also be noted that the region around 1.2 MeV is especially sensitive to subtle effects that might affect the solar neutrino spectrum, for example,

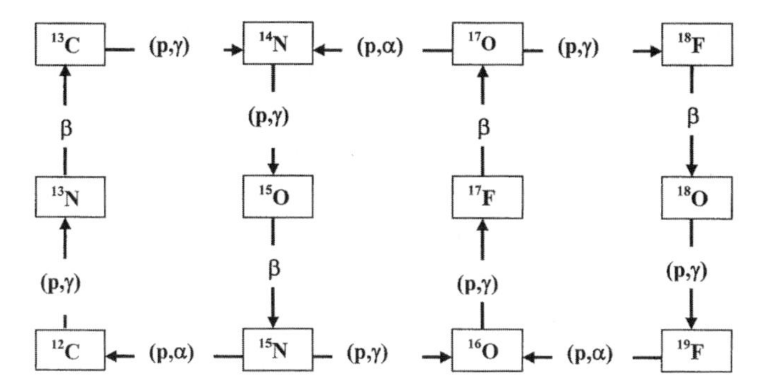

Fig. 5.10. The CNO cycles of hydrogen burning. The times between captures in these cycles are much greater than the typical times required for β-decay for the temperatures at which these cycles operate, so the decays always have time to occur.

the neutrino magnetic moment and the need to actually describe neutrinos as wave packets instead of the usual plane wave treatment described above. This makes it especially imperative to study this energy region in detail.

5.6 CNO Cycles

As discussed above, hydrogen burning in more massive stars is dominated by the CNO cycles, although these cycles do contribute a small amount of the hydrogen burning that occurs in the Sun (see sec. 5.7). These cycles are indicated in figure 5.10, in which it can be seen that the separate subcycles are closely linked.

The primary cycle, referred to as the CN cycle, operates with a ^{12}C catalyst via the following reactions:

5.6.1 $$^{12}\mathrm{C} + \mathrm{p} \rightarrow {}^{13}\mathrm{N} + \gamma, Q = 1.944\,\mathrm{MeV};$$

5.6.2 $$^{13}\mathrm{N} \rightarrow {}^{13}\mathrm{C} + \mathrm{e}^{+} + \nu_{\mathrm{e}}, Q = 2.220\,\mathrm{MeV};$$

5.6.3 $$^{13}\mathrm{C} + \mathrm{p} \rightarrow {}^{14}\mathrm{N} + \gamma, Q = 7.551\,\mathrm{MeV};$$

5.6.4 $$^{14}\mathrm{N} + \mathrm{p} \rightarrow {}^{15}\mathrm{O} + \gamma, Q = 7.297\,\mathrm{MeV};$$

5.6.5 $$^{15}\mathrm{O} \rightarrow {}^{15}\mathrm{N} + \mathrm{e}^{+} + \nu_{\mathrm{e}}, Q = 2.754\,\mathrm{MeV};$$

5.6.6 $$^{15}\mathrm{N} + \mathrm{p} \rightarrow {}^{12}\mathrm{C} + \alpha, Q = 4.965\,\mathrm{MeV}.$$

As with the pp-chains, this cycle combines four protons into an α-particle, producing two positrons, two neutrinos, and 26.73 MeV in the process. It is indicated as the left-most of the three loops in figure 5.10.

It is of interest to use this CN cycle to indicate the differential equations by which the abundances of the nuclides in a cycle such as this one are calculated. In hydrogen burning, ^{12}C can be destroyed by capturing a proton, and it can be created by a $^{15}N(p, \alpha)^{12}C$ reaction. Thus the differential equation governing the rate of change of its abundance $[^{12}C]$ is

5.6.7 $$d[^{12}C]/dt = -[^{1}H][^{12}C]\langle\sigma v\rangle_{1,12\gamma} + [^{1}H][^{15}N]\langle\sigma v\rangle_{1,15\alpha}.$$

Similarly,

5.6.8 $$d[^{13}N]/dt = -[^{13}N]/\tau_{13} + [^{12}C][^{1}H]\langle\sigma v\rangle_{1,12\gamma},$$

5.6.9 $$d[^{13}C]/dt = -[^{1}H][^{13}C]\langle\sigma v\rangle_{1,13\gamma} + [^{13}N]/\tau_{13},$$

5.6.10 $$d[^{14}N]/dt = -[^{1}H][^{14}N]\langle\sigma v\rangle_{1,14\gamma} + [^{1}H][^{13}C]\langle\sigma v\rangle_{1,13\gamma} + [^{1}H][^{17}O]\langle\sigma v\rangle_{1,17\alpha},$$

5.6.11 $$d[^{15}O]/dt = -[^{15}O]/\tau_{15} + [^{1}H][^{14}N]\langle\sigma v\rangle_{1,14\gamma},$$

5.6.12 $$d[^{15}N]/dt = -[^{1}H][^{15}N]\langle\sigma v\rangle_{1,15\alpha} - [^{1}H][^{15}N] <\sigma v>_{1,15\gamma} + [^{15}O]/\tau_{15},$$

where the subscript 1,15α (or 1,15γ) means that a proton interacted with ^{15}N in a (p, α) (or (p, γ)) reaction. Note that the equations for the rates of change of abundance of ^{14}N and ^{15}N are complicated by a third term associated with a slightly higher mass cycle, as described below. These equations are, therefore, examples of the complexity of the coupled equations that must be solved to describe stellar evolution.

The CNO cycle reactions have also undergone intense study over the past several decades. One example of such effort is a recent study of the $^{14}N(p, \gamma)^{15}O$ reaction, performed by the *LUNA* collaboration (Formicola et al. 2004) using proton beams ranging from 140 to 400 keV from their accelerator in the Laboratori Nazionali del Gran Sasso. This reaction is thought to be the slowest one in the CN cycle (see discussion below), the result of which is that its reaction rate has interesting implications to several questions in astrophysics. The Gamow window for this reaction is at 27 keV at the temperature of the Sun's core, so this facility has the capability of measuring the cross section to as low an energy as those at which they occur in stars. It is especially important to measure the cross section for this reaction to energies around the Gamow window, as it has possible contributions from a subthreshold state, that is, a state in the ^{15}O compound nucleus that is below zero proton energy, the tail of which can contribute to the cross section at low energies. Studies of this reaction prior to the one presented here had produced inconsistent results for

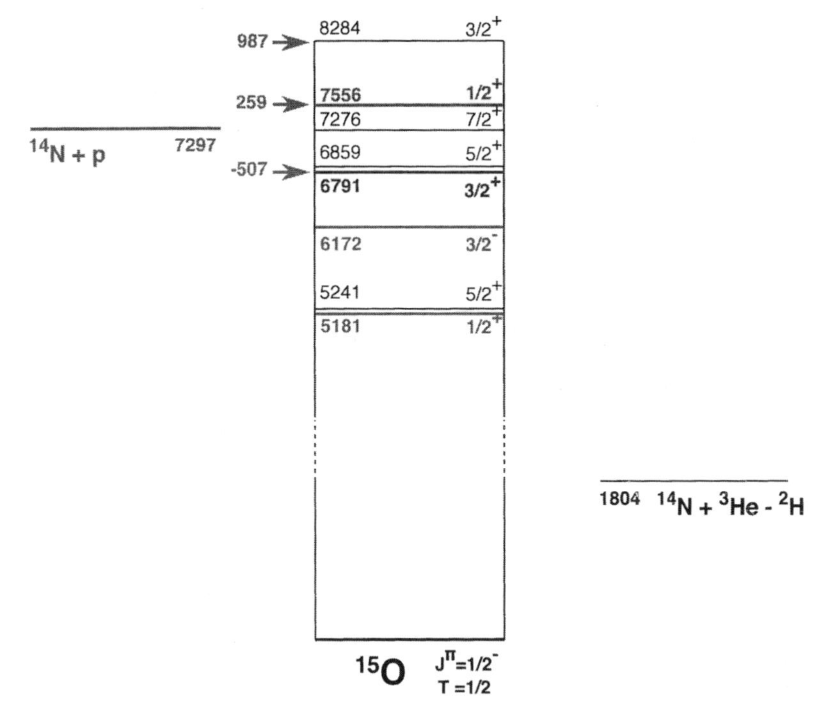

Fig. 5.11. Energy levels in ^{15}O (in keV). The resonance energies corresponding to relevant energy levels are indicated to the left (in keV), as is the ^{14}N + p threshold. From http://edocs.tu-berlin.de/diss/2004/bemmerer_daniel.pdf. Reprinted with permission from D. Bemmerer.

the extrapolated *S*-factor due to a significant extent to uncertainty about the contribution of the subthreshold state. Thus this experiment was undertaken to measure the cross section, hence astrophysical *S*-factor, to energies close to the Gamow window. ANC measurements (see sec. 3.2) had shown that there was an appreciable direct component to this reaction cross section, and known states in ^{15}O above the threshold energy (see fig. 5.11) suggested that resonant terms would also contribute to the cross section, and that was confirmed by previous measurements.

In this experiment, the capture gamma rays were detected with Ge detectors, which have an intrinsic resolution of around 2 keV. Thus the energy widths observed in the scans of the energy-dependent cross section were more a reflection of the intrinsic width of the state than of the resolution of the detector (although that had to be unfolded from the observed widths to determine the actual intrinsic widths). A high-resolution spectrum is shown in figure 5.12. There the sharpness of the gamma-ray lines can be seen. Also to be

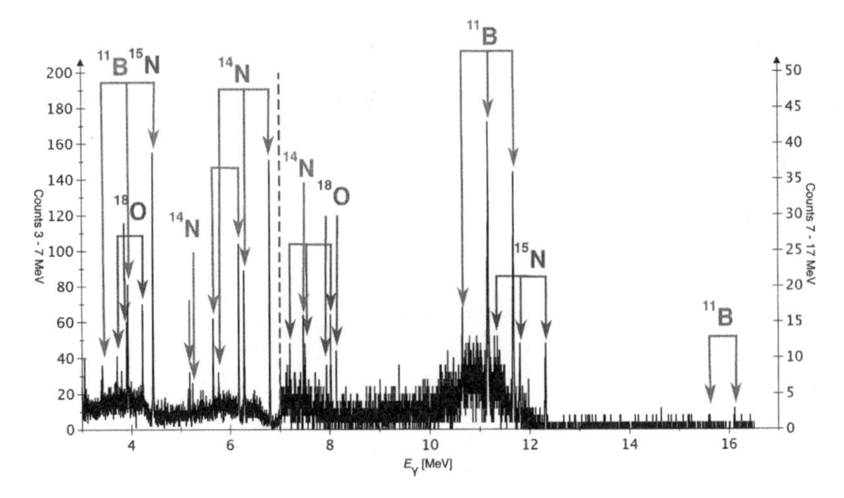

Fig. 5.12. Gamma-ray spectrum for $^{14}N(p, \gamma)^{15}O$ obtained at $E_p = 200$ keV. The identities of the observed lines, including full-energy gamma-ray peaks and first and second annihilation photon escape peaks, are indicated by the arrows. This spectrum would have been swamped by cosmic ray–induced backgrounds had the experiment been done at the Earth's surface. Lines other than those from $^{14}N(p,\gamma)^{15}N$ are beam-induced backgrounds. The ordinate on the left is valid for 3 MeV $< E_\gamma <$ 7 MeV, and that on the right is valid for 7 MeV $< E_\gamma <$ 17 MeV. (From http://edocs.tu-berlin.de/diss/2004/bemmerer_daniel.pdf. Reprinted with permission from D. Bemmerer.

noted are the triplets of lines associated with each gamma ray; this results from the fact that when high-energy gamma rays interact with matter they have a high probability of losing their energy by pair production. Another possibility for interaction of gamma rays in this energy range, as well as for those of lower energy, is Compton scattering (see, e.g., Segre 1977). However, when pair production occurs and the resulting positron annihilates, zero, one, or two of the resulting 0.511 MeV photons can escape the detector, producing the three peaks separated by 0.511 MeV. The multiple lines result from the multiple branches of the gamma rays that can result from direct capture or from cascades through the states of ^{15}O.

Figure 5.13 shows an "excitation function," that is, the *S*-factor that results from the cross section measurement over the energy spanned by the measurement, plotted with results from some of the previous measurements. Similar, often more complicated appearing, excitation functions exist for the other gamma rays that were observed. And, of course, similar data have been acquired for many of the other reactions involved in the CNO cycles.

Our basic discussion above of the reaction rates should allow us to infer which nuclide will be the most abundantly produced from the basic CN cycle. That will be the nucleus that is most slowly destroyed, as nuclei involved in the cycle will tend to accumulate at that nucleus. If we assume that resonances will not distort the picture too greatly, we can infer that the proton capture rates on ^{12}C and ^{13}C should be about equal because the Coulomb barriers will be about the same. However, the increased nuclear charge on ^{14}N will render the proton capture rate thereon to be slower because of the increased Coulomb barrier. The next reaction is $^{15}N(p, \alpha)$, which, because its rate is governed by a strong interaction, would be expected to have a much larger cross section than the (electromagnetically governed) proton capture reactions. Thus the nucleus that is destroyed least rapidly is expected, from this simple argument, to be ^{14}N; that would be expected to be the most abundant nucleus produced in CN cycle hydrogen burning.

There are two other cycles that are offshoots of the primary CN cycle. These operate through the reactions:

5.6.13 $$^{15}N + p \rightarrow {}^{16}O + \gamma, \ 12.127\,\text{MeV};$$

5.6.14 $$^{16}O + p \rightarrow {}^{17}F + \gamma, \ 0.600\,\text{MeV};$$

5.6.15 $$^{17}F \rightarrow {}^{17}O + e^+ + \nu_e, \ 2.761\,\text{MeV};$$

5.6.16 $$^{17}O + p \rightarrow {}^{14}N + \alpha, \ 1.192\,\text{MeV}.$$

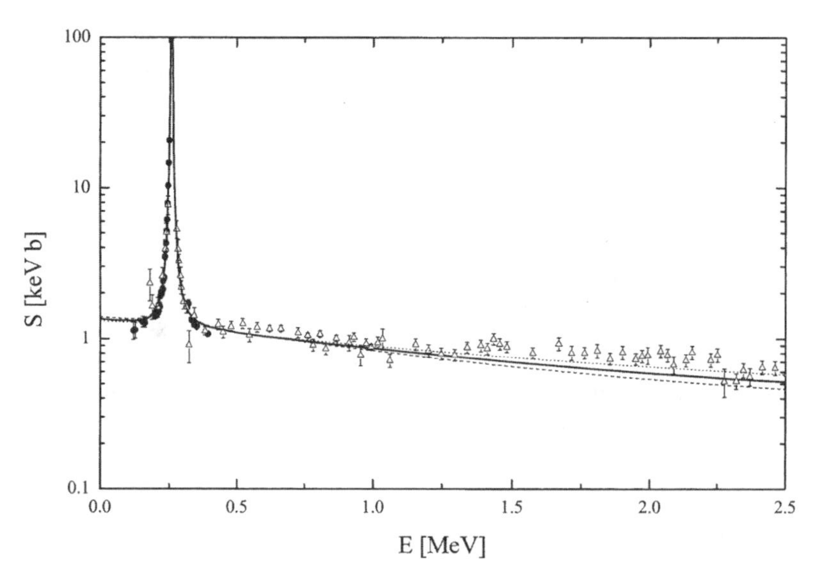

Fig. 5.13. Astrophysical *S*-factor as a function of energy for the $^{14}N(p,\gamma)^{15}O$ reaction to the 6.79 MeV level in ^{15}O. The data from present work are represented by the solid points, and those from a previous study by open triangles. From Formicola et al. (2004). Copyright 2004, with permission from Elsevier.

This secondary cycle utilizes the occasional proton capture on ^{15}O to branch out of the primary CN cycle but then returns most of the nuclei that participate in it back to the primary cycle at ^{14}N via the ^{17}O(p, α) reaction. If, however, ^{17}O captures a proton, then it can lead to a tertiary cycle through the reactions

5.6.17 $$^{17}\text{O} + \text{p} \rightarrow {}^{18}\text{F} + \gamma, 5.607\,\text{MeV};$$

5.6.18 $$^{18}\text{F} \rightarrow {}^{18}\text{O} + \text{e}^+ + \nu_\text{e}, 1.655\,\text{MeV};$$

5.6.19 $$^{18}\text{O} + \text{p} \rightarrow {}^{19}\text{F} + \gamma, 7.994\,\text{MeV};$$

5.6.20 $$^{19}\text{F} + \text{p} \rightarrow {}^{16}\text{O} + \alpha, 8.114\,\text{MeV}.$$

Of course, this discussion has assumed that the proton capture reactions proceed at sufficiently slow rates that the β-decays will always have time to occur before the next proton capture occurs. While this is generally the case in burning scenarios that occur inside stars, there are some sites that occur, for example, when matter from a star accretes onto a collapsed companion such as a white dwarf, in which that is not the case. The temperatures in such sites will be sufficiently high that the reactions can occur much more rapidly than the β-decays. We will return to this situation later on when we discuss hot hydrogen burning in chapter 8.

Since the energies of the neutrinos emitted in these three cycles are somewhat different, and are also different from those of the pp-chains (which themselves have differing neutrino energies), and the neutrinos nearly all escape from the star in which they are produced, the amount of energy deposited in the star per α-particle produced will differ slightly from one cycle to another.

As indicated above, the reactions in the pp-chains and the CNO cycles have been the focus of intense study by experimental nuclear astrophysicists for several decades. The usual situation in nuclear astrophysics, that is, that the energies at which these reactions occur in stars are below that to which present technology will allow measurements, persists here. Nonetheless, additional nuclear information—energies, spins, and parities of nuclear levels, and, often, their single-proton spectroscopic strengths—often exists. As discussed in chapter 3, this allows the existence of resonances to be inferred and often allows calculation of their contributions to the reaction rates. Thus virtually all of these reactions are fairly well understood. That does not apply, unfortunately, to the first reaction in the pp-chains, that in which two protons form a deuteron. As noted above, because it is a weak interaction, its cross section is extremely small; far too much so to be measured at the energies of interest to stellar processes. As noted in section 3.6, however, intense effort has gone into calculating it to an accuracy of a few percent, using a time-dependent perturbation approach.

5.7 The Standard Solar Model

So how well does the standard solar model predict the neutrino output of the Sun? Using as input data decades of measurements of nuclear cross sections by nuclear astrophysicists, as well as all the considerations discussed in chapter 4, the standard solar model (Bahcall and Pinsonneault 1998) predicts the structure of the Sun. In particular, the Sun is found to have an envelope in which energy transfer is convective and a core in which energy transfer is radiative (Bahcall and Pinsonneault 2001), as is discussed in section 4.5.

Remarkably, the Sun's temperature, composition, and motions deep within can be tested by measurements of helioseismology, that is, observations of sound speeds (by measuring their Doppler shifts) of standing waves observed on the surface of the Sun (Bahcall and Ulrich 1988; Bahcall and Pinsonault 2001). These result from the Sun acting like a resonant cavity, trapping acoustic waves (as well as gravity waves) in regions bounded on top by the density decrease near the surface and on the bottom by an increase in sound speed that refracts downward-propagating waves back toward the surface (http://soi.stanford.edu/results/heliowhat.html). Waves of different multipolarity, that is, wavelength (both radially and in angular variables), sample different regions of the Sun. Figure 5.14 shows one such wave. Since the sound speeds depend on the structure of the Sun, the remarkable agreement that exists between the standard solar model predictions of sound speeds and those observed is a striking confirmation of the ability of the model to describe the Sun's structure.

A major international collaboration, the Global Oscillation Network Group (GONG), has produced a vast amount of helioseismology data, as well as an excellent description of how helioseismological waves are generated in the Sun. The GONG Web site is at http://GONG.nso.edu.

It is worth considering for a moment just what the predictions of the model are. A number of details about the structure, temperature, and density of the Sun are predicted by the model. Just a few of them are shown in figure 5.15. One interesting feature to be observed in that figure is that most of the Sun's energy production occurs within the inner 20 %, by radius, and the density falls off from 140 g/cm^3 at the center to about 50 g/cm^3 at the 20 % point. The temperature falls off considerably less rapidly, dropping from 15.5 million K at the center to just under 10 million K at the 20 % radius. These results

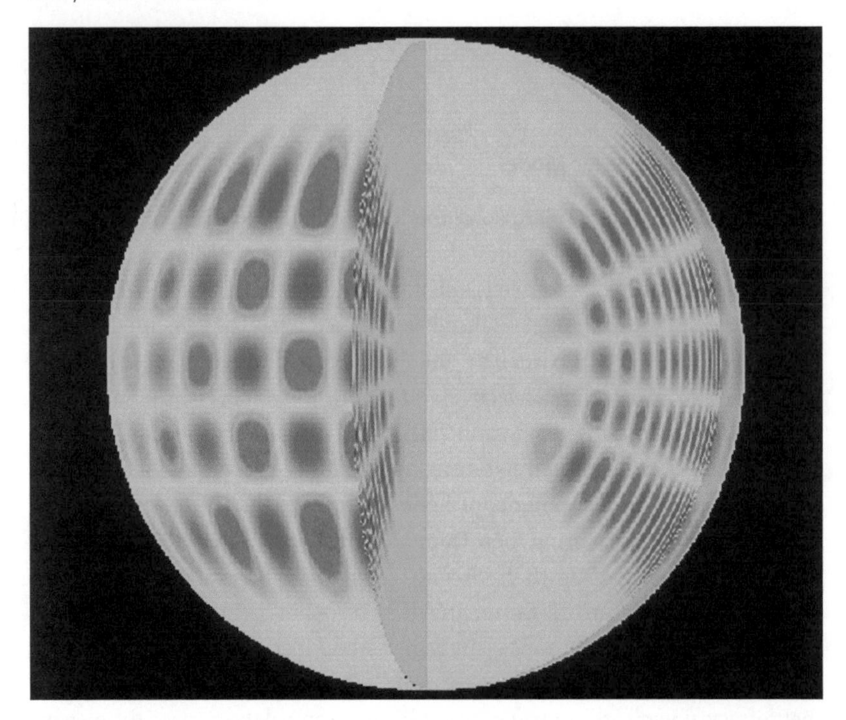

Fig. 5.14. This shows schematically a computer image of one set of acoustic standing waves of the Sun's vibrations. The radial order of the standing wave is $n = 14$, and the angular order is $m = 16$. Red and blue show element displacements of opposite sign. The frequency of oscillation of this mode has been determined to be 2935.88 ± 0.2 mHz. Reprinted with permission of the Solar Physics Group at Stanford University; from http://soi.stanford.edu/results/heliowhat.html.

are directly related to the rapidity with which the nuclear reaction rates vary with temperature, that is, a small temperature drop essentially shuts down the nuclear reactions.

Another interesting feature, that of solar neutrino production, is that the different neutrino source reactions operate at different regions in the Sun (the yields shown in fig. 5.16 are weighted by r^2). This is as would be expected, since the temperature varies with radius (see fig. 5.15) and the different reactions have different temperature dependences. Specifically, the $p + p \rightarrow d + e^+ + \nu_e$ reaction has a much flatter temperature dependence than does the $^7Be + p \rightarrow {}^8B + \gamma$ reaction. Thus, the yield from the former reaction would be expected to be distributed over a much wider radius in the Sun than would be that from the decay of ^8B. The standard solar model confirms this expectation, as can be seen in figure 5.16, which shows that the $p + p \rightarrow d + e^+ + \nu_e$ reaction

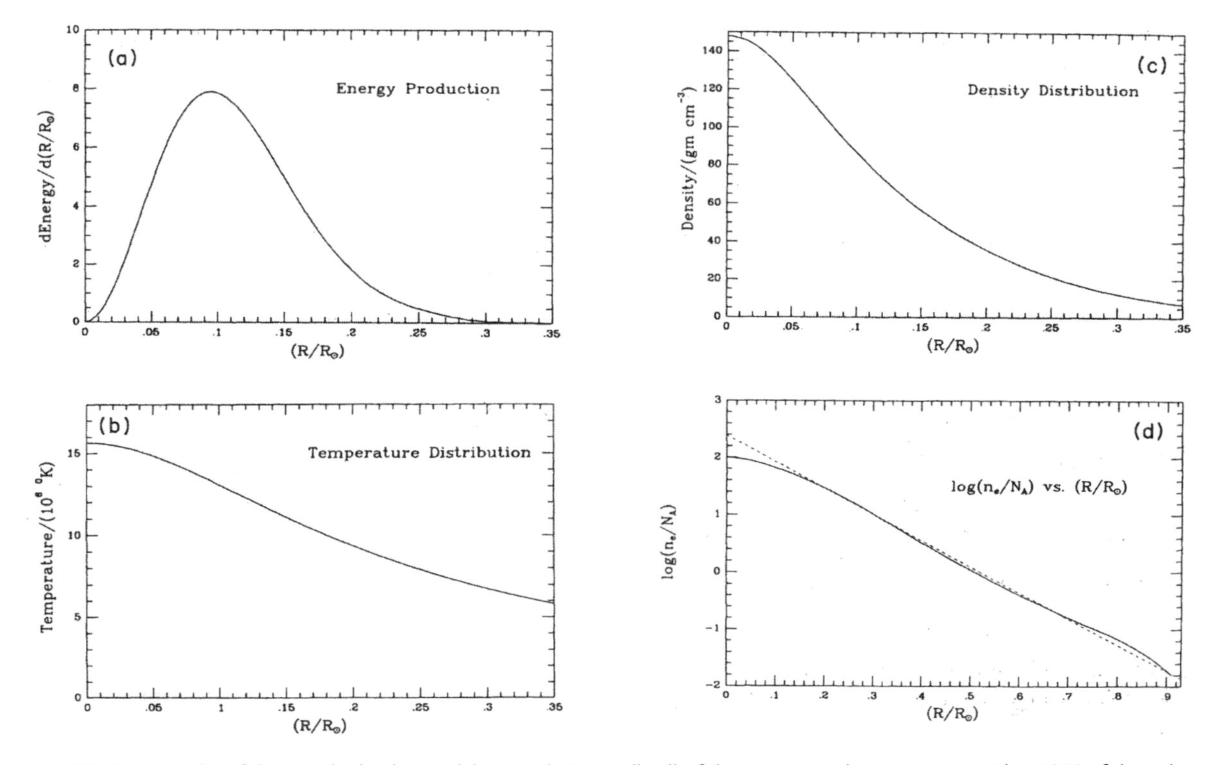

Fig. 5.15. Basic results of the standard solar model. Note that virtually all of the energy production occurs within 20 % of the solar radius, that is, within the inner 1 % of the solar volume. Reprinted with permission from Bahcall and Ulrich (1988). Copyright 1988 by the American Physical Society.

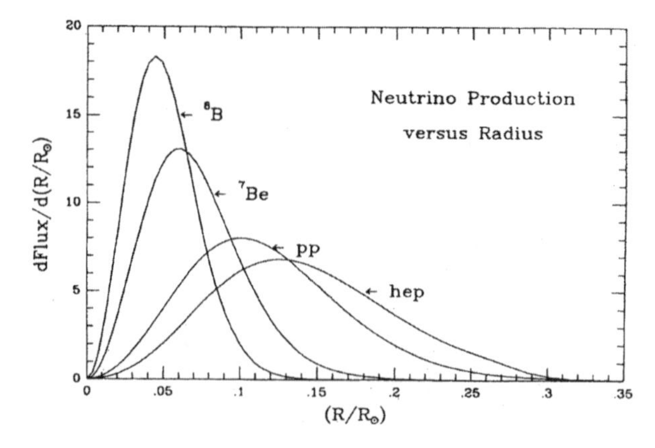

Fig. 5.16. Neutrino production as a function of radius for the different neutrino sources. The fraction of neutrinos that originates in each fraction of the solar radius is given as $[dFlux/d(r/R_{\odot})][d(R/R_{\odot})]$. Reprinted with permission fromBahcall and Ulrich (1988). Copyright 1988 by the American Physical Society.

yield is peaked at 1/10 of the solar radius, while that from the $^7Be + p \rightarrow {}^8B + \gamma$ reaction is peaked at a radius less than half that value.

The model predicts the solar neutrino fluxes from the various nuclear processes; these are indicated, along with their yields for the chlorine and gallium detectors, in Table 5.1. There it is seen that the ^8B neutrinos dominate the chlorine yield, while the pp-I neutrinos dominate the gallium yield. Of course, we know how many neutrinos per second the Sun produces, since we can easily measure its luminosity, and we know that energy is the result of creating ^4He nuclei out of four protons and that each ^4He nucleus produced requires two β-decays and hence produces two neutrinos. However, the energy distribution of the neutrinos produced is important in testing the model; the standard solar model predicts that the bulk of the solar neutrinos are produced from the $^1H + {}^1H \rightarrow {}^2H + e^+ + \nu_e$ reaction. These neutrinos are detectable by the gallium experiments, as noted above, which indicate a large deficit when compared with the predictions of the standard solar model (Abdurashitov et al. 1999; Altmann et al. 2000). The next largest contribution is from the ^7Be neutrinos, which are just barely detectable by the chlorine experiment and are not detectable at all by *Super-Kamiokande* and *SNO* because of their higher detection thresholds. The smallest solar neutrino yield from pp-chain burning is from the decay of ^8B, but this decay produces much higher energy neutrinos than do the other neutrino-producing processes. Since the neutrino interaction cross section increases with the square of its energy above threshold, a one part in 10^4 branch of the neutrino yield (see

Table 5.1 Solar Neutrino Fluxes

Predicted Yields of Various Sources of Neutrinos from the Standard Solar Model of Bahcall and Pinsonneault (1998)

Source	Flux ($cm^{-2}s^{-1}$)	Cl (SNU)	Ga (SNU)
pp	$5.94\ (1.00^{+0.01}_{-0.01})$	0.0	69.6
pep	$1.39 \times 10^{-2}\ (1.00^{+0.01}_{-0.01})$	0.2	2.8
hep	2.10×10^{-7}	0	0
^{7}Be	$4.80 \times 10^{-1}\ (1.00^{+0.00}_{-0.00})$	1.15	34.4
^{8}B	$5.15 \times 10^{-4}\ (1.00^{+0.19}_{-0.14})$	5.9	12.4
^{13}N	$6.05 \times 10^{-2}\ (1.00^{+0.19}_{-0.13})$	0.1	3.7
^{15}O	$5.32 \times 10^{-2}\ (1.00^{+0.22}_{-0.15})$	0.4	6.0
^{17}F	$6.33 \times 10^{-4}\ (1.00^{+0.12}_{-0.11})$	0.0	0.1
Total		$7.7^{+1.2}_{-1.0}$	129^{+8}_{-6}

table 5.1) is predicted to be the dominant yield in the chlorine experiment and essentially the total yield for *Super-Kamiokande* and *SNO*.

While the results of these experiments agree that there is a large ν_e deficit by the time the ^{8}B neutrinos reach the Earth, the combination of the results from *SNO* (Ahmad et al. 2001, 2002; Ahmed et al. 2004), together with those from *Super-Kamiokande* (Y. Fukuda et al. 1998; S. Fukuda et al. 2001), leave no doubt that these neutrinos do oscillate on their way from their point of production to the Earth. Furthermore, these experiments show that the number of neutrinos produced in the Sun agrees with the predictions of the standard solar model (Bahcall, Pinsonneault, and Basu 2001; Bahcall and Pinsonneault 1998), at least for the high-energy portion of the solar neutrino spectrum. The *SNO* results, of course, are crucial in this regard; they indicate that some of the solar neutrinos that reach the Earth, and that *SNO* detects, have resulted from oscillation of the e-neutrinos produced in the Sun to some other flavor. Note, however, that both the *SNO* and *Super-Kamiokande* results are necessary to produce a statistically significant result; *SNO*'s results must be combined with the electron-scattering yields from *Super-Kamiokande* because it has produced much better statistics on that type of event than has *SNO*. Nonetheless, the ability of *SNO* to detect both charged-current and neutral-current events confirms that the number of ^{8}B neutrinos produced by the Sun is as expected from the standard solar model. It will remain for subsequent experiments such as *Borexino* to test the theoretical-observational concurrence for all regions of the solar neutrino spectrum.

From the perspective of neutrino physics, the *SNO* results confirm the concept of neutrino oscillations as observed for μ-neutrinos by the *Super-Kamiokande* collaboration (Fukuda et al. 1998). Of course, the solar oscillations do not occur at the same Δm^2 and mixing angle as the μ-neutrino oscillations. These two distinct solutions are exhibited in figure 5.8.

CHAPTER 5 PROBLEMS

1. Calculate the amount of energy produced in the $p + p \rightarrow d$ reaction (eq. 5.1.1). Be sure to include the energy produced when the positron emitted in the reaction annihilates with an electron.
2. Now assume that the Sun's energy is produced entirely by $4^1H \rightarrow {}^4He$ and that 3 % of the energy is carried off in the neutrinos. From this, calculate the number of neutrinos emitted per second. What is the neutrino flux at the Earth?
3. Beginning with the differential equation for the processes that create and destroy ^3He (similar to that for ^2H in eq. 5.2.4), derive the expression for the time rate of change of the ^3He abundance in terms of the ^1H and ^4He abundances and the appropriate reactions rates.

   Answer:

$$d[^3He]/dt = [^1H][^2H]\langle\sigma v\rangle_{12,3} - (1/2)[^3He]^2\langle\sigma v\rangle_{33,4} - [^3He][^4He]\langle\sigma v\rangle_{34,7}.$$

   Assuming quasistatic equilibrium can be achieved (a questionable assumption for this nuclide, but explain why), determine an equation in the ^3He abundance (which will be quadratic in the ^3He abundance) that can be solved for the equilibrium abundance of ^3He in terms of that for ^1H, ^4He, and the relevant reaction rates. Do not assume that the $^3He + {}^4He \rightarrow {}^7Be$ reaction rate can be neglected. Based on the temperature dependence of the two reaction rates inside the square root, which rate would dominate at higher temperatures? Which would dominate at lower temperatures?
4. Derive the expression for the time rate of change of the ^7Be abundance in terms of the ^1H, ^3He, and ^4He abundances, the relevant reaction rates, and the ^7Be lifetime.

   Answer:

$$d[^7Be]/dt = [^3He][^4He]\langle\sigma v\rangle_{34,7} - [^7Be]/\tau_7 - [^7Be][^1H]\langle\sigma v\rangle_{17,8}.$$

   Then, assuming quasistatic equilibrium, derive the expression for the ^7Be abundance.
5. One of the "explanations" offered for the factor of 3 discrepancy between the predicted and observed ^8B neutrino flux was that perhaps mixing-length theory

was "just a little bit" wrong so that the Sun's core was actually convective. The resulting mixing could lower the core temperature a bit, thereby reducing the ^{8}B neutrino flux. First calculate the temperature dependence of the ^{7}Be(p, γ)^{8}B reaction, using equation 3.9.44. Then determine how much the temperature at the Sun's core would have to be lowered to produce the required factor of 3 reduction in the production of ^{8}B and hence in the ^{8}B neutrino flux. Note that this cannot really explain away the discrepancy, as the standard solar model actually describes the evolution of the Sun from its birth to the present, and if one changes mixing-length theory "just a little" one does not get a star that looks like the Sun after 4.7 Gyr. Neither can one arbitrarily reduce the temperature at the Sun's core without violating other parameters of the Sun.

6. You can run your own standard solar model calculation, using either the code EZ given, along with sample results, at http://www.kitp.ucsb.edu/~paxton/ or that given at http://chandra.as.arizona.edu/~dave/tycho-intro.html. You can check the results of your calculations by comparing to the information given in this chapter or, if you want to go into more detail, at the Web site of the late John Bahcall, http://www.sns.ias.edu/~jnb/.
7. It was stated in chapter 1 that we should be grateful that no stable mass 5 nuclei exist as the Sun would have burned up much too rapidly for us to have existed. Estimate how long it would take the Sun to burn up if ^{5}He and ^{5}Li were stable so that one could run a CN-like cycle, but with He and Li. The slowest reaction in this cycle, and thus the reaction that would determine how long a Sun with this stuff in it would last, would be (analogous to ^{14}N(p, γ)) ^{6}Li(p, γ). Assume that the temperature at the core of the Sun would be the same as it is with our Sun, estimate the relative reaction rates for those two reactions just assuming nonresonant reactions, and take the ratio (and assume that the real Sun will last 10 billion years in its hydrogen-burning phase) to estimate how long the Sun would live under these conditions.

6

ADVANCED STELLAR EVOLUTION, SUPERNOVAE, AND GAMMA-RAY BURSTERS

6.1 Introduction

In this chapter, we will discuss the successive burning stages through which massive stars proceed, each one occurring following depletion of the nuclear fuel of the previous one and following a compression to a higher density and temperature. An excellent discussion of this topic is given in the review paper by Woosley, Heger, and Weaver (2002). We will also demonstrate the value of gamma-ray astronomy in serving as a diagnostic of the completion of a massive star's life as well as the current limited knowledge of the neutrinos that are emitted from that final event, the core-collapse supernova. Since nuclear-structure physics has played a crucial role in our understanding of many of these stages of stellar evolution, and in the final event, some discussion of the relevant topics will be presented. The final section presents the state of the art, at least at the time of publication of this textbook, of our understanding of gamma-ray bursts.

6.2 The Triple-Alpha Reaction

Following the discussion of the previous chapter and the conversion of most of the massive star's hydrogen into helium, the core of the star, in order to maintain hydrostatic equilibrium, will contract. In so doing it will convert half of its gravitational potential energy into thermal energy, as prescribed by the Virial theorem, and half into radiation. This will increase the temperature of the core to the point at which the helium ashes remaining from hydrogen burning ignite. Schaller et al. (1992) performed a comprehensive study of stars ranging in mass from 0.8 to 120 $M_\odot$ ($M_\odot$ denotes 1 solar mass) and of metallicity 0.02 and 0.001. They found that helium burning occurred at around $T_9 = 0.20$ throughout these masses, ranging between a T_9 of about 0.01 and

0.03 through the helium-burning phase. The density during helium burning varies somewhat more; it was generally around 10^3 g cm^{-3}. During this phase of stellar burning, the amount of energy emitted from the core of the star and from the hydrogen-burning zone outside the helium core is so large that the resulting radiation pressure forces the periphery of the star to expand and redden, thereby producing a red giant.

Two sets of reactions dominate helium burning, although others will be discussed below that are important for other reasons. The first converts the ^{4}He into ^{12}C by the "triple-α reaction":

6.2.1 $$^4\text{He} + {}^4\text{He} \leftrightarrow {}^8\text{Be},$$

6.2.2 $$^8\text{Be} + {}^4\text{He} \leftrightarrow {}^{12}\text{C}.$$

These reactions allow nucleosynthesis to circumvent the absence of stable nuclides at mass 5 and 8 u. The process is inhibited by the very short lifetime of ^{8}Be, around 10^{-16} s, which means that virtually all of the ^{8}Be formed breaks up almost immediately into its two α-particle constituents. However, as noted in chapter 1, Salpeter, Opik, and Hoyle realized that, at the density and temperature at which helium burning occurs, there will be a buildup of a small abundance of ^{8}Be, which then allows the second reaction to proceed.

The energetics of the triple-α process are indicated in figure 6.1. There it can be seen that the ground state of ^{8}Be is unbounded by 92 keV but that the ^{8}Be that is formed can capture an α-particle through a resonance associated with the 0^+, 7.654 MeV state in ^{12}C. While this usually results in subsequent breakup into three α-particles, occasionally γ-decay to the ^{12}C(2^+, 4.43 MeV) state results. That state will γ-decay to the ground state, forming ^{12}C. It is through this process that most of the ^{12}C in the universe has been synthesized.

The differential equation for the rate of formation of ^{8}Be is given approximately by

6.2.3 $$d(^8\text{Be})/dt = (1/2)(^4\text{He})^2\langle\sigma v\rangle_{4,4} - (^8\text{Be})/\tau_8,$$

where τ_8 is the lifetime of ^{8}Be. Assuming "quasi-static equilibrium," the left-hand side of equation 6.2.3 can be assumed to be essentially zero, allowing one to obtain

6.2.4 $$(^8\text{Be}) = (1/2)\tau_8(^4\text{He})^2\langle\sigma v\rangle_{4,4}.$$

The rate of formation of $^{12}\text{C}^*$ is given by

6.2.5 $$R_{12} = (^8\text{Be})(^4\text{He})\langle\sigma v\rangle_{4,8} = (1/2)\tau_8(^4\text{He})^3\langle\sigma v\rangle_{4,4}\langle\sigma v\rangle_{4,8}.$$

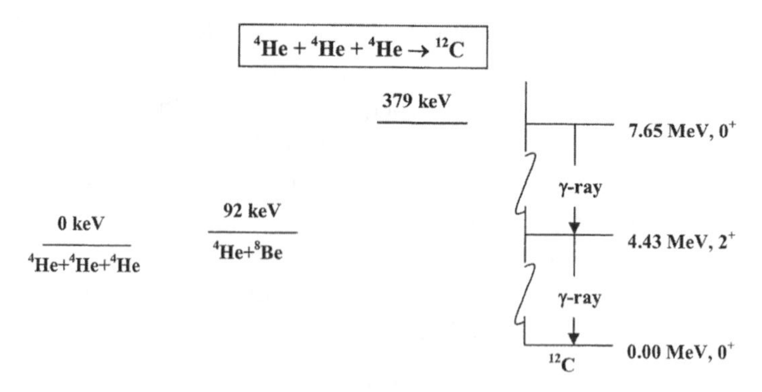

Fig. 6.1. The energetics of helium burning. The Gamow window for ^{4}He + ^{8}Be at a typical helium burning temperature of 200 × 10^6 K is at 230 ± 70 keV, which includes the 287-keV resonance associated with the 7.65-MeV level in ^{12}C.

However, both reactions are resonant, giving the following expressions for the two reaction rates:

6.2.6 $$\langle\sigma v\rangle_{4,4} = (2\pi/\mu_8 k_B T)^{3/2}\hbar^2\omega_8(\hbar/\tau_8)\exp(-E_{R8}/k_B T)$$

6.2.7 $$\langle\sigma v\rangle_{4,8} = (2\pi/\mu_{12} k_B T)^{3/2}\hbar^2\omega_{12}\Gamma_{\gamma 12}\exp(-E_{R12}/k_B T),$$

where ω is the spin statistics factor in equations 6.2.6 and 6.2.7. Equations 6.2.6 and 6.2.7 can be substituted into equation 6.2.5, along with the values for the reduced masses, to give

6.2.8 $$R_{12} = (^4\mathrm{He})^3(3^{3/2}/2)(2\pi\hbar^2/M_\alpha k_B T)^3(\Gamma_{\gamma 12})/\hbar)\exp(-Q/k_B T)$$

where $Q = 379$ keV (see fig. 6.1), the sum of the energies from the three α-particles to the ^{12}C(7.654 MeV) state. Obviously the rate for a three-body reaction cannot be measured directly. Although the rate for the triple-α reaction is now well determined, it required heroic efforts of many experimenters to establish the parameters necessary for its calculation. The rate for this process has been investigated recently (Fynbo et al. 2005), using β-decay of ^{12}N and ^{12}B to populate the states of interest in ^{12}C. The rate was found to be changed by only a small amount from the generally accepted one except at the high and low temperature extremes.

6.3 Determining the ^{12}C(α,γ)^{16}O Reaction

As helium burning begins, the primary result will be the production of ^{12}C. As its abundance builds up ^{16}O also begins to be produced. Of course, this consumes α-particles and therefore competes with the production of ^{12}C; the balance ultimately determines the $^{12}C/^{16}O$ ratio. The reaction that produces ^{16}O is the other important reaction in helium burning:

6.3.1 $$\alpha + {}^{12}C \rightarrow {}^{16}O + \gamma.$$

The energetics for this reaction are shown in figure 6.2. The center-of-mass energy at which this reaction occurs in typical helium-burning conditions is around 300 keV, which is far below the lowest energy, about 1 MeV, to which direct measurements of the cross section have been made. Unfortunately, the cross section drops by many orders of magnitude from 1 MeV to 300 keV, because of the Coulomb barrier, so indirect means are required to determine the cross section at astrophysically relevant energies. It turns out to be dominated by resonances associated with the high-energy tails of two energy levels that are actually just below the threshold for $\alpha + {}^{12}C$ at 7.16 MeV. These are the $J^{\pi} = 1^{-}$, 7.12 MeV and $J^{\pi} = 2^{+}$, 6.92 MeV states in ^{16}O. The contribution from the latter state also can interfere with a direct capture reaction background.

Experiments to determine the resonance parameters of these two states have thus far been performed only for the 7.12 Mev state. The experiment involved studying the β-decay of ^{16}N to ^{16}O in the excitation energy region of interest. Ions of ^{16}N from a low-energy beam from the RNB facility at TRIUMF (Canada) were implanted in very thin foils. The ^{16}N nuclei β-decayed to an excited state of ^{16}O, which decays to $\alpha + {}^{12}C$; these two particles were then observed in coincidence (Azuma et al. 1994). The phase space associated

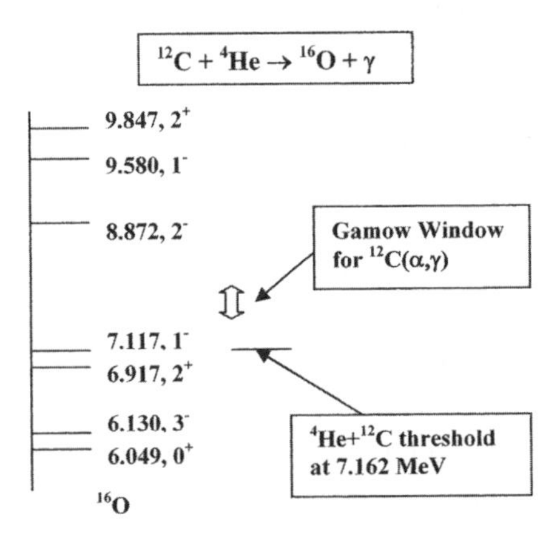

Fig. 6.2. Energetics of the the $^{12}C(\alpha, \gamma)^{16}O$ reaction. At astrophysically interesting energies, S is dominated by the high-energy tails of the resonances associated with the subthreshold states at 7.12 and 6.92 MeV and by an $E2$ direct capture component.

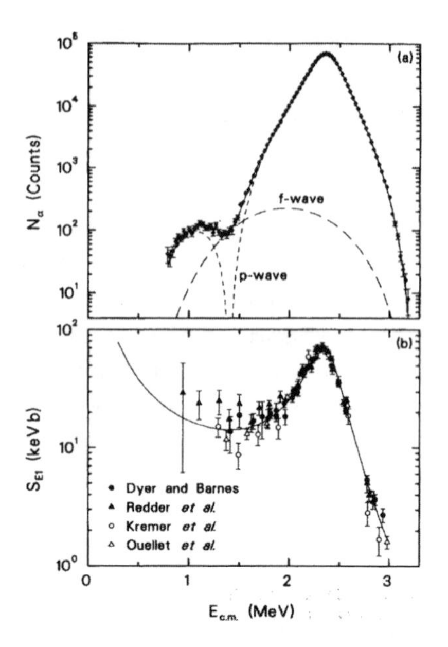

Fig. 6.3. The ^{16}N β-delayed α-spectrum, along with the fit assuming both resonant and direct reaction contributions (along with all other measured S-factors) for E1 capture for the ^{12}C(α,γ)^{16}O reaction. (Reprinted with permission from Azuma et al. (1994), in which the references indicated are defined. Copyright 1994 by the American Physical Society.

with β-decay greatly enhances the region in which the high-energy tail of the 7.12 MeV state contributes to α + ^{12}C compared with the gamma-ray yield that would be produced in the ^{12}C(α, γ)^{16}O reaction to that same energy region. The analysis required to extract the nuclear spectroscopic information from the β-decay spectrum to yield the required ^{12}C(α, γ)^{16}O information involves extremely complicated and detailed nuclear physics (Buchmann et al. 1996). However, the ^{16}N β-decay experiment has determined one of the components of the low-energy *S*-factor, probably the most important component, for this reaction over the energy region for which it occurs in helium burning. This result, together with those from experiments in which the *S*-factor was measured directly at higher energies, is shown in figure 6.3. Unfortunately, ^{16}N does not β-decay to the ^{16}O(2+, 6.92 MeV) state, so other means will have to be used to determine the contributions to the cross section from that state.

One such experiment is ^{12}C(α, α)^{12}C elastic scattering, which has significant sensitivity to the resonance parameters of interest to the ^{12}C(α, γ)^{16}O experiment. Although the elastic scattering does not provide all the information one might wish to calculate the resulting *S*-factor, it surely does provide some of that information, for example, the total width and energy of the resonance and the partial width of the entrance channel.

The elastic scattering experiment had been done some years back (Plaga et al. 1987), but analysis of those data (Buchmann et al. 1996) strongly suggested that they contained systematic errors, possibly in the normalization of the data. Thus the experiment was redone (Tischhauser et al. 2002), this time with detectors at many scattering angles simultaneously. This allowed treating the data as ratios of the elastic scattering yields at different angles, a result which is independent of that (often difficult to obtain, especially at low energy) cross section normalization.

Figure 6.4, from Tischauser et al. (2002), shows the ratio of the yields at two angles. There it can be seen that the dramatic fluctuations in the data allow a very sensitive test of the resonance dependent fit to the ratio data. The nuclear-structure results of this experiment were found to have fairly large error bars, so that they actually did agree with most of the experimental results previously obtained, in some cases, in experiments that disagreed among themselves. However, this study did demonstrate the power of taking the data ratios when a more accurate normalization is required than the data can provide.

Another study was performed from a completely different perspective to study the effect of the $^{12}C(\alpha, \gamma)^{16}O$ cross section on nucleosynthesis. This was

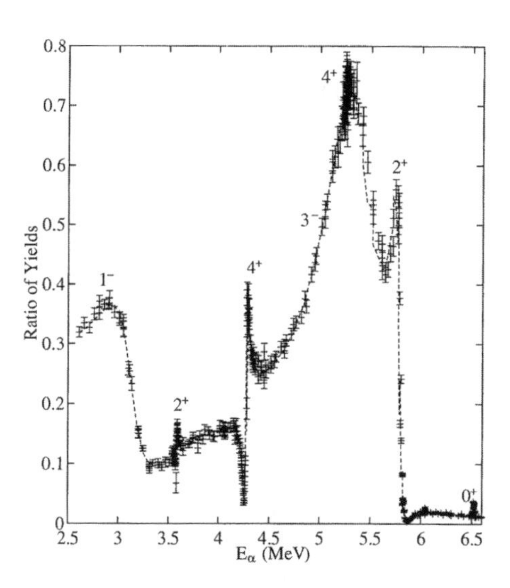

Fig. 6.4. Excitation curve of the yield ratio for θ_{Lab}= 84° and 58.9° and best fit with a specific choice of reaction model parameters. The errors shown are statistical only. Reprinted with permission from Tischauser et al. (2002). Copyright 2002 by the American Physical Society.

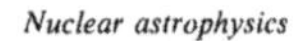

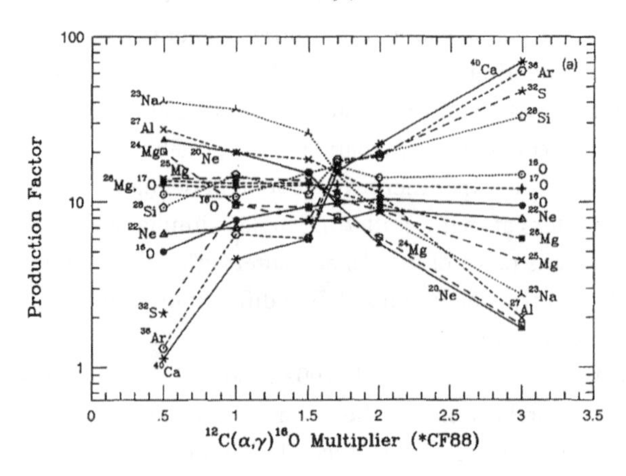

Fig. 6.5. Yields of nuclides relative to solar abundance as a function of the $^{12}C(\alpha,\gamma)^{16}O$ reaction cross section. As can be seen, the abundances of the nuclides vary wildly with this reaction rate, but remarkably all seem to fall into about the same production factor with a single value of the cross section. From Weaver and Woosley (1993). Copyright 1993, with permission from Elsevier.

done by Weaver and Woosley (1993); they found that, when the magnitude of that cross section was varied to observe its impact on the theoretically predicted nucleosynthesis of the medium-mass nuclei, one saw a rather dramatic effect. As can be seen in figure 6.5, although the abundance ratios of these nuclei were extremely sensitive to the value of that cross section, a single value did produce the same abundance ratios of nearly all of those nuclei as is found for the solar abundances, and that value is well within the existing error bars for the best current experimentally determined cross section (Azuma et al. 1994).

Other Reactions of Helium Burning

Additional α-particle radiative capture reactions have been found to have extremely small cross sections, so they are not thought to be very important in consuming helium. However, they are important in forming some of the nuclides that are important, for example, to the s-process (see chap. 7). Specifically, the nucleus ^{22}Ne, thought to be a source of neutrons for the s-process, is produced by the helium-burning reaction sequence

6.3.2 $$^{14}N(\alpha,\gamma)^{18}F(\beta^+)^{18}O(\alpha,\gamma)^{22}Ne.$$

The nucleus ^{13}C is also thought to be a possible neutron source, although it would be produced in a hydrogen-burning shell and then mixed into the helium-burning shell. The (α, n) reactions on ^{13}C and ^{22}Ne in the helium-burning region would produce the requisite s-process neutrons, although they do so at different temperatures and in different regions. These reactions, and the regimes in which each is important, are discussed in the section in chapter 7 on the s-process.

The buildup of neutron excess is important to the considerations of nucleosynthesis in stars and to considerations of what supernovae can expel into the interstellar medium, so it is something that will be tracked through the various stages of stellar burning. Some excess neutrons are developed from the production of ^{22}Ne in helium burning, so defining η, the neutron excess, as

6.3.3 $$\eta = [\text{neutrons} - \text{protons}]/[\text{neutrons} + \text{protons}],$$

(where the sum is over free neutrons and protons as well as neutrons and protons bound in nuclei) then η achieves a value of 0.002 in the helium-burning region (Woosley 1998).

6.4 Advanced Stages of Stellar Burning

Toward the end of a massive star's helium-burning phase, the conversion of its helium to carbon and oxygen will eliminate most of that nuclear fuel in its core. Gravity will again prevail, and the core will contract, again converting some of its gravitational potential energy into thermal energy. If the star exceeds 8 $M_{\odot}$ the resulting increase in temperature will ignite the next phase of burning, carbon burning. Each successive phase will consume the remaining nuclear fuel that is most reactive; that is determined by both the Coulomb barrier and the binding energy. In this context, we will see that oxygen is especially tightly bound, so it violates the natural progression of fuels toward heavier nuclei and, therefore, nuclei of greater charge.

The stages of stellar burning through which a star ultimately passes, as well as the details of each of the stages and its final state, are determined primarily by its initial mass. A very massive star may undergo all the stages of stellar burning described below, ultimately achieving core collapse and exploding as a supernova. The remnant of such a scenario might be either a neutron star or a black hole. A much less massive star might go through some of the stages of stellar burning and then end as either a carbon-oxygen (if $< 8 M_{\odot}$) or an oxygen-magnesium (if slightly more massive) white dwarf. It is currently thought that all stars $< 8 M_{\odot}$ lose their envelopes and become white dwarfs (see, e.g., Woosley 1998). If such a star then accreted some matter from a companion star, the result could be a detonation/deflagration when it exceeded its Chandrasekar (maximum) mass. This would also produce a type Ia supernova but would leave no remnant at all.

In addition to the mass dependence, there are other caveats and uncertainties in our understanding of stellar evolution. For example, the above scenarios are not independent of the mass loss that a star has undergone during its evolution; this can also affect its final state. In addition, if rotation is sufficiently strong, it can halt the contraction of the star. Finally, the treatment of convection has major uncertainties associated with it. Keeping these caveats in mind, we will nonetheless proceed in the sections that follow to present the state of the art in stellar evolution calculations in indicating the characteristics of its different stages.

Before continuing the story of stellar evolution through to collapse, it is worth reminding the reader that the description of the nucleosynthesis that all the stars that have ever existed have contributed to the abundances of the interstellar medium is the subject matter of galactic chemical evolution (GCE). That involves several considerations. First, the newly synthesized nuclides must be expelled into the interstellar medium by the star. Those nuclides that end up in white dwarfs, neutron stars, or black holes will not contribute to the observed abundances. Second, less massive stars are more abundant, so although more massive stars may synthesize many more new nuclei per star, there are fewer of them. Thus the massive stars may not contribute as much to the interstellar medium as might be expected (see, e.g., Miller and Scalo 1979). Finally, less massive stars live longer than more massive stars, so this must also be taken into account. GCE was discussed in some detail in section 4.8.

Carbon Burning

Because carbon and oxygen are the ashes of helium burning, and oxygen is unusually tightly bound, carbon burning will be the stage that follows helium burning. However, if a star initially has a mass of less than about 8 $M_\odot$ it will not, after consuming its helium, be able to generate a sufficiently high core temperature, $T_9 = 0.8$ (Schaller et al. 1992), to ignite its carbon. What occurs is that the degeneracy pressure halts the contraction before that temperature is reached. In this case, the ashes of its helium-burning phase will constitute its core forever, barring some event such as accretion that could disturb the core. It will then exist as a carbon-oxygen white dwarf, a 1.1 $M_\odot$ stellar cinder that is supported by electron degeneracy pressure. If, however, the mass does exceed 8 $M_\odot$ then degeneracy pressure is a smaller fraction of the total pressure and the requisite temperature of $T_9 = 0.8$ for carbon burning will be achieved, with the density being roughly 10^5 g cm^{-3} (Schaller et al. 1992).

The reactions that are involved in this phase are

$$^{12}\mathrm{C} + {}^{12}\mathrm{C} \rightarrow {}^{20}\mathrm{Ne} + {}^{4}\mathrm{He}, Q = 4.62\,\mathrm{MeV}, \tag{6.4.1}$$

$$\rightarrow {}^{23}\mathrm{Na} + \mathrm{p}, Q = 2.24\,\mathrm{MeV}, \tag{6.4.2}$$

$$\rightarrow {}^{23}\mathrm{Mg} + \mathrm{n}, Q = -2.60\,\mathrm{MeV}. \tag{6.4.3}$$

Although the direct fusion of two ^{12}C nuclei would produce ^{24}Mg, production of that final nucleus is not highly probable; those in equation 6.4.1–6.4.3 are much more so. Note that the reaction indicated in equation 6.4.3 is endothermic. However, the temperature is sufficiently high in the carbon-burning phase that this reaction can proceed, especially because a neutron (which does not have to overcome the Coulomb barrier) is emitted. Note also that the nuclei ^{23}Mg and ^{23}Na are not tightly bound, so they will probably be further processed, tending to result in ^{20}Ne, although some of those nuclei may also remain as ashes of carbon burning. In addition to the above reactions, some ^{4}He can be captured on ^{20}Ne to synthesize some ^{24}Mg.

The carbon-burning phase lasts for roughly 1000 years (Weaver, Zimmerman, and Woosley 1978), depending on the mass of the star and its metallicity, during which time it produces ^{20}Ne to add to the preexisting core oxygen. This will serve as the fuel for the next stage of stellar evolution, Ne burning. The geography of carbon burning can be quite complex, occurring in several possible zones of some massive stars and being virtually circumvented altogether in some others (a 25 $M_\odot$ star; Woosley 1998), in which stellar evolution moves directly from helium burning to oxygen-magnesium burning. Since some neutron-rich nuclides are produced in carbon burning, specifically ^{23}Na and ^{23}Mg (after it undergoes β-decay), a small neutron excess will develop. Neutrinos are also produced, primarily from e$^+$-e$^-$ annihilation, which goes to two photons nearly every time it occurs, but can go to two neutrinos about once in 10^{19} times, which carry away significant amounts of energy from the core of the star.

Neon Burning

Following carbon burning, a massive star's core will contract to a density of about 10^7 g cm^{-3} and will achieve a temperature of around $T_9 = 1.4$ (Weaver, Zimmerman, and Woosley 1978). This will be sufficient to ignite the neon-burning phase, which can last for roughly a few years. It is characterized by the reactions

$$\gamma + {}^{20}\mathrm{Ne} \rightarrow {}^{4}\mathrm{He} + {}^{16}\mathrm{O}, \quad (6.4.4)$$

$$^{4}\mathrm{He} + {}^{20}\mathrm{Ne} \rightarrow {}^{24}\mathrm{Mg} + \gamma. \quad (6.4.5)$$

Thus some of the ^{20}Ne is destroyed by photonuclear reactions to produce the α-particles needed to boost other ^{20}Ne nuclei to ^{24}Mg. Although there will be other reactions that can occur, for example, ^{20}Ne(α,p)^{23}Na, the resulting nuclei will be sufficiently loosely bound that they will tend to be processed to one of the more tightly bound α-nuclei, ^{16}O, ^{20}Ne, ^{24}Mg, ^{28}Si, and so forth. In this case, the primary nuclei produced will be ^{16}O and ^{24}Mg. Nonetheless, small abundances of other nuclei will remain as the ashes from neon burning.

Since the net result of these reactions is

$$^{20}\mathrm{Ne} + {}^{20}\mathrm{Ne} \rightarrow {}^{16}\mathrm{O} + {}^{24}\mathrm{Mg}, \quad (6.4.6)$$

the evolution of the compositions in neon burning is given by

$$d(^{16}\mathrm{O})/dt = d(^{24}\mathrm{Mg})/dt, \quad (6.4.7)$$

$$= -(1/2)d(^{20}\mathrm{Ne})/dt. \quad (6.4.8)$$

Some ^{28}Si is also made in neon burning. If the mass of the initial star was 8–9 $M_\odot$ it will probably be insufficient for it to follow through all the phases of stellar evolution (Maeder and Maynet 1989); the neon-burning phase would be its final stage. In this event, the star will end up as an oxygen-magnesium white dwarf, which will have a mass of 1.1–1.4 $M_\odot$.

Oxygen Burning

A star with a mass of 10–11 $M_\odot$ may have its core supported to some extent by electron degeneracy pressure. Indeed, the details of stellar evolution for such stars are complicated (Maeder and Maynet 1989). However, if the initial mass of the star is in excess of this it will follow through all of the successive stages of evolution to formation of an iron-nickel core. In this case, the next stage will be oxygen burning.

Following neon burning, the core of the star will contract somewhat more to achieve a temperature of $T_9 = 2$, and a density of a few times 10^7 g cm^{-3} (Weaver, Zimmerman, and Woosley 1978). The nuclei that exist at the inception of oxygen burning are the neon-burning ashes: ^{16}O, ^{24}Mg, and ^{28}Si, and traces of other Mg and Si isotopes along with some Al, P, and S (Woosley 1986). The next nucleus that can undergo reactions is, because of its low

Coulomb barrier, oxygen. Complete fusion of two ^{16}O nuclei would produce ^{32}S, but direct formation of that nucleus is not highly probable, as the states that are formed decay into other nuclei. The reactions that characterize this phase of burning are, therefore

6.4.9 $$^{16}O+^{16}O \rightarrow {}^{28}Si+^{4}He, Q = 9.59\,MeV$$

6.4.10 $$\rightarrow {}^{31}P + p, Q = 7.68\,MeV$$

6.4.11 $$\rightarrow {}^{31}S + n, Q = 1.50\,MeV$$

6.4.12 $$\rightarrow {}^{30}P + d, Q = -2.403\,MeV$$

6.4.13 $$\rightarrow \text{many many others.}$$

Woosley and Hoffman (1992) found that 90 % of the mass of the stellar core is in ^{28}Si and ^{32}S at the end of oxygen burning. However, there are also produced many nuclei for which $N > Z$ (Woosley 1986) in processes such as

6.4.14 $$^{33}S + e^{-} \rightarrow {}^{33}P + \nu_e,$$

6.4.15 $$^{30}P + e^{+} \rightarrow {}^{30}S + \nu_e,$$

6.4.16 $$^{35}Cl + e^{-} \rightarrow {}^{35}S + \nu_e,$$

6.4.17 $$^{37}Ar + e^{-} \rightarrow {}^{37}Cl + \nu_e.$$

The high neutron excess in these nuclei can produce an η in excess of 1 % from oxygen burning, which is well in excess of its value in normal matter (Woosley and Hoffman 1992). Because these nuclei have achieved very nonsolar abundances, they must not escape into the interstellar medium if modern stellar evolution calculations are to make any sense at all; they must ultimately be swallowed by the collapsed core, neutron star or black hole, that is the endpoint of the star's evolution.

An interesting development that begins to appear during oxygen burning is quasiequilibrium clusters (Woosley, Arnett, and Clayton 1972; Woosley 1986) of nuclei the abundances of which are maintained in approximate equilibrium under exchange of light particles. For example, near the end of oxygen burning, $^{28}Si(n,\gamma)^{29}Si$ is in equilibrium with $^{29}Si(\gamma,n)$ and $^{29}Si(p,\gamma)^{30}P$ is balanced by $^{30}P(\gamma,p)^{29}Si$, so that ^{28}Si, ^{29}Si, and ^{30}P are in equilibrium. A similar set of reactions maintains $^{34,35}S$ and $^{35,36}Cl$ in equilibrium. As the temperature increases, these small clusters grow to include more nuclei and also begin to

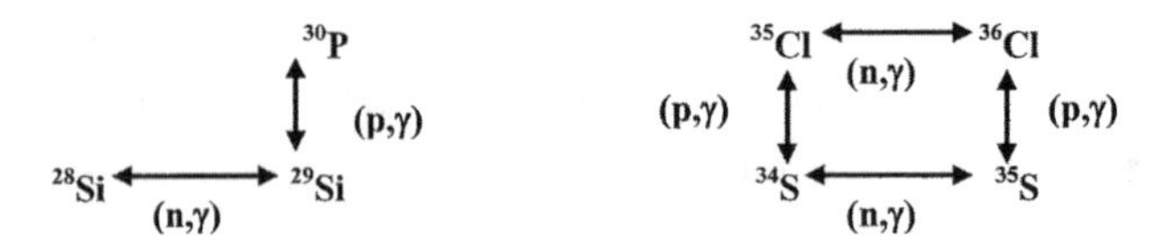

Fig. 6.6. Nuclides in equilibrium at the start of oxygen burning.

merge with each other (Woosley, Arnett, and Clayton 1972; Woosley 1986). Some of these are indicated in figure 6.6.

The dominant source of neutrinos during oxygen burning is, as with carbon burning, e^+-e^- annihilation, although other neutrino processes (see secs. 6.6 and 6.9) begin to be important. These will be sufficient to carry away a considerable amount of energy from the core of the star during this burning stage.

6.5 Silicon Burning and Nuclear Statistical Equilibrium

Following oxygen burning, the core of the star contracts further, increasing its density to 10^8 g cm^{-3}, and achieving a temperature of around $T_9 = 3.5$ (Weaver, Zimmerman, and Woosley 1978). At this temperature, some of the ^{28}Si ashes formed in the previous burning stage undergo photonuclear processes such as

6.5.1 $$^{28}\mathrm{Si} + \gamma \rightarrow {}^{24}\mathrm{Mg} + \alpha,$$

6.5.2 $$^{24}\mathrm{Mg} + \gamma \rightarrow {}^{20}\mathrm{Ne} + \alpha,$$

and so forth, ultimately producing seven α-particles from each of the ^{28}Si nuclei that began this process. The resulting α-particles can be captured on other ^{28}Si nuclei to produce ^{32}S and higher mass nuclei, but additional nuclear reactions result in an extremely complex network involving proton, neutron, and α-particle photonuclear processes as well as their inverses (Woosley 1986). The light particles produced can add to the quasi-equilibrium group nuclei above ^{28}Si, gradually pushing them to more massive, and more tightly bound nuclei, eventually reaching the Fe-Ni peak (actually, ^{64}Ni has the largest binding energy per nucleon of any nucleus).

The quasi-equilibrium group during silicon burning includes all nuclei heavier than ^{24}Mg initially, but at later times it includes all nuclei more massive than ^{16}O (Woosley 1986). The abundances of all the nuclides in this group at a given temperature and density are most sensitive to the abundance of ^{28}Si, that is, how much of the silicon has been burned. As its abundance decreases, the mean mass of the quasi-equilibrium group increases.

The neutron excess plays an important role in determining the nuclei that will be synthesized; if it is large, these will primarily involve nuclei to the neutron-rich side of the $N = Z$ line. Note, however, that η does not change rapidly; its changes occur on the timescales required for typical β-decays. If $\eta < 0.01$, the primary product will be ^{56}Ni, but that will shift to ^{54}Fe if η is larger, for example, > 0.02 (Woosley 1986), and to ^{64}Ni if η is even larger still. However, silicon burning may happen too quickly for most β-decays to occur; the typical amount of time for the core of a star to undergo silicon burning is 1 day (Weaver, Zimmerman, and Woosley 1978)!

The last reaction to come into equilibrium is the triple-α reaction, which eventually occurs at a rate that balances photodisintegration of carbon. At that time, all strong and electromagnetic reactions occur at rates that are balanced by their inverse reactions. Because of the tiny cross section for neutrinos to interact, weak interactions, however, are not in equilibrium. Nonetheless, this nuclear statistical equilibrium (NSE) allows us to describe the abundances of all the nuclides involved in a consistent set of equations. NSE is governed by the Saha equation, adapted to the nuclear case by Hoyle (1946) (and, e.g., Woosley and Hoffman 1992) as

6.5.3
$$Y(^{A}Z) = C(^{A}Z, \rho, T_9)\, Y_n^N Y_p^Z,$$

6.5.4
$$C(^{A}Z, \rho, T_9) = (\rho N_A)^{A-1} C(^{A}Z, T_9),$$

where

6.5.5
$$C(^{A}Z, T_9) = G(^{A}Z, T_9)\, A^{3/2} 2^{-A} \Theta^{1-A} \exp[BE(^{A}Z)/k_B T].$$

In equation 6.5.5, $\Theta = 5.943 \times 10^{13}\ T_9{}^{3/2}$, $Y(^{A}Z)$ is the number fraction of the nucleus with mass number A, charge number Z, and neutron number N, $Y_{n(p)}$ = neutron (proton) mass fraction, and $G(^{A}Z, T_9)$ = the nuclear partition function. This will produce a distribution involving many nuclei. But note that, because of the exponential of the binding energy in the expression for $C(^{A}Z, T_9)$, the nuclei produced will have a tendency toward maximal binding energy, hence toward the stable nuclei. While that feature might suggest that the α-nuclei would be produced, it is mitigated by the dependence in NSE on the neutron number (note the proton and neutron mass fractions in eq. 6.5.3) and the need for N to exceed Z for maximum nuclear stability for higher mass nuclei. Indeed, the nuclei favored by NSE will shift around somewhat depending on the relative numbers of protons and neutrons. The conditions and results of NSE have been studied in detail by Woosley and Hoffman (1992), in what they called the α-process. They note that the limitation to synthesizing heavier nuclei that are usually thought to be possible in NSE is not the Coulomb barrier but photodisintegration of the nuclei that are too proton rich to have a maximal binding energy.

The ashes of silicon burning are extremely tightly bound nuclei; any increase in the temperature of the core of the star produced by subsequent collapse cannot produce another burning stage that can stabilize itself. When the core mass exceeds that which can be supported by electron degeneracy, it begins its final collapse phase. Then, for stars with masses $<15 M_\odot$ the dominant process will be electron captures, which remove some of the electrons that support the core and also emit neutrinos, thereby further facilitating collapse. In more massive stars, the dominant process is photonuclear reactions, which absorb some of the energy of the core so that pressure does not increase in concert with gravity. Once either of these effects begins, collapse follows rapidly. However, other effects can also be important. The material in the core as it begins its final collapse is likely to be very neutron rich, and these nuclei will trap neutrinos inside the star's core. Furthermore, energy can be stored in the excited states of the nuclei involved (Bethe 1990). This allows the star to remain relatively cool until it reaches nuclear density and, therefore, for the nuclei to avoid destruction by photonuclear reactions until that time. Thus the details of this collapse process involve many facets. For many more details on the stages of massive star evolution and collapse, the reader is referred to Woosley (1986) and Arnett (1996).

At this stage our massive star has assumed an onionskin structure, with the most advanced stages of stellar evolution in the center and less advanced stages coming into play with increasing radius. This structure, along with the fuels being consumed at each stage, is indicated schematically in figure 6.7.

6.6 The Collapse Phase, Creation of a Neutron Star, and Neutrino Cooling

When the core of a massive star has completed its silicon-burning phase, turning its core into iron group nuclei, no subsequent compression can be arrested by a new nuclear fuel, as the iron group nuclei are the mostly tightly bound. Thus formation of any heavier nuclides will necessarily be endothermic, which means that the next compression phase cannot be halted until the core of the star achieves something close to nuclear matter density.

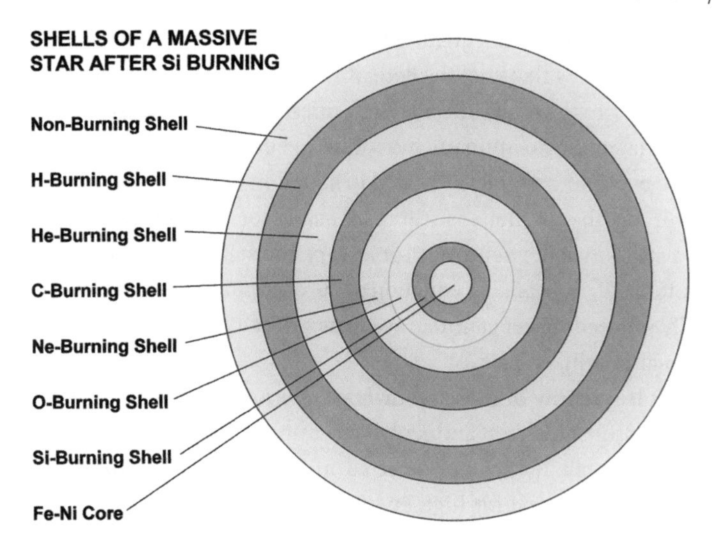

Fig. 6.7. The burning stages of a well evolved massive star, assumed to be at least 10 solar masses, showing its burning layers, in the usual "onion skin" form, just before it undergoes its final stage of core collapse.

The collapse will ultimately produce temperatures high enough to reduce all the preexisting nuclei in the core to α-particles and neutrons and, subsequently, into the constituent nucleons. The subject of collapse through this "neutronization" and the subsequent formation of the neutron star has been covered thoroughly (Bethe 1990; Burrows 1990; Langanke and Martinez-Pinedo 2003). The density will increase rapidly to several times nuclear density, at which point the core will undergo bounce and, as has been thought for several decades, will emit a shock wave that will expel the outer stellar layers. Additional nucleosynthesis will occur in the layers heated by this shock (via the rp-, γ-, and ν-processes). The energy contained in the core is enormous, of order 10^{53} ergs. The core cools within a few seconds by emitting neutrinos by the processes

6.6.1 $$\mathrm{p} + \mathrm{e}^- \rightarrow \mathrm{n} + \nu_\mathrm{e},$$

6.6.2 $$\mathrm{n} + \mathrm{e}^+ \rightarrow \mathrm{p} + \bar{\nu}_\mathrm{e},$$

6.6.3 $$\gamma \rightarrow \mathrm{e}^- + \mathrm{e}^+,$$

6.6.4 $$\mathrm{e}^+ + e^- \rightarrow \gamma + \gamma,$$

6.6.5 $$\rightarrow \nu_i + \bar{\nu}_\mathrm{i},$$

6.6.6 $$A^* \rightarrow A + \nu_i + \bar{\nu}_i.$$

The reactions indicated in equations 6.6.1 and 6.6.2 can operate as a cycle that produces neutrinos that cool the core; it is called an "URCA process," named after a casino (there are significant similarities!). The reaction in equation 6.6.5 indicates that neutrino-antineutrino pairs can be created by e^+e^- annihilation. While the two-photon mode indicated in equation 6.6.4 is favored by 19 orders of magnitude, the $\nu\bar{\nu}$ mode is still responsible for most of the cooling of the neutron star, since the neutrinos can escape the star much more readily than can photons. As indicated in equation 6.6.6, decay of an excited nucleus, while generally proceeding via gamma-ray emission, can also proceed by emitting a neutrino-antineutrino pair.

That at least some of these neutrinos are emitted was confirmed in 1987, when a handful of electron antineutrinos emitted from SN1987A (Hirata et al. 1987; Bionta et al. 1987) were detected in two large underground detectors that had been designed for high-energy physics experiments to observe proton decays, the Kamiokande and Irvine-Michigan-Brookhaven detectors. The reaction by which the neutrinos were detected is

6.6.7
$$\bar{\nu}_e + p \rightarrow e^+ + n.$$

The signature of the supernova neutrinos was the Cherenkov light created by the positrons recoiling through the water of the two detectors. As noted in chapter 5, neutrino cross sections are extremely small, but the fact that the energies of neutrinos from supernovae are considerably higher than those from the Sun, and that this reaction proceeds via a charged-current interaction, aids in their detection.

This supernova occurred in the Large Magellanic Cloud, which is not actually in our Galaxy but was close enough to be observed. The neutrinos were distributed over about 10 s, as shown in figure 6.8, with the bulk of them, and certainly those at highest energy, arriving within 2 or 3 s. While an extraordinary number of papers were written on the distribution of the neutrinos observed, the fact that there were only about 20 events is probably responsible for most of the effects that were noted. The upshot of these 20 events, however, is that the standard model of stellar collapse was qualitatively confirmed. The events also whetted the appetites of astrophysicists for an observation of a much larger number of events and possibly for observation of more than electron antineutrinos from the next nearby supernova.

We will return below to a discussion of the basic features of the supernova neutrino distributions and the means by which they might be observed. The neutrinos are also responsible for a significant amount of nucleosynthesis. Although we will await a detailed discussion of their effects until chapter 7, in

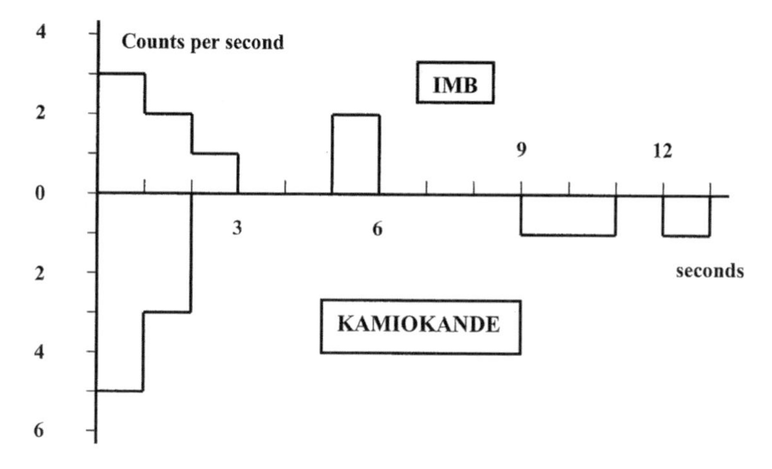

Fig. 6.8. The neutrino distribution from SN 1987A, as seen by the *Kamiokande* and *Irvine-Michigan-Brookhaven* collaborations.

which we discuss the processes of heavy-element nucleosynthesis, it is worth mentioning some of the basic features in which neutrinos play a role at this point. Certainly they can play a role in what has been called the ν-process, in which neutrino spallation reactions are responsible for synthesizing some of the less abundant nuclides from more abundant ones that are nearby in mass. However, they may also play roles in the r-process, both as a means to expedite that process and as a possible hindrance. The effects of the neutrinos on this latter process are so subtle that much work is still needed to understand them fully.

6.7 Classifications of Supernovae

While we have followed the evolution of a massive star through to its formation of a final collapsed object in the preceding sections, that scenario is not the only possible source of supernovae. Furthermore, the classifications depend on whether the description is from an astronomical or astrophysical perspective. Thus it is worth considering the specifics of at least some other supernova types. The main ones will be discussed; there are subclasses that are often sufficiently similar to the ones presented below that they will not be discussed separately.

In section 6.6, we described a type II, or core-collapse, supernova, that is, one that results from a star that is sufficiently massive ($M > 10M_\odot$) that it will explode as a natural consequence of its successive stages of gravitational contraction and collapse. This type of supernova is characterized by astronomers as one that has a high abundance of hydrogen in its observable spectrum. By contrast, as noted above, type Ia supernovae result from accretion onto a white dwarf to the extent that an instability occurs as a result of the accumulation beyond a critical mass. Hydrogen is not expected to be, nor is it, a part of the spectra of such stars, and it is this feature that is used by astronomers to characterize this type of supernova. Type II supernovae are powered by the energy resulting from its gravitational collapse, while type Ia supernovae are driven by thermonuclear explosions. However, curiously, the explosion energy in each case is about the same: of order 10^{51} ergs.

The light curve of the prototypical type II supernova (see, e.g., Weaver and Woosley 1980; Arnett 1996) rises sharply initially as the expanding shock wave hits the periphery of the star. The periphery is in the extended state that accompanies its helium-burning (red giant) phase, which is the only observable precursor to the supernova. The energy output decreases quickly and then becomes a broad bump or a plateau that can last for 2–3 months and that is powered by cooling from the shock-heated stellar material. The hydrogen periphery of such a star is opaque at this time, so this energy is trapped and can escape only slowly. Beyond the plateau period the light curve falls off with the characteristic 80-day half-life from the decay of the ^{56}Ni and, subsequently, ^{56}Co, the ^{56}Ni having been produced as the endpoint of the stellar nucleosynthesis just prior to the explosion. A typical light curve from a type II supernova is shown as the upper curve throughout most of the postexplosion time period in figure 6.9.

As noted above, type Ia supernovae are thought to result from accretion from one member of a binary system onto the other, which is a white dwarf (Arnett 1996). The star that is initially the more massive will go through its stages of stellar evolution and, assuming it has the appropriate initial mass, will end up as either a carbon-oxygen or an oxygen-magnesium white dwarf. As the companion star completes its hydrogen-burning phase and becomes a red giant, the matter from its periphery will be attracted by the gravitational potential of the white dwarf, achieving a high temperature as it falls onto its surface. The white dwarf is supported by electron degeneracy, so the thermonuclear reactions that will occur can produce an enormous amount of energy without affecting the pressure. The result will be thermonuclear runaway. If the mass accretion is large enough that the composite object exceeds its Chandrasekar mass, the maximum mass for a star that is supported by electron degeneracy pressure, or if sufficient electron captures occur to reduce appreciably the electron degeneracy pressure, then the entire star will

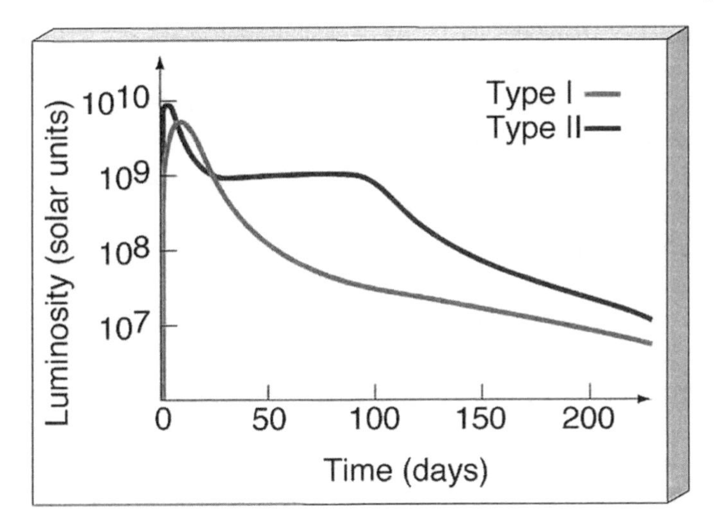

Fig. 6.9. The idealized light curves for a "classical" type II supernova, that is, for one that has a significant hydrogen envelope, and for a type Ia supernova. From Chaisson and McMillan (2002). Reprinted by permission of Pearson Education, Inc. Upper Saddle River, NJ.

generally explode and be ejected into the interstellar medium. In rare cases, however, a neutron star remnant may result (Woosley and Baron 1992).

Since type Ia supernovae have no outer shell of material to trap the radiation from the core of the star, the light curve from these supernovae will look different from those from type II supernovae (see, e.g., Arnett 1996). However, the burning is thermonuclear, and does progress via NSE, so it will also produce a great deal of ^{56}Ni. The radioactive decays of this and of its daughter, ^{56}Co, power the light curve, giving it the characteristic roughly 80-day falloff, as shown in figure 6.9. However, the early time features of the type II supernovae light curves will be missing, as will the plateau.

Two more types of stars that would be expected to have similar stages of evolution to those of a type II supernova are type Ib and type Ic supernovae. These appear to be the result of massive stars that have undergone mass loss. Thus they are produced by the core-collapse processes of type II supernovae but do not have the outer shell of material to trap the energy of the explosion. Woosley, Langer, and Weaver (1993) studied such effects in massive stars of 35–85 $M_\odot$. They found that mass loss that could shed most of the mass of the stars occurred during the helium-burning phase, producing a presupernova core of about 5 $M_\odot$. They also note that such mass loss depends critically on stellar metallicity. By the time the core helium has been exhausted, all but the 35 $M_\odot$ star had completely shed their hydrogen envelopes, while that star had a residual

2 $M_\odot$ envelope. From this point it was found that the 35 $M_\odot$ star would evolve as would a star that had not undergone mass loss, ultimately to produce a type II supernova. The others evolved somewhat differently from their non-mass-loss counterparts, but still ultimately became core-collapse supernovae.

However, because this type of star will have shed its outer envelope as a stellar wind, it will present a very different appearance to astronomers from that of a type II supernova. It was found (Woosley, Langer, and Weaver 1993) that the final stage would proceed in NSE but that the final stages of the more massive stars would differ somewhat from that of a "standard" type II supernova. They would evolve to ^{56}Ni cores and would produce primarily the α-nuclei along the way. Their light curves would have the characteristic 80-day half-life produced by the $^{56}\mathrm{Ni} \rightarrow {}^{56}\mathrm{Co} \rightarrow {}^{56}\mathrm{Fe}$ decays seen in type II supernovae. However, these more massive stars have a different astronomical classification as a result of having very little hydrogen in their spectrum and because they do not have much of an envelope to trap the radiation being emitted from the decay of their ^{56}Ni, so they will not exhibit the plateau that characterizes a classical type II supernova. These stars are designated as Wolf-Rayet stars; they may exhibit strong N or C features in their spectra, depending on the extent to which their outer shells have been stripped away, prior to their achieving supernova status.

In a subsequent study, Woosley, Langer, and Weaver (1995) concluded that the progenitors of type Ib supernovae might have been stars with less initial mass than in the 1993 study but that the mass loss might have occurred as a result of mass transfer within a close binary system.

One longstanding issue with core-collapse supernovae involves understanding the explosion mechanism. The energy output of a supernova is prodigious; only a small fraction of that energy needs to be coupled to the outer reaches of the star in order to make it explode. All supernovae were initially thought to result from a thermonuclear explosion (Fowler and Hoyle 1964). While this still prevails as the explanation for type Ia supernovae, it is not thought to power type II supernovae. About the same time, Colgate and White (1966) suggested that neutrinos could provide the necessary momentum to the outer shells of the star to expel them into the interstellar medium, but this mechanism has been found not to be robust. However, it was revived when neutral weak currents were discovered (Wilson 1974), but even their addition failed to provide a compelling explosion mechanism.

It has been observed that at least some of the explosions are asymmetric; in some cases, the neutron star has been observed moving at fairly high speed away from the center of the explosion, and neutron stars moving at fairly high velocity with respect to the Galactic disk, possibly representing old core-collapse supernovae, are well-known objects. For a review of issues related to the explosions, see Arnett (1996).

Recent work by Blondin et al. (2003) and by Burrows et al. (2006) has focused on instabilities as the explosion driver. Burrows et al. (2006) produced a model, using an 11 $M_\odot$ progenitor in which infalling matter from accretion streams begins, after about 200 ms, to generate-oscillations-in the core that ultimately translate into acoustic waves that transfer their energy to the outer reaches of the star with high efficiency. The power from this core oscillator appears to be quite sufficient to drive the explosion of the star (unlike that of the neutrinos), which if it turns out to be a general feature of core-collapse supernovae, solves a longstanding problem. In addition, asymmetric explosions, which have been observed as a product of many core-collapse supernovae, appear to be a natural consequence of this model.

6.8 Nuclear Structure and the Core-Collapse Process

In section 6.5, we discussed how the core collapsed by capturing electrons. How the core gets to that state, however, is interesting, and nuclear physics experiments have played a significant role in our understanding thereof. It is thought (Bethe et al. 1979) that the collapse must occur at a low entropy so that preexisting heavy nuclides, especially the iron group nuclei formed in the silicon-burning phase, will not be destroyed too early in the collapse process. These nuclei are needed to trap the neutrinos (large nuclei have large neutrino scattering cross sections due to coherence effects) that will be emitted as neutronization of the core proceeds so that ultimately the core temperature will become sufficiently high to reduce all the nuclei to (mostly) neutrons. The neutronization proceeds by electron captures on the iron group nuclei via Gamow-Teller (GT) transitions to nuclei with successively lower nuclear charges. This process has long been described by simplified shell-model descriptions of the nuclei involved, which assumed that many of the potential levels in the final state nuclei to which transitions might proceed would be Pauli blocked. However, residual interactions can unblock these shells somewhat; shell-model Monte Carlo calculations (Koonin, Dean, and Langanke 1997) have shown that this can occur.

Fortunately, these states in at least some of the nuclei of interest can be studied, and the strengths of the GT transitions can be determined, via charge exchange reactions such as (n,p) and (t,^{3}He); these reactions would be expected to populate the states of interest strongly. The results can then

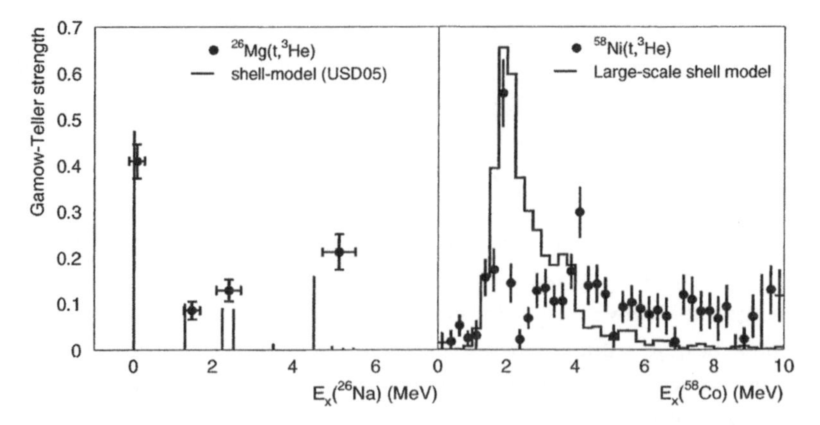

Fig. 6.10. Results for the $^{26}Mg(t, {}^{3}He)$ and $^{58}Ni(t,{}^{3}He)^{58}Co$ reactions at 112 MeV/nucleon: The main components of the GT strength, both from the reactions studied and from shell-model calculations, are as indicated. The experimental-theoretical comparison is seen to be quite good. Reprinted with permission from R. Zegers et al. 2005.

be used to provide some indication of the capability of the theoretical models to be extrapolated to the nuclei that cannot be studied. A recent study by Zegers et al. (2005) is prototypical of such experiments. This study was of the $^{26}Mg(t,{}^{3}He)$ and $^{58}Ni(t,{}^{3}He)^{58}Co$ reactions, which were performed at the *National Superconducting Cyclotron Laboratory* of Michigan State University (see sec. 2.12.2) with a secondary triton beam. The experiments measured cross sections at very forward angles (the transition is expected to peak at 0°) and that cross section is proportional to the GT strength B(GT) as

6.8.1 $$d\sigma/d\Omega(q = 0) = KN_D|J_{\sigma\tau}|^2 B(\text{GT}).$$

In equation 6.8.1, ($q = 0$) implies that the relationship holds at zero momentum transfer, K is a kinematical factor, N_D is the "distortion factor", and $|J_{\sigma\tau}|^2$ (see sec. 3.2 for some information on direct reactions) is the volume integral of the nucleon-nucleus interaction.

The results from these experiments are shown in figure 6.10. There the GT strength functions measured in the two reactions are compared to shell-model results. This clearly shows that the measured and predicted GT strengths are in good agreement, both in strength and distribution, especially for the lowest excitation energy groups (those seen to the left in both figures). These results have been found to be at least qualitatively consistent with those of Hagemanns et al. (2004, 2005), taken with a (d,^{2}He) reaction. The results from the (t,^{3}He) experiment, however, are more straightforward to interpret, as the unboundedness of ^{2}He can complicate the analysis of those data.

6.9 Supernova Neutrino Distributions and Supernova Neutrino Detection

As noted in section 6.6, neutrinos and antineutrinos of all flavors are produced in the hot core of a collapsing star. Subsequent scattering produces the neutrino distributions that, when fit with Fermi-Dirac distributions that include a chemical potential, comprise the standard model of supernova neutrino production (McLaughlin et al. 1999; Qian et al. 1993). The equation for such a distribution is given by

6.9.1 $$f_\nu = \left[T_\nu^3 F_2(\eta)\right]^{-1} E_\nu^2[\exp(E_\nu/T_\nu - \eta) + 1]^{-1},$$

where η is the chemical potential divided by T_ν and $F_2(\eta)$ is a normalization factor. T_ν is the temperature at the neutrinosphere. The standard model predicts that $\langle E(\nu_x)\rangle = 25$ MeV, $\langle E(\nu_e)\rangle = 11$ MeV, and $\langle E(\bar{\nu}_e)\rangle = 16$ MeV (although inclusion of different neutrino effects will produce different energies). Here ν_x refers to ν_μ, $\bar{\nu}_\mu$, ν_τ, and $\bar{\nu}_\tau$, which are grouped together because these flavors all interact only through the neutral-weak interaction, so they are expected to have identical energy distributions. However, these are only the features of the neutrinos as they are emitted from their neutrinospheres, that is, their surfaces of last scatter. The results from *Super-Kamiokande* and *Sudbury Neutrino Observatory* (*SNO*) on MSW neutrino oscillations (see chap. 5) determine that many of the (higher energy) ν_xs produced in supernovae will transform to ν_es of the same energy and vice versa. The effects of oscillations have been studied in detail by Dighe and Smirnov (2000). Because of oscillations, in order to sample the energy distributions of the ν_xs at their neutrinospheres, one must detect the ν_es that arrive at Earth. In addition, subtle effects (Raffelt 2001), such as "neutrino bremsstrahlung," "inelastic scattering," and "spectral pinching," would also affect the spectra.

The effect of spectral pinching is illustrated in figure 6.11. There it is seen that, because smaller neutrinospheres will be at higher temperatures than larger radius neutrinospheres, one would expect that higher energy neutrinos would be emitted from the former. However, this effect is mitigated by the fact that higher energy neutrinos have larger interaction cross sections, extending somewhat their zeroth-order neutrinospheres. Conversely, the smaller cross sections of the lower energy neutrinos will tend to shift their zeroth-order neutrinospheres inward somewhat. This will tend to "pinch" the tails of the neutrino spectra toward the mean energy (compare the darker line in fig. 6.11 to the lighter line). This effect is simulated analytically by the nonzero chemical potential added to the Fermi-Dirac distribution that would be used in the most naive model.

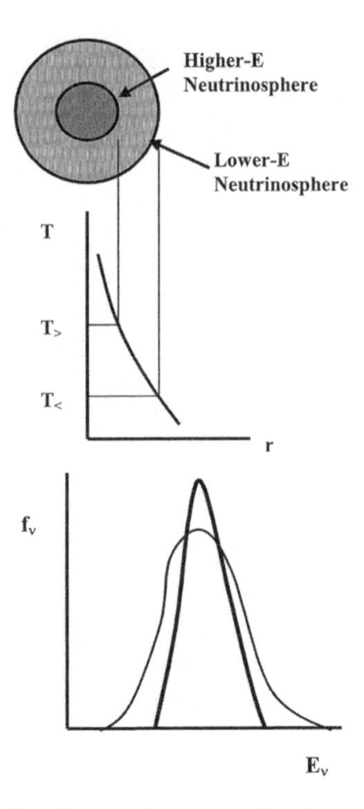

Fig. 6.11. Effects of spectral pinching on the neutrino distributions, as discussed in the text.

The combined time and energy distributions might also indicate the presence of convection and rotation (Burrows 1990; Mezzacappa et al. 1998; Totani et al. 1998); convection, in particular, might be expected to mix the hotter material from deep within the core with the cooler material nearer the neutrinosphere, thereby hardening the spectrum of neutrinos emanating from the neutrinosphere.

Detection of the neutrinos from a supernova involves some interesting nuclear physics. The events observed by *Super-Kamiokande* (see sec. 2.10.1) will be predominantly due to the interactions

6.9.2 $$\bar{\nu}_e + p \rightarrow e^+ + n,$$

and

6.9.3 $$\nu_i + e^- \rightarrow \nu_i + e^-.$$

The yield from the former reaction is much larger than that from the latter, but the latter has the advantage that, since electrons are light, their direction of recoil will allow pointing back to the direction from which the neutrinos came, thus providing some indication of the location of the supernova. The Cherenkov radiation that results from the energy loss of the e^+ from the first reaction allows determination of the energy spectrum of the $\bar{\nu}_e$s. In addition, a much smaller number of neutral-current events will be detected from inelastic excitation of the ^{16}O to states that are above proton and neutron emission thresholds (Langanke, Vogel, and Kolbe 1996). The resulting ^{15}N and ^{15}O nuclei will occasionally produce detectable deexcitation gamma rays. The threshold for this process is relatively high so its yield will have particular sensitivity to the high-energy tail of the ν_xs, and to any processes that can affect it. Of course, the high threshold also reduces the resulting yield.

SNO (see sec. 2.10.2) operates with both light water and heavy water. Its light water will produce events from the reactions indicated in equations 6.9.2 and 6.9.3. However, its heavy water will also produce events from the charged-current reactions

6.9.4 $$\bar{\nu}_e + d \rightarrow e^+ + n + n,$$

and

6.9.5 $$\nu_e + d \rightarrow e^- + p + p,$$

and from the neutral-current reaction

6.9.6 $$\nu_i + d \rightarrow \nu_i + n + p.$$

The charged-current reactions produce energetic leptons that can be observed via their Cherenkov light, whereas the neutral-current interactions can be observed by detecting the neutrons they produce via several techniques (see sec. 5.4), depending on the state of evolution of the detector when the supernova neutrinos arrive at Earth.

KamLAND (see http://www.awa.tohoku.ac.jp/html/KamLAND/) also detects $\bar{\nu}_e$s via the reaction indicated in equation 6.9.2 and all neutrinos from that in 6.9.3. However, it will also observe interactions of the $\bar{\nu}_e$s and ν_es on ^{12}C that populate the ground states of ^{12}N and ^{12}B. *KamLAND* will see the resulting βs from the decay of those unstable nuclei in (very) slow coincidence with the βs from the charged-current neutrino-induced interaction. Neutral-current interactions will proceed primarily through the 15.11 MeV state in ^{12}C, which will produce a sharp gamma-ray peak.

A particularly interesting possibility for converting supernova neutrinos into detectable signals is through the use of lead. This could involve just detection of the neutrons emitted either through charged-current interactions or neutral-current interactions (Zach et al. 2002) or might also utilize some lead-containing compound that is soluble in water such as lead perchlorate (Elliott 2000). In the former case, lead sheets could be interspersed with some sort of scintillator material, and in the latter, the events could be detected from the Cherenkov light produced either by recoiling leptons or the gamma rays resulting from neutron captures. Lead has been found to be an extremely efficient converter of supernova neutrinos to neutrons (Zach et al. 2002), given expected neutrino-lead cross sections (Kolbe and Langanke 2001; Volpe et al. 2002; Engel, McLaughlin, and Volpe 2002), reaching mean values as large as 10^{-40} cm^2 for the charged-current cross sections at some of the energies. Lead produces events from neutral-current interactions:

6.9.7 $$\nu_i + {}^{208}\mathrm{Pb} \rightarrow \nu_i + {}^{208}\mathrm{Pb}^* \rightarrow {}^{208}\mathrm{Pb} + \gamma, \text{ or}$$

6.9.8 $$\rightarrow {}^{207}\mathrm{Pb} + \mathrm{n}, \text{ or}$$

6.9.9 $$\rightarrow {}^{206}\mathrm{Pb} + 2\mathrm{n}.$$

Lead's thresholds for neutral-current neutron-emitting reactions are 7.37 MeV for one-neutron events and 14.11 MeV for two-neutron events. These events merely indicate that an interaction with a neutrino has occurred; because there is an undetected outgoing neutrino, it is not possible to determine the energy of the incident neutrino.

However, lead can also produce neutrons from charged-current interactions via

6.9.10 $$\nu_e + {}^{208}\mathrm{Pb} \rightarrow e^- + {}^{208}\mathrm{Bi}^* \rightarrow {}^{208}\mathrm{Bi} + \gamma, \text{ or}$$

6.9.11 $$\rightarrow {}^{207}\mathrm{Bi} + \mathrm{n}, \text{ or}$$

6.9.12 $$\rightarrow {}^{206}\mathrm{Bi} + 2\mathrm{n},$$

with a threshold of 9.26 MeV for one-neutron and 17.35 MeV for two-neutron events. The charged-current interactions have a larger cross section than do the neutral-current interactions, so even though the former reactions have a higher threshold energy, they will dominate the yield, all other things being equal.

The charged-current events have the potential for measuring the energy spectrum of the electron neutrinos, if the detector has the capability of detecting the energy of the electron emitted and the number of neutrons emitted. The energy of each electron produced will be equal to the energy of the incident neutrino minus the energy required to excite the state populated in ^{208}Bi; these

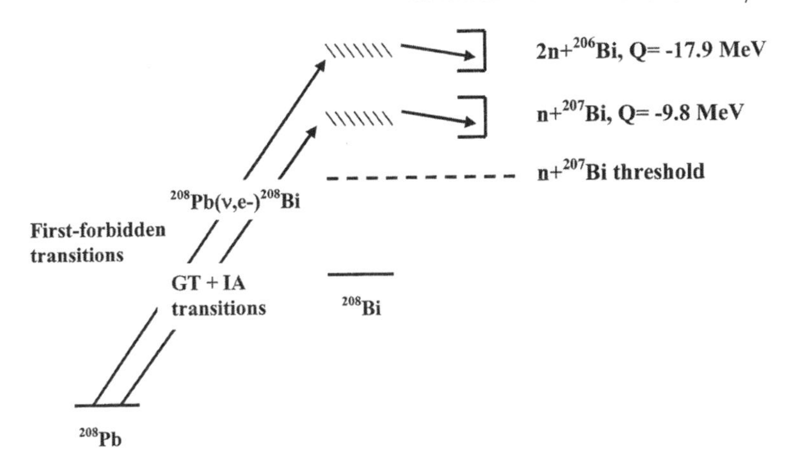

Fig. 6.12. Energetics of the charged-current reactions on ^{208}Pb. The transitions are dominated by clustered states indicated as the cross-hatched regions.

states are clustered, either at an excitation energy of around 15 MeV (the GT resonance states and the isobaric analog state) or around 21 MeV (the states populated by first forbidden transitions). The (fairly localized) GT resonance and the isobaric analog state will decay by emitting a single neutron. The states populated by first forbidden transitions tend to decay by two-neutron emission. This situation is indicated in figure 6.12. Some of the states that would be populated by supernova neutrinos, especially the GT resonance and isobaric analog state, have been studied via direct reactions. Specifically, the ^{208}Pb(^{3}He,t)^{208}Bi reaction was studied by Akimune et al. (1995), using the accelerator and ancillary facilities at the Research Center for Nuclear Physics at Osaka University. Similar studies have been done (Flanders et al. 1989) with the ^{208}Pb(p,n)^{208}Bi reaction, which was studied at the Indiana University Cyclotron Facility. Of course, these nuclear reactions would be expected to have a somewhat different selectivity than would the neutrino-induced reactions, so their application to neutrino detectors must be viewed with some caution.

Supernova neutrino luminosities might also signal the formation of a black hole should such an event result from stellar collapse. If that were delayed by even a fraction of a second from the time the first neutrinos were emitted, an abrupt termination of the neutrino luminosities would be observed (Burrows 1990). However (Baumgarte et al. 1996), since the ν_x*s* interact differently with matter than do the ν_es and $\bar{\nu}_e$s, their neutrinospheres would probably not be at the same distance from the center of the collapsing star. Thus termination time differences of $\sim$1 ms might be expected (Beacom, Boyd, and Mezzacappa

2000, 2001; Baumgarte et al. 1996) for the different flavors, a timing capability that could be achieved by *Super-Kamiokande*. However, the need to measure time differences of that magnitude necessitates a close supernova in order to achieve the necessary statistics. Nonetheless, the sudden termination of the neutrino spectra would be a stunning confirmation of the formation of a black hole.

It should be noted that these distributions discussed above may not reflect the distributions that would be observed. Takahashi et al. (2001) and Fuller and Qian (2005) have suggested that maximal mixing could occur deep within the star and that the result would be a considerably more mixed spectrum than might be expected from the above discussion. This effect, if it does occur, would be expected to produce very similar distributions of the different neutrino species. Further theoretical work may help to resolve the issue, but it may also require the observation of supernova neutrinos and probably both ν_es and $\bar{\nu}_e$s to resolve these issues.

An important aspect of supernova neutrino detection is the frequency with which Galactic core-collapse supernovae occur. Uncertainties arise because the observational record reflects only those supernovae that were close enough to Earth that they were not obscured by dust in the Galaxy. The situation was reviewed recently (Beacom, Boyd, and Mezzacappa 2001); the Galactic supernova rate seems to be 3–10 per century, depending on the details of the estimate. The ratio of black hole to neutron star formation was also recently estimated (Beacom, Boyd, and Mezzacappa 2001) to range from 1:4 to as high as 9:1. Despite the wide range, these values do suggest that black hole formation is at least not a much rarer event than neutron star formation.

6.10 Neutron Star Structure and Strange Quark Matter

A subject of great current interest is that of the structure of a neutron star. Detailed models suggest that a crust develops at the surface of the neutron star, with nuclear reactions at extremely high density and neutron richness, so-called pycnonuclear reactions, transforming the crust just below the surface into truly exotic nuclei and crystalline structures. As the density increases and the temperature decreases (energy is radiated away), the Gamow peak is increased in height and shifted toward a lower energy (Harrison 1964). Ultimately, this effect will shift the center of the Gamow peak across the origin so that the reactions are now sensitive to the high-energy tail of the peak. The reaction probability now depends primarily on very low energy particles.

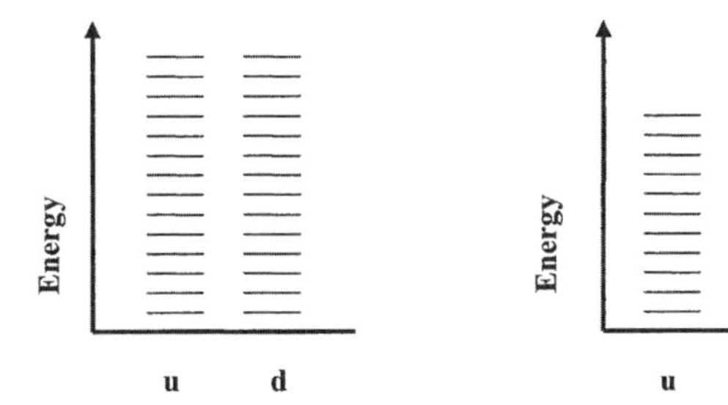

Fig. 6.13. Energy levels of the quarks in ordinary matter (*left*), which is made up of up (u) and down (d) quarks, and strange quark matter (*right*), which is made up of u, d, and strange (s) quarks. The total energy of the ordinary matter can be seen to be higher than that of the strange quark matter, as the highest energy states occupied by the up and down quarks in ordinary matter are above those of the strange quarks in the strange quark matter. The s quarks begin at a higher energy than do the u and d quarks because of the larger s-quark mass.

The structures below the crustal surface are also unlike anything that occurs anywhere else in nature. Indeed, Ravenhall, Pethick, and Wilson (1983), and more recently Watanabe et al. (2005), have discussed the transitions to unusual shapes that nuclei would be expected to undergo as the density of their environment increases while core collapse proceeds. These shapes can be spherical in one phase, tubular in another, and planar in yet another, leading to their characterization as "nuclear pasta."

Below this region is thought to occur a (mostly) neutron superfluid, and below that, a region in which the constituents may no longer be neutrons and protons, but up, down, and strange quarks. Indeed, it has been suggested (Witten 1984) that this may be the "ground state" of nuclear matter, but the strange quark matter objects, SQMs, can only be achieved under the extreme conditions that exist in neutron stars. The reason that this may be a more tightly bound state than ordinary nuclear matter is that the three constituent quark flavors allow the quark matter fluid to circumvent the Pauli principle to some extent by converting the highest energy up and down quarks to strange quarks at the bottom of their energy distribution, thus lowering the total energy of the system (see fig. 6.13). However, the strange quark is thought to have a considerably higher mass in such an SQM fluid (Berger and Jaffe 1987, 1991), perhaps 250 MeV compared with the 5 or so MeV of the up and down quarks, so such matter could not be formed if the SQMs were too light (Berger and Jaffe 1987, 1991; Takahashi and Boyd 1988). In addition, since ordinary nuclei were formed following the big bang, it would be difficult to make such SQMs except in the environment of a neutron star.

However, if SQMs were to escape into the interstellar medium, they would have a most unusual appearance (Witten 1984): massive "nuclei" with anomalously low nuclear charge, such as ^{65}He (Takahashi and Boyd 1988), or even nuclei with masses of thousands of atomic mass units, but with low charges. Although detection of anomalous events has occasionally raised excitement as to the possible discovery of SQMs (Saito et al. 1990; Boyd and Saito 1993), more deliberate searches for them have not confirmed their existence (Vandegriff et al. 1996; Lu et al. 2005), and some doubts have been raised that they could exist at all (Madsen 1988).

Another approach to searching for SQMs is to attempt observations of neutron stars that would showcase the differences between ordinary neutron stars and SQM stars. If SQM matter is the true ground state of nuclear matter, then neutron stars would be more tightly bound than they would be if they consisted only of protons and neutrons. This would allow them to be slightly denser objects and therefore to spin more rapidly than would be the case for nuclear matter. Extensive theoretical effort has gone into identifying and quantifying signatures of SQM stars (see, for example, Alford et al. 2005). Many of these have focused on pulsars. Since pulsars are thought to be the result of rapidly spinning neutron stars, searches for "millisecond pulsars" have been ongoing for many years to see if such objects exist, with the belief that this rotation frequency would not be possible with ordinary matter but might be possible with SQM stars.

A recent observation of a neutron star that appeared to be too small to be composed of protons and neutrons illustrates the excitement that a quark matter star discovery would generate. Observations (Slane, Helfand, and Murray 2002) of 3C 58, thought to be a neutron star, with the X-ray satellite *CHANDRA* suggested that the emission from this star was far too low for it to be characteristic of neutron star cooling. The cooling curve would be blackbody in shape but would not have the magnitude expected for an object the size of a neutron star. This led those authors to suggest that this observation required exotic cooling processes in the neutron star core. This was taken a step further by Drake et al. (2002), who suggested that this object might be smaller than a conventional neutron star by about a factor of at least 2 in radius, suggesting a neutron-star-like object, but one that would be more compressed than a neutron star. They hypothesized that the object might be a quark matter star. However, it was subsequently pointed out by Walter and Lattimer (2002) that the distance to this object might well be incorrect and that if a new value were imposed for that distance, the cooling rate would be quite consistent with a compact object of the usual size of a neutron star.

Another suggestion for observing SQM stars, by Lin et al. (2006), would be to observe the transition from normal matter to quark matter by the gravitational waves such a transition would emit. The flurry of activity around 3C 58, as well as the ongoing activity in this area, shows the interest in determining if quark matter can exist inside a neutron star. There is no doubt that discovery of a definitive means for detecting such matter would have a profound effect on our understanding of neutron stars, and this remains an active area of research.

6.11 Gamma-Ray Astronomy

Although some of the preceding discussion has focused on detection of neutrinos as a diagnostic tool for supernovae, gamma-ray astronomy can also provide significant information on supernovae by studying their remnants. For example, ^{44}Ti is thought to be made in the relatively low-density region outside the just-collapsed core (see, e.g., Timmes et al. 1996). This region undergoes α-rich freezeout, with the abundances of the nuclides being dictated, at least initially, by NSE. ^{44}Ti has special significance because of observations of Dupraz et al. (1997), using the *Compton Gamma Ray Observatory*, of ^{44}Ca deexcitation gamma rays from Cas A that are initiated by the β-decay of ^{44}Ti. More recently, evidence for ^{44}Ti decay was observed with *INTEGRAL* (see Renaud et al. 2006 and sec. 2.6.5). The decay of ^{44}Ti proceeds as

6.11.1 $$^{44}\mathrm{Ti}(T_{1/2} = 60\,\mathrm{years}) - \beta \rightarrow {}^{44}\mathrm{Sc}(T_{1/2} = 4.0\,\mathrm{hours}) - \beta \rightarrow {}^{44}\mathrm{Ca}^*.$$

The first excited state of ^{44}Ca is indicated as $^{44}\mathrm{Ca}^*$. Once populated by the β-decay of ^{44}Sc, it immediately decays to the ^{44}Ca ground state, producing a 1.157 MeV gamma ray that can be observed by spaceborne gamma ray detectors. Cas A is thought to have exploded as a core-collapse supernova more than 300 years ago. (See, for example, Fesen et al. 2006 for discussion of this determination.) The amount of ^{44}Ti suggested by the observations could either be on the edge of agreement with stellar collapse models, if the half-life for ^{44}Ti were at the upper end of the range of values resulting from years of measurements, or badly out of agreement with the models if the shorter value were the correct one. The half-lives were grouped into values either around 45 years or around 60 years. New measurements of that half-life (Gorres et al. 1998; Ahmed et al. 1998; Norman et al. 1998) determined that the correct value is the larger one; the weighted average from Ahmed et al. (1998) and Gorres et al. (1998) is 59.2 $\pm$ 0.6 years, leaving the stellar-collapse model in acceptable, albeit somewhat tenuous, agreement with the observations (Timmes et al. 1996).

A possible solution to this potential problem was offered by Motizuki et al. (1999), who observed that ^{44}Ti is a nucleus that undergoes electron capture (see sec. 3.5), so that if the ^{44}Ti produced in the supernova existed for some time in a highly ionized state, its apparent lifetime would be extended and would therefore make it appear that the amount of ^{44}Ti currently observed from Cas A was the result of larger ^{44}Ti production. The highly ionized state could be the result of the shock wave that produced the Cas A ejecta. Another possible solution was provided by a measurement of the ^{40}Ca(α,γ)^{44}Ti cross section (Nassar et al. 2006). Use of an intense ^{40}Ca beam from the ATLAS linear accelerator of the Argonne National Laboratory in this experiment allowed direct detection of the recoiling ^{44}Ti ions. The result obtained from this experiment indicated that the cross section that had been used to predict the ^{44}Ti yield from the supernovae was lower than the actual result. The ^{44}Ti yield that would be produced with the new cross section would be higher by about a factor of 2 than that predicted previously, alleviating the theoretical-observational conflict to a significant extent.

Another important product of core-collapse supernovae, indeed of all supernovae, is ^{56}Ni; up to several tenths of a solar mass are produced in the region outside the core by explosive nucleosynthesis in supernovae and so will be ejected when a supernova explodes. ^{56}Ni β-decays to ^{56}Co ($T_{1/2} = 6$ days) and then to ^{56}Fe ($T_{1/2} = 77$ days). The light curves from supernovae are observed to decay, after an initial period that depends on the type of supernova, with about an 80-day half-life (as noted in sec. 6.7, some light trapping occurs in type II supernovae); they are indeed powered by the decay of the ^{56}Ni produced.

The nucleosynthesis that ultimately results from a core-collapse supernova involves several processes subsequent to those discussed thus far. These include the s- and r-processes, the "slow" and "rapid" neutron capture processes, among others, which are discussed in detail in chapter 7. The nucleosynthesis output of a type II supernova has been studied in detail by Woosley and Weaver (1995); the interested reader is directed to that paper.

6.12 Gamma-Ray Bursts

This chapter closes with a discussion of gamma-ray bursts (GRBs). These were discovered accidentally in the 1960s by a military satellite that was launched to monitor compliance with the Nuclear Test Ban Treaty. These satellites were sensitive to gamma rays, which would be the signatures of nuclear blasts in the Earth's atmosphere. Since that time, the systematics of GRBs have been

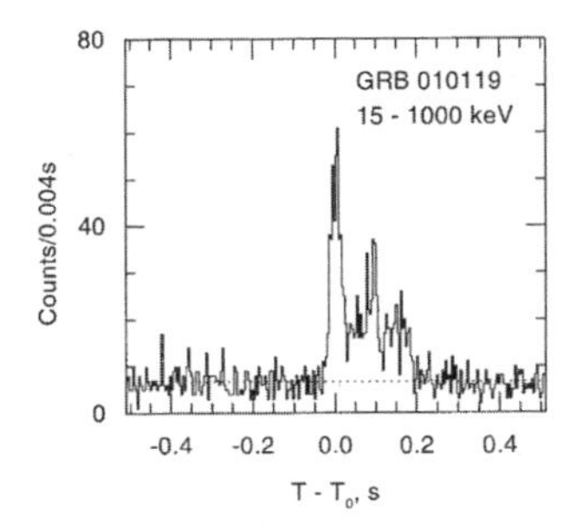

Fig. 6.14. Light curves of two GRBs, the first, GRB 010119, in the short-duration category, and the second, GRB 920622, in the long-duration category. The second one was observed by three different detectors, so was seen in three energy ranges. As can be seen, GRBs can have profoundly different features; this can apply even to GRBs within the same category. From Hurley et al. (2002) for the first and Greiner et al. (1995) for the second.

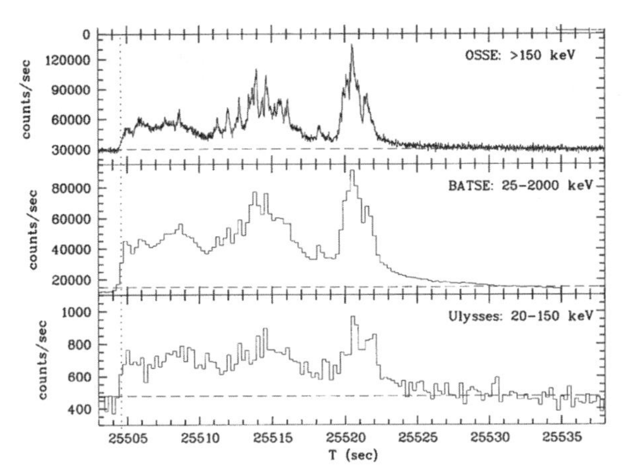

greatly refined (Kouveliotou 1997), and the determination that they are both very frequent and extragalactic has led to a flurry of theoretical activity to explain how any object could produce such prodigious amounts of energy over such a short timescale. Since the reigning model at present for one class of GRBs seems to involve a core-collapse supernova, it is natural to include this discussion in this chapter. Figure 6.14 shows the observed light curves from two GRBs, one in the short-duration category, defined as lasting less than 2 s, and the other in the long-duration category, those lasting for more than 2 s. This time cutoff seems at present to provide the demarcation between the two classes of GRBs.

The first GRB was detected by the *Vela* (satellite) and dates to July 1967. However, because of security concerns, it was not until 1973 that Klebesadel, Strong, and Olson (1973) published their discovery. That first paper reported 16 GRBs that occurred between July 1969 and July 1972.

Early gamma-ray detectors were not capable of localizing the source of the gamma rays they observed with much precision. Triangulating the detections from several different widely spaced gamma-ray detectors could produce some

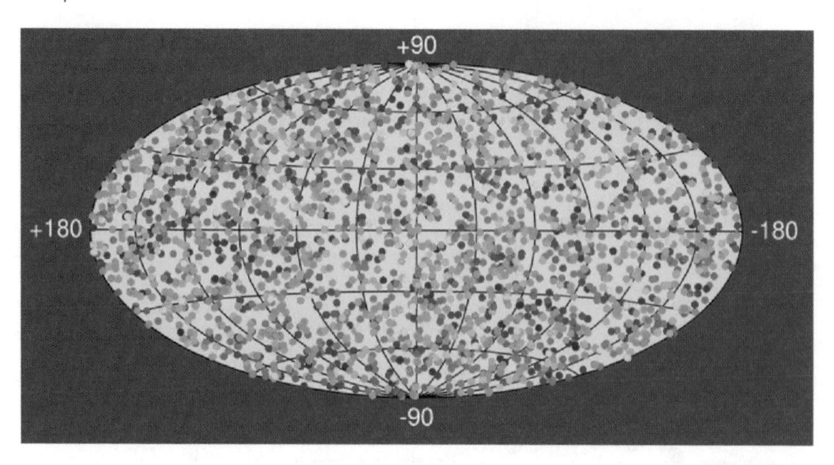

Fig. 6.15. All-sky map of the locations of GRBs detected by BATSE. Notice they are distributed quite uniformly over the sky; they certainly are not constrained to the plane of the Galaxy. From http://f64.nsstc.nasa.gov/batse/grb/skymap/.

spatial localization, but this approach did not provide a measurement of the distance to the bursts. This lack of distance information led to two very different classes of theories about where GRBs come from. One was that GRBs occur at cosmological distances (that is, hundreds of millions to billions of light-years away). However, most astronomers favored a scenario in which GRBs occurred much closer to Earth since this would make the GRB's energy output easier to explain. One such theory was that GRBs originate from neutron stars within the Milky Way.

The *BATSE* satellite, which was part of the *Compton Gamma-Ray Observatory* (see sec. 2.7.4), was designed to measure this distribution to "confirm" the Galactic-origin hypothesis. If that were the case, the GRBs would be confined to the Galactic plane. However, the GRB distribution that *BATSE* observed was clearly uniform over the entire sky, as is shown in figure 6.15.

While these observations make it unlikely that GRBs originated in neutron stars in the Milky Way, without GRB distance measurements it was impossible to determine that definitively. Unfortunately, *BATSE*'s spatial localization was inadequate to identify the source of the GRBs sufficiently well to allow follow-up observations on observed GRBs with optical and X-ray telescopes, which could have provided the distance measurement.

BeppoSAX and the Gamma-Ray Burst Afterglow

However, *BeppoSAX*, an Italian-Dutch X-ray satellite (see sec. 2.6.3), which was designed to look at lower energy electromagnetic radiation, X-rays, could achieve the needed spatial resolution; roughly 0.01°, to provide coordinates for optical observations. In 1997, it detected X-ray "afterglow," the residual radiation from a GRB, which dies away with a longer timescale than does the burst itself. For the first time astronomers could now be sure the GRB and the afterglow that they observed at the *BeppoSAX*-specified coordinates were coming from the same object. This afterglow is radiation emitted after the initial burst of gamma rays, possibly produced by the interaction of the radiation from the initial burst, possibly from a jet, with surrounding gas. Afterglows can usually be seen in the X-rays, optical, and radio wavelengths. Indeed, much of what is known about GRBs, certainly their redshifts, comes from afterglow observations.

The redshift measurements allowed a determination of the distance to at least some GRBs. The result is that all such GRBs are at cosmological distances. The most distant occurred at a redshift of about 4.5, which translates to a distance of a few billion light-years. An afterglow was observed by *BeppoSAX* of GRB 970228 (year = 1997, month = February, day = 28) 8 hours after the burst and again 3 days after the burst. Over that time span, the afterglow faded considerably but was still unmistakably detectable. GRB 970228 also heralded the observation of afterglow in optical spectra, which was first observed by the *William Herschel Telescope*, in the Canary Islands. This optical afterglow could also have had important consequences in identifying the source of the GRBs. However, the optical spectrum of this GRB was too faint for a spectroscopic measurement that would have provided the redshift.

However, on May 8, 1997, a GRB provided both an X-ray and an optical counterpart, and the *Keck Observatory* (see sec. 2.5.1) was able to measure a spectrum. This determined the redshift to have a z of 0.835, suggesting a distance of several Gpc, a cosmological distance. Of course, this traded one problem for another: how to understand what kind of an object could produce the energy observed for the GRBs.

Subsequent optical observations of GRB afterglows confirmed the cosmological distances for GRBs and provided some likely candidates for their host galaxies. Within about 4 weeks of GRB 970228, the *Hubble Space Telescope* (*HST*; see sec. 2.5.5) observed the sky surrounding the burst. The image in figure 6.16 shows the *HST* view of the rapidly fading optical afterglow of the burst, seen as the bright spot just below the center of the image. Notice the dimmer, extended object below and to the right of the bright afterglow. That appears to be the host galaxy of GRB 970228. Observations of host galaxies are important, as the characteristics of host galaxies might give clues about the nature of the burst.

What are GRBs? While this has been a subject of intense debate among astrophysicists, there now seems to be some convergence of thought as to their nature. Before describing these, let us review the facts associated with GRBs:

- GRBs are distributed uniformly over the sky.
- Their detection rate would be about one per day if we could observe all directions all the time from Earth.
- GRBs are of two classes, roughly divided into those lasting less than 2 s and those lasting more than 2 s.
- The longer bursts are at cosmological distances, and recent results (2005) suggest that the distances to the shorter ones are also cosmological.
- Longer burst GRBs emit more than 10^{53} ergs of energy, about what a core-collapse supernova emits, if that energy were emitted uniformly in all directions. Recent results (2005) suggest that the shorter burst GRBs emit less energy.
- However, the energy from the longer burst GRBs is apparently beamed; indeed, that may be essential to explain the energy output. GRBs do seem to be highly kinetic objects; possibly consistent with asymmetric explosions.

Astrophysicists now generally believe that the longer duration GRBs are explainable by "hypernova" or "collapsar" models (Woosley 1993; Paczynski 1998; Woosley 1998; Fryer, Woosley, and Heger 2001); the timescales, energy outputs, and afterglows, with beaming, are consistent with those observed. Note that without beaming, the GRB's output would represent an incredible energy source; it would be equivalent to the rest mass of several Suns.

What is a hypernova? A spectacular event occurred on March 29, 2003, which provided striking confirmation and understanding of this model. This was detected by the High-Energy Transient Explorer-II (HETE-II). This observation was particularly special to the study of GRBs because it was considerably closer than other GRBs observed and thus provided much more observational detail. It was explainable by the model of Woosley (1993) and of others as a very massive star, greater than 40 $M_{\odot}$ that could produce an anomalously energetic (compared to a typical core-collapse supernova) event when it explodes. This could be related to type Ic supernovae; such a supernova is a core-collapse supernova but results from a massive star that has shed at least its hydrogen, and possibly its helium, shells before it went into its final collapse phase. However, hypernovae must have some special features in addition to those of a typical type Ic supernova.

Fig. 6.16. *Hubble Space Telescope* pictures of GRB 990123 about 4 weeks after the GRB was observed by *BeppoSAX*. The bright yellow spot is apparently the GRB, while the extended emission below and to the right may be from the host galaxy. Reprinted courtesy of STScI, from the *HST* Web site: http://hubblesite.org/newscenter/newsdesk/archive/releases/1997/10/image/b.

The phenomena associated with the March 29, 2003, GRB, GRB 030329, have been explained in the context of the following model (Hjorth et al. 2005). Just prior to explosion the star probably had about 10 $M_{\odot}$ in the remaining star. When the core underwent its final collapse, a black hole formed near its center, which, assuming the star had been rapidly rotating, would have been surrounded by a disk of accreted matter. These conditions—the black hole and the rapid rotation—provided the conditions for creation of a jet, which was launched within a few seconds of the creation of the black hole. Because the matter was concentrated in the accretion disk, the jet, which would have been emitted normal to the plane of the accretion disk, was able to punch through the outer material of the star and blow off the typical material, most notably ^{56}Ni, that is produced in a supernova. Its radioactive decay powers the light curve observed for type Ic supernovae, which decays with a characteristic time associated with the radionuclides produced and which is seen in at least some GRBs. However, because the jet would have originated near the core of the star, considerably more ^{56}Ni might be emitted than would be the case for an ordinary core-collapse supernova. The jet produced might be expected to result from an asymmetric explosion. As the jet shot beyond the star, finally impacting the material that was previously shed from the outer shells of the massive star, it would be expected to produce a detectable afterglow. Observation of the afterglow in the optical part of the spectrum provided the redshift of this object; it was found to be 2.6 million light-years away (http://www.eso.org/outreach/press-rel/pr-2003/pr-30-03.html).

Another important clue to the nature of hypernovae comes from a recent study by Mazzali et al. (2005). They studied optical spectra of a number of known type Ic supernovae for some signs that they might have been correlated with a GRB. These studies used the 8.3 m Japanese *Subaru* telescope (Kawabata et al. 2004) and the *Keck-I* telescope (see sec. 2.5). One supernova remnant, SN 2003jd, was found to be anomalous in several ways. First, it was an unusually bright supernova remnant, putting it in a class with the supernova remnants associated with other GRBs. Second, its remnant seems to have contained considerably more ^{56}Ni than would a typical supernova. Third, its emission lines suggested that some of them were being affected by motion. While the lines from ^{56}Ni decay were observed to be singly peaked, those from oxygen were double peaked, suggesting that the oxygen lines might be Doppler shifted from their motion both toward and away from Earth from being contained in an accretion ring rotating about the central core. The oxygen would be expected to be produced farther from the core of the star than the Ni would be from standard supernova models and very general physics considerations, so this would be expected to be the case for a hypernova. Furthermore, the shapes of the emission lines were used to constrain the geometry of the hypothesized jet-disk system.

However, there are some missing GRB/hypernova signatures as well. This supernova was not detected as a GRB, nor was it observed in the X-ray spectrum. However, Mazzali et al. (2005) addressed this question by performing some simulations that would test how readily the object could be observed at those wavelengths as a function of several of the variables, most notably the direction of the jet, which is associated with the asymmetric explosion. They concluded that there was a large parameter space such that if the object exploded with those orientation parameters it would not produce sufficiently strong signals to be observed at those wavelengths. Furthermore, the parameter space was consistent with the geometry inferred from the emission lines. As they conclude, "absence of evidence is not necessarily evidence of absence." Their interpretation of their observation is that SN 2003jd might well have been a hypernova, that is, a GRB, but its jets were not pointed at Earth. This observation would certainly be consistent with the suggestion that the prodigious energy output of GRBs must result from beaming. Although this does seem to be a plausible explanation, the absence of all the indicators for a GRB leaves the conclusion uncertain.

What is the nature of the shorter duration GRBs? In late 2005, several papers were published dealing with results from two shorter burst GRBs (Fox et al. 2005; Gehrels et al. 2005; Villasenor et al. 2005), GRB 050509 and GRB 050709, with redshifts, respectively, of $z = 0.225$ and $z = 0.160$. The key to these results was an ability to perform observations at longer wavelengths than those of the GRB; they utilized the three cameras of *Swift* (see sec. 2.7.6) and HETE + follow-up by X-ray and optical observatories to determine the redshifts and obtain additional data. *Swift* has also observed several other shorter burst GRBs. In both of the above cases, the optical light curve, which was weak to nonexistent, ruled out the possibility that the GRB was a hypernova, thereby establishing that the shorter burst GRBs are of a completely different nature from the longer burst ones. These observations suggest that these GRBs would be consistent with mergers of a neutron star and a black hole or two neutron stars (see, for example, Belczynski et al. 2006; Faber et al. 2006).

One fascinating aspect of these shorter burst GRBs is that, if they are, as hypothesized, mergers of a neutron star with a black hole or of two neutron stars, they would be expected to generate very strong gravitational wave signals. Unless they were quite close, these would still not be strong enough to be detected by Laser Interferometer Gravitational-Wave Observatory (LIGO) (see sec. 2.11), but the next stage of gravitational wave detectors, Advanced LIGO, should be capable of detecting many such events. Such an observation would provide striking confirmation of this basic picture of the mechanism by which the shorter burst GRBs are produced.

CHAPTER 6 PROBLEMS

1. You can see what happens in the collapse of a massive star by going to the Web site http://www.jinaweb.org/html/movies.html. This shows three supernova movies: 1) a radiation-diffusion/hydrodynamic simulation of the early development in two dimensions of the explosion of the core of a massive star 2) the collapse of a rapidly rotating massive star core, and 3) a representative two-dimensional

multi-group, flux-limited diffusion rad/hydro simulation of the collapse of a rotating Chandrasekhar core."

2. You can also see the development of the r-process abundances from the same Web site, http://www.jinaweb.org/html/movies.html. This shows how the r-process synthesizes the heavy elements.
3. Derive equation 6.2.8, the rate at which the triple-α reaction proceeds.
4. Determine numerical expressions for the (temperature-dependent) values for E_o and ΔE_o for ^{8}Be + ^{4}He. What would be the values for these two parameters at a typical temperature for helium burning of $T_9 = 0.2$?
5. Run your own alpha-rich freeze-out calculation. Go to the Web site http://nucleo.ces.clemson.edu/home/online_tools/three_phase_ism/. Follow the instructions, and try some different initial conditions to see how high in mass the products of alpha-rich freeze-out can be made to go. Test the hypothesis that the entropy needs to be above a certain value so that the production of ^{12}C, and therefore all heavier nuclides, is limited, thereby allowing production of nuclei up to mass 90 or so.
6. Assume that a supernova exploded a distance R from Earth, and that it emitted e-, μ-, and τ-neutrinos with masses (energies) $m_e(E_e)$, $m_\mu(E_\mu)$, and $m_\tau(E_\tau)$
   a. Assuming the masses of the mass eigenstates are much less than their energies, derive an approximate formula for the difference in arrival time of the neutrinos of mass m_1 and m_2 as a function of their masses and energies and of R.
   b. If R is 10 kpc (3×10^{19} cm), the energies of both neutrino species are 10 MeV, and $m_1 = 0$ and $m_2 = 1$ eV, what will be the difference in arrival times between the neutrinos of mass m_1 and of m_2?
7. The ground state of ^{8}Be is unbound by 92 keV. However, in high-density stellar environments, the electron density can be sufficiently high that they will neutralize part of the Coulomb energy in ^{8}Be and so make the ground state bound. So,
   a. Assume a plausible size for ^{8}Be and estimate what electron density, hence what density of ^{4}He, would be required to make the ^{8}Be ground state bound, that is, to reduce the 92 keV to zero.
   b. Explain in words what effects you would have to consider to estimate the production rate of ^{12}C as a function of the density up to the density you determined in part a.
   c. Estimate the ratio of the ^{12}C production rate at half the density you determined from part 1 compared with the lower density production rate of normal helium burning, taking into account the factors you discussed in part b.

7

PRODUCTION OF THE ABUNDANT HEAVY NUCLIDES

7.1 Introduction

The abundant heavy nuclides are, for the most part, produced by processes involving neutron capture on preformed seed nuclei. This is necessitated primarily by the fact that the temperatures required to produce sufficient energy for charged particles to overcome the Coulomb barriers of the heavy nuclides would be so high that they would, instead, destroy those seed nuclei through photonuclear reactions. Two types of neutron-capture processes are generally thought to occur: the s-process, or slow process, and the r-process, or rapid process. Here the "slow" and "rapid" refer to the times between successive neutron captures compared with the typical β-decay times of the nuclei through which these processes proceed. Each produces about half the nuclides more massive than iron, but the distributions of the nuclides they produce are qualitatively different, as can be seen in figure 7.1. The s-process produces sharp abundance peaks at 88, 138, and 208 u, whereas the r-process produces much broader peaks at 130 and 195 u, and possibly at 110 u as well. Clearly, any successful description of these two processes must describe both the locations of the peaks and their widths.

In the description of the s-process, the β-decays nearly always have time to occur before another neutron capture occurs, forcing it to follow a path that runs through either stable nuclei or nuclei that are one neutron removed from stability. The r-process pathway, by contrast, results from many successive neutron captures, driving the nuclei along its path far beyond, from 10 to 30 neutrons, the neutron-rich side of stability. The trajectories of these two processes through at least part of the chart of the nuclides are shown in figures 7.2 and 7.7. The r-process is seen to pause at the neutron closed shells (see sec. 3.2); this will ultimately result in the elevated abundances in the decay

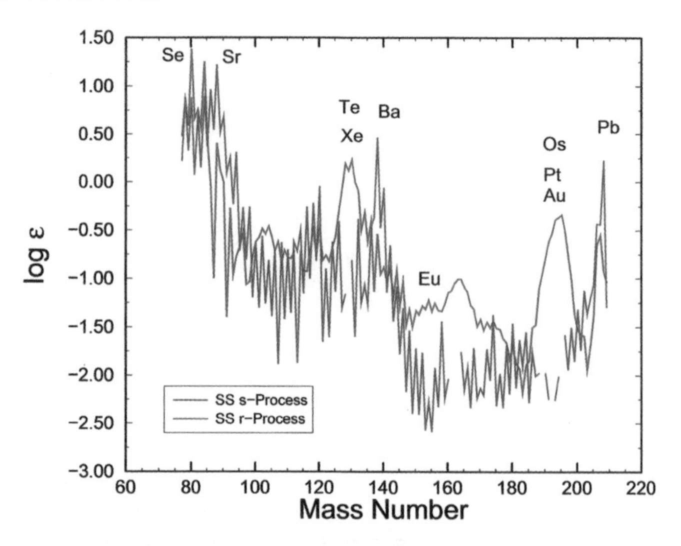

Fig. 7.1. Abundances of the nuclides produced by the s- and r-processes. Plotted values are $12 + \log_{10}$ of the abundance relative to hydrogen. The sharp abundance peaks at 88 (Sr), 138 (Ba), and 208 (Pb) u are due to the s-process, and the broader peaks at 130 (Te, Xe) and 195 (Os, Pt, Au) u are due to the r-process. From Cowan and Thielemann (2004).

products of those nuclides. The s-process also pauses at the neutron shells, although the effect is much less dramatic.

Because the nuclides through which the s-process passes are stable or nearly so, their properties and the reaction cross sections on them have been studied in the laboratory. These determine the path of the s-process, so its path, and hence its abundances, are generally calculable with good accuracy. The same is not true of the r-process; except for those nuclides that are produced only by the r-process, its abundances are determined by subtracting the s-process abundances from the total abundances. These procedures produce the abundances indicated in figure 7.1. Note, however, that these are solar system abundances, to be distinguished from the cosmic abundances presented in chapter 1. Of course, the two abundance data sets are similar.

There are also some other processes, especially those that appear to be required to produce the very neutron-poor nuclides, the p-process nuclei. These do not involve neutron radiative capture directly, although they do, in many cases, require the products of the s- or r-processes as seed nuclei. In some cases, these processes involve neutrino-induced reactions. Some of these are dealt with later in this chapter, and some in the next chapter.

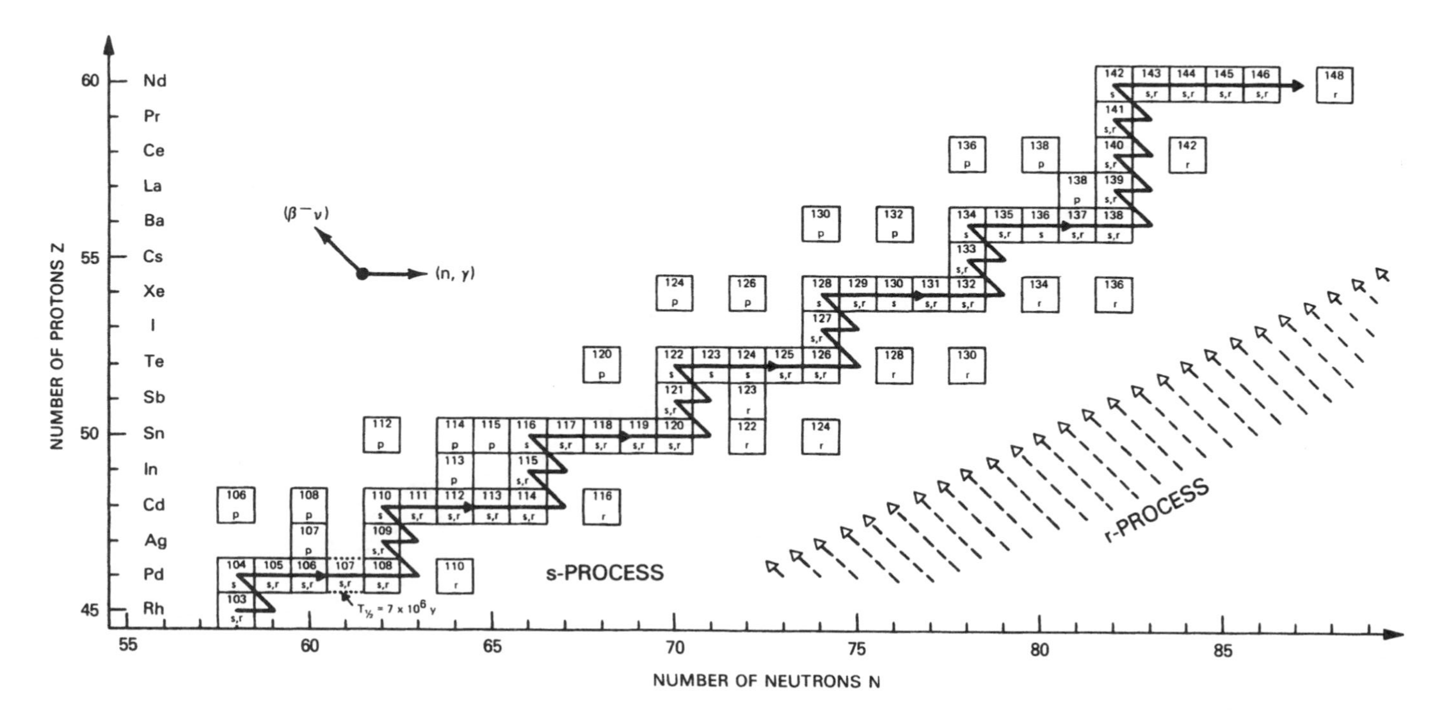

Fig. 7.2. A portion of the chart of the nuclides, showing the path of the s-process. Some of the nuclei, for example, ^{128,130}Xe and ^{134,136}Ba, are shielded from r-process production by the presence of stable isobars, so are s-process-only nuclides. Note also that some nuclides, for example, ^{124}Xe, ^{126}Xe, and ^{138}La, cannot be formed by either the s- or the r-processes; they are p-process nuclides (see chap. 8). From Rolfs and Rodney (1988). Courtesy of C. Rolfs.

7.2 The s-Process

The path of the s-process is indicated in figure 7.2, which illustrates the point, made above, that it tracks the nuclei that are either stable or one neutron beyond the neutron-rich edge of stability. The s-process is thought to occur during helium burning. It therefore occurs off the main sequence of the Hertzsprung-Russell (HR) diagram (see chap. 1), along the asymptotic giant branch (AGB). Thus the stars that undergo this type of nucleosynthesis are referred to as AGB stars. The details of this process, especially the hydrodynamics, are both important and complex, and they have implications for the stellar structures involved and the nucleosynthesis that can occur. The source of the neutrons for this process has long been thought to come from either the $^{13}C(\alpha, n)^{16}O$ reaction or the $^{22}Ne(\alpha, n)^{25}Mg$ reaction, both of which have positive Q-values and either of which could in principle provide the necessary neutron density, albeit at quite different temperatures. However, the sources of those nuclei are necessarily quite different. All the CNO nuclei produced by any preceding hydrogen burning, including the ^{12}C produced in a helium-burning region and dredged up into a hydrogen-burning region, would have been converted to ^{14}N. Helium burning will then produce a large abundance of ^{22}Ne through successive α-particle captures (and one β-decay) on those ^{14}N nuclei. Thus, ^{13}C would have to be produced in a hydrogen-burning region outside of the helium-burning core of a red giant and then mixed down into the helium-burning region, to be resupplied through the period of time during which it must provide the neutrons for the s-process. Comparison of higher mass and lower mass abundances produced in the s-process suggests (see discussion by Smith, in Wallerstein 1997) that the dominant source is ^{13}C, but it has been found to be difficult in stellar evolution calculations to mix the ^{13}C produced in the hydrogen-burning shell outside the helium-burning region sufficiently to produce the required ^{13}C abundance (see, e.g., Busso, Gallino, and Wasserburg 1999; Lugaro et al. 2003; Wasserburg et al. 2005). The (α, n) reaction on ^{13}C occurs at a lower temperature than that on ^{22}Ne because of its lower Coulomb barrier. It may well be that both sources contribute, although probably in different s-process temperature regimes. In addition, another process, the "weak s-process," occurs in massive stars. All of these will be discussed below.

Hydrodynamical Aspects of the s-Process

The stellar-burning scenarios that describe the s-process are both complex and beautiful, and demonstrate a synergy between observation and theory. Once a significant fraction of a star's hydrogen has been consumed it will leave the main sequence and evolve into one of several modes, depending on its mass and metallicity. There can be four major dredge-up scenarios, events in which material is dredged up from deeper zones within the star to the surface, which a single star may experience. These will be described below, drawing from descriptions of these processes in the section by Iben in Wallerstein (1997) and in Straniero, Gallino, and Cristallo (2005).

The first dredge-up phase occurs after the hydrogen is exhausted over a substantial fraction of the star's interior, about 10 % of the mass for less massive stars and somewhat more for more massive ones, as the star reaches its red giant phase for the first time. As the core contracts and heats up, the periphery of the star expands, increasing the opacity of the region and steepening the temperature profile until turbulent convection is forced to carry the outward flux of energy emerging from the hydrogen-burning shell that surrounds the core. The base of the convective envelope moves inward (in mass; it is often more convenient to plot "radius" in terms of the mass included within a given radius) eventually getting to a region where preexisting ^{12}C was converted to ^{13}C during main sequence burning, then into a region where the ^{12}C was converted to ^{14}N (see chap. 5). Since the entire outer portion of the star is convective, these abundance changes are reflected in the surfaces of stars that are undergoing these processes; they can be observed!

In stars less massive than 2.3 $M_\odot$ the helium-rich core becomes electron degenerate before the helium is ignited. As the core grows in mass, gravitational potential energy is converted into thermal energy, and, when the mass of the core reaches $\sim$0.5 $M_\odot$, helium is ignited and burns its way in a series of flashes into the center of the star, lifting the degeneracy. The star continues to burn its helium into carbon and oxygen under nondegenerate conditions. For more massive stars the helium ignites before the electron degeneracy is established.

The second dredge-up occurs for all intermediate mass stars with mass $M > 3.5 M_\odot$ before they ascend the AGB (of the HR diagram; see secs. 1.6 and 9.5) phase. All intermediate-mass AGB stars as well as low-mass AGB stars will then undergo a thermal pulsing–AGB phase, in which the He shell at the top of the C-O core and the H in the outer convective shell alternately burn, and the He-shell convectively penetrates the envelope.

The third dredge-up (TDU) phase is probably the most interesting one from the viewpoint of nuclear astrophysics, as it produces much of the ^{12}C, and most of the s-process elements, in the universe, with the exception of those predominantly produced in weak s-processing in massive stars. Approximately

97% of the stars that can leave the main sequence in less than the age of the universe become AGB stars and experience thermal pulses due to helium shell flashes (Schwarzschild and Harm 1965; Weigert 1966), which periodically interrupt their normal hydrogen burning. However, the TDU phase will occur only in stars with envelopes more massive than about 0.5 $M_{\odot}$. Thus the lowest mass stars, for example, the Sun when it becomes an AGB star, will undergo thermal pulses but will not undergo TDU, so will not expel anything into the interstellar medium. For the stars that do undergo TDU, the newly synthesized nuclides are returned to the interstellar medium following the TDU phase by a stellar wind phase, in which the AGB star will eventually lose all its envelope and be converted to a planetary nebula (Iben, in Wallerstein 1997). This results from several complicated factors that are peculiar to this particular phase of a star's life (see Iben, in Wallerstein 1997).

These stars produce thermonuclear shell flashes from the following mechanism. During quiescent hydrogen burning, the resulting helium is deposited in a growing helium shell above the CO (or possibly ONe) core but at a rate that is insufficient to maintain quiescent helium burning in that shell. As the helium builds up it is compressed and heated, and eventually the helium ignites. Helium burning is proportional to the 40th power of the temperature, so nuclear energy is released very rapidly as it begins, and a thermonuclear runaway occurs. It is eventually damped out by expansion and cooling. The stars in this phase of their evolution can undergo many such shell flashes; we return to this point below.

However, this generates a convective zone nearly throughout the helium shell, but not quite to the hydrogen shell (Iben 1976, 1977). The ^{14}N synthesized earlier is converted almost entirely to ^{22}Ne (see chap. 6). In principle, this could produce a high neutron flux from ^{22}Ne$(\alpha, n)^{25}$Mg with which to drive the s-process. However, this source requires higher temperatures, around 350×10^6 K. Thus it appears that the source of the neutrons is more likely to be from ^{13}C via the ^{13}C$(\alpha, n)^{16}$O reaction. The means by which this reaction is activated is not fully understood but does require diffusion of a few hydrogen atoms during a TDU episode, when the hydrogen-rich envelope and the helium-rich intershell come into contact and the hydrogen-shell is temporarily inactive, forming a "proton pocket." At reignition of hydrogen burning, these few protons will be captured by the abundant ^{12}C nuclei present in the helium-burning intershell, thus creating (following β-decay) a ^{13}C pocket. As noted by Gallino et al. (1998), a sharp H-He instability exists between the bottom of the convective envelope and the top of the intershell region. It would be unlikely if this instability did not result in some diffusion of the H into the ^{12}C-rich zone to produce the ^{13}N and, subsequently, ^{13}C needed to run the s-process.

However, it is a challenge to theorists to describe this situation; this intershell is so thin that the pressure changes little across it. Thus it does not really respond as an ideal gas, which ultimately produces the instability. As Gallino et al. (1998) note, "a hydrodynamical treatment is required, and this constitutes a major challenge in s-process nucleosynthesis studies." Nonetheless, Gallino et al. conclude that the $^{13}C(\alpha, n)^{16}O$ reaction must be the source of the s-process neutrons; the $^{22}Ne(\alpha, n)^{25}Mg$ reaction simply cannot produce the necessary reaction rate at the temperature at which the H-rich pockets exist.

This situation was studied by Lattanzio and Lugaro (2005); they observe that the "largest unknown in [this scenario] is the mechanism that causes the protons to be mixed into the carbon-enriched region." They discuss four mechanisms by which this might occur: semiconvection, convective overshoot, gravity waves, and rotation. Although some of these mechanisms seem promising, their details remain to be worked out (Lugaro et al. 2003; Lattanzio and Lugaro 2005). Lugaro et al. (2003) did work out some of these details, with a parameterization of some of the physics that is not well understood at present. This represents the state of the art in such calculations. However, their study is strongly indicative of the extent to which the results from the s-process models depend in the extreme on the details of the calculation. Lugaro et al. studied the results from three of the standard computer codes that are used in s-process calculations. Some studies performed in the past had found a large amount of TDU at solar metallicity only for intermediate-mass stars (Iben 1975), while others find TDU to occur in lower mass stars ($M < 5\,M_{\odot}$), but only for low metallicities: as low as 1/200 of solar (Iben and Renzini 1982). In yet other calculations (Lattanzio 1989; Straniero et al. 1997), TDU was found to occur in stars with initial masses as low as 1.5 $M_{\odot}$ and solar metallicity.

Mass loss during the AGB phase is also a problem for which the codes used to describe it show little consistency. This is responsible for seeding the interstellar medium with the products of s-process nucleosynthesis. The main mechanism for this, thought to be radiation pressure on the grains formed in the peripheries of such stars (Lugaro et al. 2003), is usually treated in the codes in a somewhat ad hoc way, with a free parameter that depends on the luminosity, radius, and mass of the star being used to characterize the mass loss. In some of the codes, however, the calculations terminate after some number of thermal pulses because of convergence problems resulting from the fact that the envelope has disappeared, while in others the envelope remains. Bernatowicz et al. (2005) considered the properties of presolar graphite grains as deduced in laboratory studies to gain insight into the processes of the mass loss models used to describe their production.

One very important feature of the proton, or ^{13}C, pocket is that ^{13}C(p,γ) reactions will convert the ^{13}C to ^{14}N. This is significant because ^{14}N is a "neutron poison," that is, it can consume the neutrons that would fuel the s-process via the ^{14}N(n,p)^{14}C reaction, which has a positive Q-value and a large cross section. Thus the ^{12}C that is mixed into the proton pocket to make the ^{13}C must come from the helium-burning region, not an external hydrogen-burning region; in that case the CNO cycle would have converted most of its CNO elements to ^{14}N, as discussed above and in chapter 5. Furthermore, this appears to depend critically on the amount of hydrogen that gets mixed into the helium-burning region; for $\mathrm{H}/^{12}\mathrm{C} < 1/10$, the number of ^{13}C nuclei appears to be determined by the number of initial protons. For greater proton abundance, a ^{14}N pocket also emerges, until for $\mathrm{H}/^{12}\mathrm{C} > 1$, a ^{14}N pocket emerges next to the ^{13}C pocket (Lugaro et al. 2003). In this situation, most of the ^{12}C is converted directly to ^{14}N. Thus the details of the mixing that produces the pocket appear to be somewhat critical.

Finally, a number of observed stars can only be understood in terms of a final helium shell flash which a post-AGB star may experience after it has ceased to burn hydrogen; this is the fourth dredge-up phase. When the flash occurs the mass of the helium layer below the hydrogen-helium interface is slightly smaller than that necessary for initiating a flash on the AGB "because the matter in the helium layer is partially degenerate and is heated by adiabatic contraction when the hydrogen-burning shell loses its power" (Iben, in Wallerstein 1997). In the final flash, the outer edge of the convective shell forced by helium burning extends into the hydrogen-rich region of the star. The products of the resulting nucleosynthesis appear on the surface of the stars that undergo this fourth dredge-up phase. A fairly small fraction of AGB stars actually experience this final phase.

The level of detail that is described in the above discussion is intended to illustrate to students of nuclear astrophysics the extent to which astrophysicists have studied the internal processes of stars that are required to perform the nucleosynthesis that is of interest to nuclear astrophysics. The nuclear physics inputs to these calculations are the results of laboratory studies that have been performed over the past several decades, but the effort that has gone into developing the hydrodynamics and mixing theory that must also be included is extraordinary.

7.3 Nuclear Aspects of the s-Process

The nuclear physics aspects of the s-process should be extremely well described by the vast numbers of neutron-capture reaction studies (see article

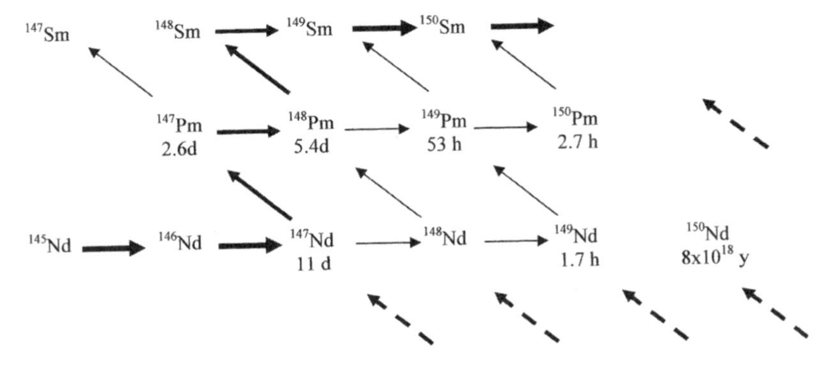

Fig. 7.3. The s- and r-process flow through the mass 145 u to 150 u nuclides. Unstable nuclides have their half-lives indicated, and the relative strength of the s-process flow is indicated by the thickness of the arrows. Dashed arrows indicate r-process production.

by Kappeler in Wallerstein 1997) that have been performed on the nuclei that are on the s-process path. Despite some significant discrepancies in the results of different groups performing these measurements, the qualitative, and semiquantitative, properties of the s-process are well established, and the abundances of the s-process nuclides are also well determined. Indeed, one can study the s-process in semi-isolation by looking at nuclides that are shielded from r-process production by stable nuclei. A case in point is indicated in figure 7.3, which shows the path of the s-process through the tellurium, xenon, and barium isotopes. ^{128,130}Xe and ^{134,136}Ba are seen to be shielded by stable, or very long-lived, isobars; these allow the separation of s-process production from that of the r-process.

Since the s-process proceeds for the most part through nuclei that β-decay rapidly compared with the time between successive neutron captures, abundances of the s-process nuclei can be compared through a simple expression. We begin by writing the equation for the rate of change of the abundance of a sequence of isotopes in an s-process environment:

7.3.1 $$dN(A, Z)/dt = n_n[N(A-1, Z)\langle\sigma v\rangle_{A-1} - N(A, Z)\langle\sigma v\rangle_A],$$

where n_n is the density of neutrons. Since β-decays are assumed to have time to occur before successive reactions in the s-process, the nuclei involved are identified by their mass number A. If the processing times that characterize the s-process are sufficiently long, the flow into and out of $N(A, Z)$ will just balance each other, so the left-hand side of this equation can be taken to be zero. This allows a simple test of our understanding of the s-process, assuming that there is no branching from β-decay:

$$N(A-1, Z)/N(A, Z) = \langle\sigma v\rangle_A/\langle\sigma v\rangle_{A-1}. \tag{7.3.2}$$

This expression can be further simplified, as neutron-capture cross sections vary approximately inversely with velocity (see sec. 3.13), so $\langle\sigma v\rangle$ is constant with energy. This allows the Maxwellian averaged cross sections, $\sigma_A = \langle\sigma v\rangle_A/v_T$, to be measured at just a single energy, characterized by velocity v_T. Then, when the velocity factors are cancelled out of the above equation, it becomes

$$\sigma_{A-1}(n, \gamma)\, N(A-1, Z) = \sigma_A(n, \gamma)\, N(A, Z), \tag{7.3.3}$$

an expression that is expected to hold throughout sequences of nuclides that are produced only by s-process nucleosynthesis. This test has been applied to a number of isotopic sequences and found to apply reasonably well (see article by Kaeppeler in Wallerstein 1997) in mass regions between neutron closed shells. However, these simple relationships break down at the closed shells, and they are often subverted by subtleties of the nucleosynthesis.

One example of both a test of this concept and of the subtleties involved is given by the Sm isotopes. Both ^{148}Sm and ^{150}Sm are shielded by the stable nuclides ^{148}Nd and ^{150}Nd from r-process production, so are s-process-only isotopes. However, several complications present themselves in testing the abundance times mean cross section comparison. The situation for the nuclei in this mass region is shown in figure 7.3. There it can be seen that a branch point could occur at ^{148}Pm that can subvert the simplest form of the test. Another complication was explained by Lesko et al. (1989), who showed that an isomeric state of ^{148}Pm would have a strong effect on the branching because its half-life of 41.3 days was much larger than that of the ground state (5.27 days) and that the two states would be in thermal equilibrium for temperatures of $T_9 > 0.15$. They were also able to use this situation to determine that the neutron density in the s-process environment in which these nuclides were synthesized was roughly 3×10^8 cm^{-3}. The situation was summarized by Toukan et al. (1995). It should also be noted that meteoritic samples (see sec. 7.4) might be expected to provide a simplified test of the s-process, as such samples might be expected to isolate the s-process from the r-process abundances.

There are also other s-process branches that can be used to determine the neutron density in a typical s-process environment. As can be seen in figure 7.4, such a branch occurs at ^{85}Kr. In a lower neutron density environment, the β-decay of ^{85}Kr would be favored over the ^{85}Kr(n, γ) reaction, whereas the

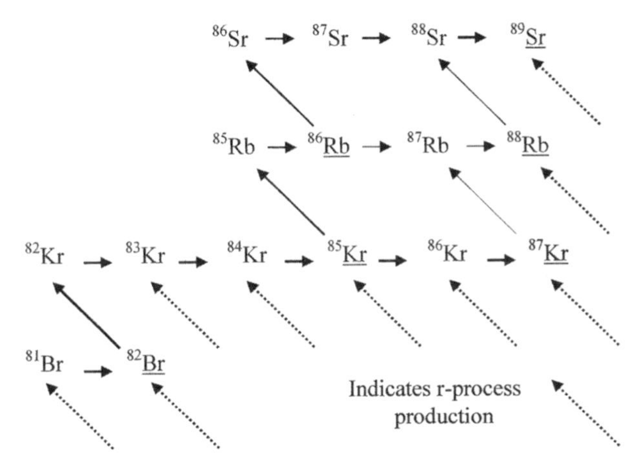

Fig. 7.4. The path of the s-process through the region around Rb. The branch at ^{85}Kr ($T_{1/2}$ = 10.76 years) can lead to synthesis of both ^{85}Rb (72.17 %) and ^{87}Rb (27.83 %). Their relative abundances can thus serve as a diagnostic of the conditions of the s-process. Radioactive nuclei are underlined. Relative strengths of s-process flow are indicated by the widths of the lines.

reverse is true in a higher neutron density environment. Thus the abundances of the nuclei in this mass region can be used to infer the neutron density during s-processing. However, this analysis is complicated by the presence of the isomer, ^{85m}Kr, which decays directly to ^{85}Rb and is fed with about 50 % probability in the neutron radiative capture on ^{84}Kr (see Kaeppeler et al. 1990). The isomer has a much different decay half-life than does the ground state: 4.48 hours compared with 3917 days. According to Kaeppeler et al. (1990), this test not only determines the neutron density during the s-process, but also strongly suggests that the s-process is pulsed, as would be the case if it proceeds as discussed in section 7.2. Although the neutron density determined from these tests depends somewhat on the specific test, this test and others (see the article by Smith in Wallerstein 1997) suggest that the s-process neutron density is in the 10^7 to 10^{10} cm^{-3} range.

Another s-process site, referred to as the "weak s-process," is thought (Raiteri et al. 1991) to occur in massive stars near the end of He burning. In this scenario, preexisting ^{14}N is rapidly converted to ^{22}Ne by a series of α-particle captures and β-decays, following which the ^{22}Ne(α, n)^{25}Mg reaction produces a source of neutrons. Preexisting seeds of iron and more massive stars can then capture the neutrons. The products of this form of nucleosynthesis can span the entire s-process mass range but are thought to be concentrated in the lighter nuclides. A detailed analysis of this situation was performed by Prantzos, Hashimoto, and Nomoto (1990). In this study a galactic chemical evolution model, in which a wide range of massive stars was invoked, appropriately weighted with a realistic initial mass function, was used to observe the products of this type of nucleosynthesis; this was found to be, as noted above, the lighter s-process nuclides, $65 < A < 90$. The efficiency of this s-process was found to be limited by the neutron poisons that would be peculiar to the nuclides that would exist in the regions in which this process took place, and this would depend strongly on the metallicity of the star. Specifically, even extremely low neutron-capture cross sections, for example, that for $^{12}C(n, \gamma)$, could be important in metal-poor stars. At higher metallicities, ^{25}Mg becomes the dominant neutron poison.

Analytic Solution to the s-Process Abundances

An elegant analytic solution to the abundances produced by the conventional s-process was worked out by Clayton (1983). Although modern network calculations long ago superceded this analysis, there is some value in seeing the forms the solutions take from this analysis because of the general features they exhibit. This approach utilized Laplace transforms, a mathematical technique that converts differential equations into algebraic equations that often can then be solved more easily than the original set of equations.

The Laplace transform of a function $f(t)$ is given by

7.3.4
$$F(s) = \int_0^\infty e^{-st} f(t)dt,$$

and the derivative of $f(t)$ by

7.3.5
$$\int_0^\infty e^{-st} df(t)/dt\, dt = sF(s) - F(0).$$

As noted above the s-process abundances form a set of coupled equations, where the $N(A)$ are the abundances:

7.3.6
$$dN_A/dt = n_n(t)[N_{A-1}\langle \sigma v\rangle_{A-1} - N_A\langle \sigma v\rangle_A],$$

where the possible time dependence of the neutron flux is now indicated explicitly. Assuming the neutron cross sections are all determined at a single energy, or velocity v_T (corresponding to the thermal energy $k_B T$, at which the cross sections are measured), this equation can be recast as

7.3.7
$$dN_A/dt = v_T n_n(t)[-\sigma_A(k_B T) N_A(t) + \sigma_{A-1}(k_B T) N_A(t)].$$

Defining the "neutron exposure" τ by

7.3.8
$$d\tau = v_T\, n_n(t)dt, \text{ and } \tau = v_T \int n_n(t)dt,$$

the differential equation becomes

7.3.9
$$dN_A(\tau)/d\tau = -\sigma_A N_A(\tau) + \sigma_{A-1} N_{A-1}(\tau).$$

Because ^{56}Fe is so abundant, and is the termination point of stellar nucleosynthesis (see sec. 6.4), it is reasonable to assume that the s-process begins with ^{56}Fe, that is, that the ^{56}Fe seeds the entire s-process. At the other end, the heaviest stable nucleus is ^{209}Bi, so it is also reasonable to assume that the s-process terminates there. It should be noted, however, that s-process nuclides from previous generations of s-processing will also serve as seeds, so this analysis must be viewed as an approximation, albeit a good one, given the large abundance of ^{56}Fe. With that in mind, it will be assumed that $N_A(0) = 0$ for all masses other than for $A = 56$ at the beginning of the s-process. Finally, mass 206 is a special case, because neutrons captured on ^{209}Bi produce ^{210}Bi, which then undergoes β- and then α-emission to give ^{206}Pb. Thus the equations for the entire s-process are given as:

7.3.10
$$dN_{56}/d\tau = -\sigma_{56} N_{56},$$

7.3.11
$$dN_A/d\tau = -\sigma_A N_A + \sigma_{A-1} N_{A-1}, 57 \leq A \leq 209, \text{ but } A \neq 206$$

7.3.12
$$dN_{206}/d\tau = -\sigma_{206} N_{206} + \sigma_{205} N_{205} + \sigma_{209} N_{209}.$$

Although the general solution to these equations is a bit tricky, solving those from $A = 56$ to $A = 205$ is straightforward. If we perform the Laplace transforms on each of the equations we obtain [where $F_A(s)$ is the transform of $N(A)$]

7.3.13
$$s F_{56}(s) = -\sigma_{56} F_{56}(s) + N_{56}(0),$$

7.3.14
$$s F_{57}(s) = -\sigma_{57} F_{57}(s) + \sigma_{56} F_{56}(s),$$

7.3.15
$$s F_A(s) = -\sigma_A F_A(s) + \sigma_{A-1} F_{A-1}(s).$$

The solution to this set of equations is

7.3.16
$$F_A(s) = N_{56}(0)[\sigma_{A-1}\sigma_{A-2} \ldots \sigma_{57}\sigma_{56}]/[(s + \sigma_A)(s + \sigma_{A-1}) \ldots (s + \sigma_{57})(s + \sigma_{56})].$$

Rearranging this a bit gives

7.3.17
$$\underline{F}_A(s) = \sigma_A F_A(s)/N_{56}(0) = [(1 + s/\sigma_A)(1 + s/\sigma_{A-1}) \ldots (1 + s/\sigma_{56})]^{-1}.$$

The solution to the original differential equations can be obtained by integrating the above equation in the complex plane:

7.3.18
$$f_A(\tau) = (2\pi i)^{-1} \int_{-i\infty}^{i\infty} e^{s\tau} \underline{F}_A(s)\, ds,$$

7.3.19
$$= \sum_{i=56}^{A} C_{Ai} \exp(-\sigma_i \tau),$$

7.3.20
$$= \sigma_A N_A(\tau)/ N_{56}(0),$$

where

7.3.21
$$C_{Ai} = [\sigma_{56}\, \sigma_{57} \ldots \sigma_{A-1}\sigma_A]/[(\sigma_A - \sigma_i)(\sigma_{A-1} - \sigma_i) \ldots (\sigma_{57} - \sigma_i)(\sigma_{56} - \sigma_i)],$$

and where the factor $1/[(\sigma_i - \sigma_i)]$ is omitted from C_{Ai}. Thus, finally,

7.3.22
$$N_A(\tau) = [N_{56}(0)/\sigma_A] \sum_{i=56}^{A} C_{Ai} \exp(-\sigma_i \tau).$$

The problem with this approach is clear; many of the cross sections are so very nearly equal that the differences between cross sections in the denominator will make the solutions unstable to small measurement uncertainties. Thus this approach has its practical difficulties. However, f_A, which is proportional to $N_A(\tau)$, can be seen to be related to all the cross sections that drive the s-process through the nuclides lighter than A. This makes sense, as the nuclides are created one neutron (and with some β-decays) at a time, so all the lighter nuclides must be created in order to synthesize the heavier ones. This was an automatic result of the assumption that the only nonzero initial abundance was that for ^{56}Fe. This in itself makes the model unrealistic; it is now well established that the s-process does depend on previous episodes of nucleosynthesis, that is, the initial abundances will generally all be nonzero, probably even for the first episode of s-processing. This is because the r-process (see sec. 7.5) will have already been busy synthesizing heavy nuclides well in advance of the first episode of s-processing.

An interesting situation to study is that in which all of the neutron capture cross sections are assumed to be equal and a solution is sought for equations 7.3.10–7.3.12. This solution is

7.3.23
$$f_A(\tau) = [\sigma/(A-1)!](\sigma\tau)^{A-1} e^{-\sigma\tau} = N_A.$$

This result is a Poisson-like distribution in A with a maximum at $A_{max} = \sigma\tau + 1$. Thus, the peak of the distribution shifts with τ, the neutron exposure;

either increased time or neutron density will push the peak of the distribution to higher mass. Despite the simplicity of the equal cross section approximation, additional features can be inferred. Because of the stochastic nature of the neutron capture process with time, some nuclei manage to capture more neutrons than average while others capture fewer. This gives rise to a spread in the peak, as is predicted by the simple model. In addition, one can also note that exceptions to this basic simplicity will occur at the neutron closed shells, at which the distributions will accumulate, causing s-process peaks. All these features are certainly consistent with what would be expected from the most basic considerations of the s-process.

7.4 Abundance Information from Meteorites

A striking way of inferring direct information about the processes of nucleosynthesis in stars is that of isotopic analysis of presolar meteoritic grains. This subject was introduced briefly in chapter 1 and will be referred to in other places when the results it has produced provide important information regarding some process of nucleosynthesis. To date, the detailed isotopic information from presolar meteoritic grains has perhaps yielded most insight into the s-process, so it is appropriate to present a discussion of its general features and capabilities at this time. However, meteorites have yielded information about many topics in astrophysics, for example, conditions of the solar nebula and chronology of parent body formation.

Basically this area of study impacts nuclear astrophysics because the carbonaceous meteorites of a specific type contain micron- or submicron-sized inclusions, or grains, that appear to have been formed from localized regions of the ejecta of stars. These truly are the grains of stardust, and the abundances of the isotopes found in those inclusions must therefore reflect the processes of nucleosynthesis that dominated the material in that local portion of the ejecta, providing a direct test of our understanding of those localized nucleosynthesis processes.

As material is ejected from a star into the interstellar medium (ISM), either by stellar winds or by stellar explosions, it contributes to the composition of the ISM. The ISM is generally assumed to be well mixed, so it should be, and does appear to be, fairly homogeneous. However, grains can apparently be formed soon after material is ejected from stars. A small fraction may survive their long trek through the ISM and thus can escape the homogenization of the ISM. In this way, the grains maintain the composition of the localized region of space in which they formed. These grains can be trapped in meteorites that

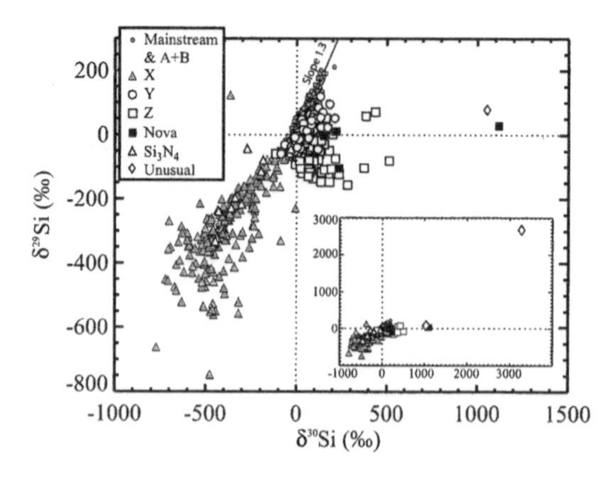

Fig. 7.5. Silicon isotopic ratios of presolar SiC and Si_3N_4 grains in meteorites. Data are from Hoppe and Ott (1997) and references therein, and unpublished results from Nittler (2003). The grains have been divided into subgroups on the basis of their Si, C, and N isotopic ratios; 90% of the grains belong to the mainstream group. Dashed lines indicate solar isotopic ratios. The mainstream grains form a line with slope 1.3, in contrast to the line of slope ~0.5, which is what would be expected for evolution in an AGB star (Nittler 2003). Nonetheless, by comparing the grain abundances with observations of stars, it is inferred that the mainstream grains originate in AGB stars. The insert shows these data on a contracted scale to show the one data point with a huge silicon isotopic anomaly; this is thought to have originated from a supernova (Amari, Zinner, and Lewis 1999). Reprinted from Nittler (2003). Copyright 2003 with permission from Elsevier.

never achieve a sufficiently high temperature (greater than about 1000°C) to fully melt, enough of which fall to Earth to be collected and analyzed with modern laboratory instrumentation.

Several excellent review articles exist on this subject (Nittler 2003; Zinner 1998; Clayton and Nittler 2004). A recent one, by Wasserburg et al. (2005), not only describes in detail the isotopic anomalies observed in meteorities but also gives an excellent review of the s-process. Of interest is the process by which the microscopic inclusions are analyzed. Fortunately, the nuclei of interest are often included in materials that are highly resistant to strong acids (Nittler 2003), for example, SiC; nanodiamonds; graphite; the refractory oxide grains corundum, spinel, and hibonite; silicates; and Al_2O_3, so much of the meteorite can be dissolved and the remaining inclusions can be separated out. The tiny inclusions can then be analyzed by ablation, followed by mass analysis via secondary ion mass spectrometry (SIMS), to determine their chemical and isotopic content. Recent advances in this area include resonance ionization mass spectrometry (RIMS) and new ion probes (nanoSIMS) that can localize their region of analysis to 50–200 nm (Nittler 2003).

An example of the sort of information that can be obtained is shown in figure 7.5, in which the results of isotopic variations of ^{29}Si versus those in ^{30}Si from many analyses of SiC grains are shown. Indicated also on the plot are some of the sources that are believed to have produced the grains in each region of the plot, along with the expected region or behavior of the abundance dependence from that source. In this figure the data are expressed as "delta ratios," that is, per mil deviations from the solar abundances, defined as

7.4.1 $$\delta^{i}\mathrm{Si} = 10^3[(^{i}\mathrm{Si}/^{28}\mathrm{Si})_{\mathrm{grain}}/(^{i}\mathrm{Si}/^{28}\mathrm{Si})_{\odot} - 1].$$

This time-honored method of representing the "isotopic anomalies" observed in the meteorites has served this research community well in presenting the anomalous abundance results that have resulted from the unmixed environments that are presented for us by these grains.

An even greater separation of meteoritic origins is seen in the comparison of the nitrogen isotopic anomalies with those from carbon, shown in figure 7.6. In this figure, what are presented are ratios of the isotopic abundances, which are seen to vary by huge factors. As in the previous figure, the mainstream grains are thought to originate in AGB stars. However, in these analyses the grain type seems to play a role. Presolar graphite data points seem to lie predominantly along the abscissa, whereas the others are spread throughout a much larger portion of the graph. However, the similarity of the nitrogen isotopic abundances to solar may simply reflect contamination (Nittler 2003).

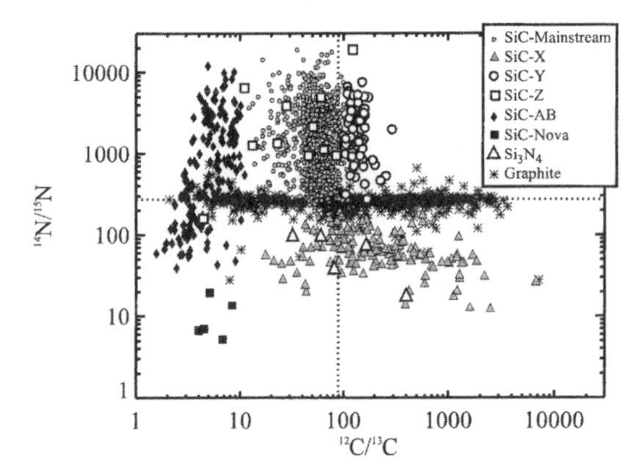

Fig. 7.6. Carbon and nitrogen isotopic ratios of presolar SiC, Si_3N_4, and graphite grains. The graphite data are from Hoppe et al. (1995). Other data are from Hoppe and Ott (1997) and references therein. Note that the abundance ratios displayed are huge; the data shown are actual ratios rather than the delta ratios shown in fig. 7.5. Reprinted from Nittler (2003). Copyright 2003 with permission from Elsevier.

These abundance ratios do indicate an interesting separation between the isotopic distributions of grains apparently produced in different sites.

The *Y* and *Z* SiC grains indicated in the two figures differ from the mainstream SiC grains studied in that they originate from AGB stars of lower inferred metallicity; the *Y* grains are about half solar, and the *Z* grains are about one-third solar. This metallicity dependence of this group of grains also strengthens the argument that these grains have originated from AGB star burning. *X* grains are thought to have originated in supernovae (Nittler 2003).

Messenger, Keller, and Lauretta (2005) provided recent confirmation that some terrestrial grains might be found that would have isotopic anomalies so unusual that they must surely have been produced by core-collapse supernovae. They identified one grain that had an $^{18}O/^{16}O$ abundance ratio that was 13 times solar, and a ^{17}O to ^{16}O ratio that was one-third solar, strongly suggesting that the material in this grain was produced in a very different environment than that in which most of the grains analyzed by these authors and those quoted by Nittler (2003) were produced.

Isotopic anomalies have now been observed in a wide variety of elements, ranging from noble gases to highly refractory elements. Most of these data, together with stellar data, confirm the AGB star origin of most of the grains, and some even provide details of the different processes that can occur in AGB stars. In some cases (see, e.g., Amari et al. 2001) correlated information about isotopic abundances in several elements can provide additional detail about the origin of the grain. Some additional results of the meteoritic analysis technique will be presented in several places in subsequent chapters.

A rather different use of isotopic anomalies is their application to possible detection of signatures of close core-collapse supernovae from the distant past. The basic concept was first suggested by Schramm and Wasserburg (1970). More recently Ellis, Fields, and Schramm (1996) and Fields, Hochmuth, and Ellis (2005) discussed the production of fairly long-lived radioisotopes in supernovae and the possibility that the wind produced by the supernova explosion might send some of them to Earth, where they might be found in terrestrial repositories, for example, glacial ice. Since the glaciers (at least until recently) added a layer of ice each year, the arriving radionuclides would be isolated in depth in the glacial ice, and the depth would indicate the time at which the supernova detritus arrived. Obviously, the supernovae would have to be reasonably close, less than 100 pc, for this effect to be observed. Note that if the supernova was too close it would be hazardous to human life; this distance has been estimated by Ruderman (1974) to be 10 pc or less.

The radionuclides so produced could either be produced within the supernova, as would be the case for ^{26}Al ($T_{1/2} = 7.17 \times 10^5$ years), ^{53}Mn ($T_{1/2} =$

3.74×10^6 years), and ^{60}Fe ($T_{1/2} = 1.5 \times 10^6$ years), or by interaction of the light particles produced in the supernova wind, the protons and α-particles, with abundant nuclides in the Earth's atmosphere. These interactions on ^{12}C and ^{40}Ar would produce ^{10}Be and ^{36}Cl, respectively, by spallation reactions.

This basic idea was used by Knie et al. (1999, 2004) to search for ^{60}Fe in deep ocean sediments, using accelerator mass spectroscopy to search for the very low abundance nuclides of interest in a background of the stable isotopes of the same element. They did identify ^{60}Fe signals well in excess of the backgrounds at between 2.4 and 3.2 Myr ago. This prompted a reanalysis of the possible output of a supernova, by Fields, Hochmuch, and Ellis (2005), to reevaluate their supernova output estimates; they concluded that, in addition to ^{60}Fe, ^{26}Al and ^{53}Mn were good candidates for these searches.

Another study, that by Cole et al. (2006), was aimed at the ice core samples suggested originally as repositories of supernova signatures by Ellis, Fields, and Schramm (1996) and Fields, Hochmuth, and Ellis (2005). This utilized samples from the Guliya glacier from the Qinghai-Tibet plateau in China (Thompson et al. 1997) to search for possible inclusions of ^{26}Al at the times suggested by observed increases in ^{10}Be and ^{36}Cl fluxes (Raisbeck et al. 1987; Thompson et al. 1997) at 35 ky and in the ^{10}Be flux at 60 ky. Grains at depths from times other than those were also sampled to determine "background" levels. Detection of ^{26}Al would have confirmed the supernova origin of the ^{10}Be and ^{36}Cl enhancements observed in the earlier studies. Although myriads of grains were found, none containing ^{26}Al was detected. Thus the search switched to that of anomalies in the O abundance ratios; again no dramatic confirmations of supernova activity were detected.

7.5 The r-Process

7.5.1 General Features of the r-Process

The basic requirements of the r-process are dictated by its fairly sharp abundance peaks at masses of 130 and 195 u (see fig. 7.1), another broad peak at 160 u, and the fact that it must synthesize all the nuclei more massive than ^{209}Bi. An additional constraint has been provided by astronomical data. Observations (Cowan et al. 1997) have shown that the r-process abundances seen in extremely low-metallicity stars appear to be in the same ratios as in stars of much higher metallicity, for example, the Sun, as is seen in figure 7.8. This suggests that the r-process must be independent of any preexisting seed nuclei, that is, it must be "primary."

Thus to summarize the basic constraints on the r-process in terms of what they impose:

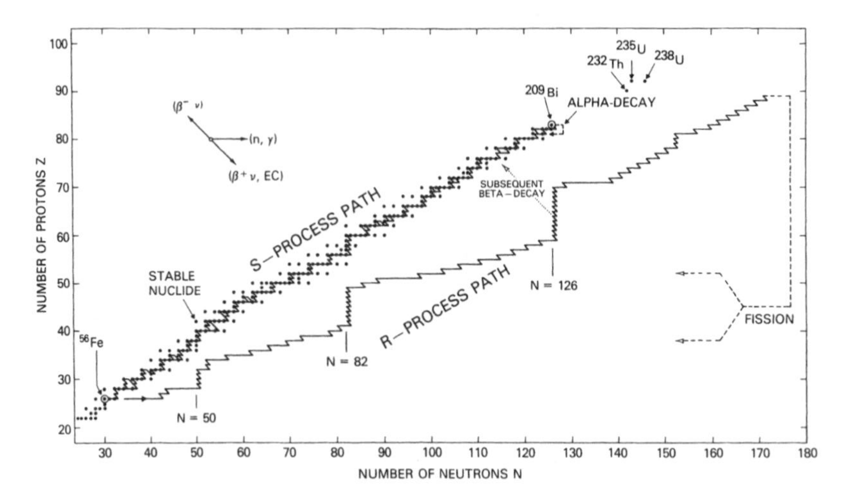

Fig. 7.7. Neutron capture path for the r-process through a portion of the chart of the nuclides. The nuclides in the band from ^{56}Fe to ^{209}Bi are stable or very long-lived nuclides. Thus the r-process path is seen to proceed 10–30 neutrons beyond stability until β-delayed fission and neutron induced fission occur (Thielemann, Metzinger, and Klapdor 1983). The r-process path was computed (Seeger et al. 1965) for the conditions $T_9 = 1.0$ and neutron density $= 10^{24}$. Adapted from Rolfs and Rodney (1988). Courtesy of C. Rolfs.

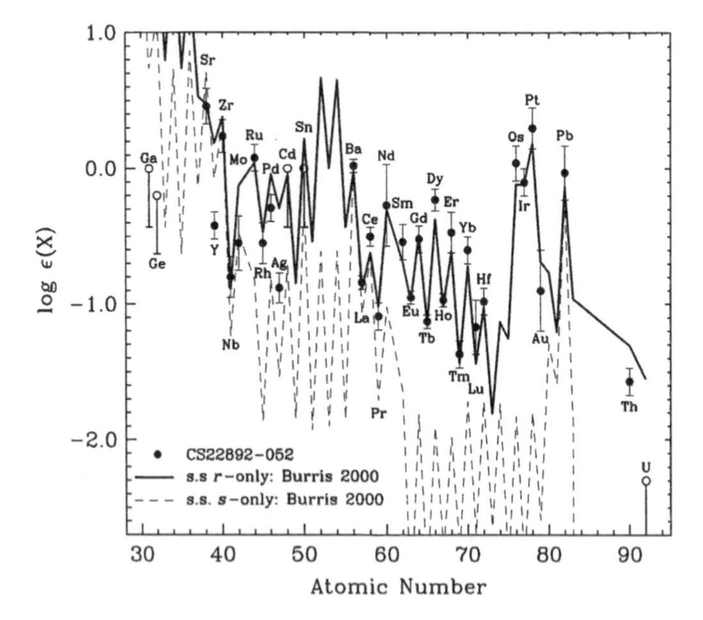

Fig. 7.8. Abundances of r-process elements seen in the metal-poor star CS 22892-052. The line through the data points is not a fit, but rather represents the solar abundances divided by about a factor of 40. The dashed line indicates the (scaled) solar s-process abundances. From Sneden et al. (2002).

1. It must produce abundance peaks at 130 and 195 u. This requires that the progenitor nuclei of the stable nuclei at those mass peaks be "waiting points" for the r-process.
2. It is primary, so it must reassemble its seed nuclei for each r-process occurrence.
3. It must operate on a timescale of seconds to produce all the nuclei beyond ^{209}Bi. This is necessitated by the need for it to proceed on a timescale shorter than the shortest β-decays that can produce a nucleus that would have a large (n, α) cross section.

The abundance peaks at 130 and 195 u have long been interpreted as representing the passage of the r-process through neutron closed-shell nuclei having $N = 82$ and 126, but with N/Z ratios much larger than those of any stable nuclei. Indeed, the r-process is thought to proceed along a fuzzy and time- (or temperature-) dependent path through nuclei approaching the neutron drip line (the lines through either the most neutron-rich or proton-rich nuclei that undergo β-decay; nuclei beyond the neutron or proton drip lines decay by either neutron or proton emission) early on and for which the Q-value for a (γ, n) reaction is around -2 MeV at the end of the r-process; these latter nuclei occur roughly 10–20 neutrons beyond the neutron-rich edge of stability (but not beyond the neutron drip line). The seed nuclei in the hot r-process environment will capture neutrons until (γ, n) and (n, γ) reactions are more or less in equilibrium, at which point they must await a β-decay before proceeding. Thus, at the r-process path away from the closed shells, the rates of those two processes must be equal, that is,

7.5.1
$$r[(\mathrm{n}, \gamma)] \approx r[(\gamma, \mathrm{n})],$$

which, using results from chapter 3, gives (eq. 3.9.39)

7.5.2
$$\left[(2\pi\mu k_B T)^{3/2}/h^3\right][G_1 G_2/G_{12}]\exp(-Q/k_B T) = \langle\sigma_{HF} v\rangle,$$

where the G_i are the nuclear partition functions and the statistical model cross section is often used to describe the neutron-capture reaction, that is (eq. 3.11.4),

7.5.3
$$\sigma_{HF} = (\pi/k^2)\eta T_1(\ell_1) T_2(\ell_2) / \sum_i \{(2J+1)/[(2j_i+1)(2J_i+1)]\} T_i(\ell_i).$$

where σ_{HF} is the Hauser-Feshbach (statistical model; see sec. 3.11) cross section for this process.

Between the closed shells, when a nucleus becomes too neutron rich, another neutron capture cannot readily occur, so the r-process will not proceed until a β-decay occurs. These shift the original nuclei in the direction of stability, that is, toward a new set of nuclei with more negative (γ, n) Q-values, and the neutron captures begin anew until the new (γ, n) ↔ (n, γ) balance point is again reached and another β-decay is required for further progress. The β-decays for these nuclei will be rapid; the extreme neutron richness will make available many proton levels to which the decay can go, with the attendant large likelihood of a low level of forbiddenness (see sec. 3.5) for the decay.

However, a very different scenario occurs at the neutron closed-shell nuclei. There an additional captured neutron will produce a nucleus that will have a much lower binding energy for the last neutron, so that a (γ, n) reaction is much more likely to occur in the high-temperature environment in which the r-process is thought to occur than would a subsequent neutron capture. In addition, the neutron-capture cross sections are extremely small at the neutron closed-shell nuclei, as can be seen in figure 3.10, further decreasing the probability of a neutron capture on these nuclei. Finally, the neutron closed-shell nuclei tend to be longer lived than their neighbors. Thus abundance piles up at the neutron closed-shell nuclei, each nucleus needing to await a β-decay before it can capture another neutron. Note, however, that this scenario assumes that the same neutron closed shells exist in nuclei 10–20 nucleons beyond the neutron-rich edge of stability as exist for stable nuclides. This has been questioned, both in the context of nuclear structure and in its impact on the r-process, and studied extensively both experimentally (see sec. 7.5.4) and theoretically (see, e.g., Haensel and Zdunik 1989; Dobaczewski et al. 1996; Pearson, Nayak and Goriely 1996; Lalazissis et al. 1998; Chen et al. 1995; Dillmann et al. 2003).

A simple analysis illustrates the dependence of the r-process abundances on the half-lives of the neutron closed-shell nuclei. Although in general nuclides can change to other nuclides either by neutron capture or by β-decay, in the r-process environment the neutron captures occur so rapidly, thereby driving the elements to the maximum neutron richness that they can achieve at the temperature at which the r-process occurs, that it is appropriate to describe the abundances of the nuclei $N(Z, A)$ along the r-process pathway only by their charge Z. Thus

$$dN_Z/dt = \lambda_{Z-1} N_{Z-1} - \lambda_Z N_Z, \tag{7.5.4}$$

where λ_Z = the rate at which nucleus (Z, A), with abundance N_Z, transforms to nucleus $(Z + 1, A)$ so, since that occurs by β-decay,

$$\lambda_Z N_Z = N_Z/\tau_Z. \tag{7.5.5}$$

However, as noted above, if (Z, A) is a neutron closed-shell nucleus, it will have a very small neutron-capture cross section (see fig. 3.10), and, furthermore, should it capture a neutron, would produce a nucleus that would be very likely to undergo a photoneutron reaction. Thus (Z, A) will remain as that nucleus until it β-decays, at which point it will immediately capture another neutron to make nucleus $(Z + 1, A + 1)$ but keeping the same neutron number. This process may continue through several neutron closed shell nuclei. At each successive nucleus, the Q-value for a (γ, n) reaction following a neutron capture will become more negative, and the neutron-capture cross section will increase, until finally a nucleus is reached which has the same neutron closed shell, but for which neutron capture becomes possible. Then the process of many neutron captures, with an occasional β-decay, will take over until the next neutron closed shell is reached.

At the neutron closed shells, we can establish a simple relationship between the abundance and the lifetime. There the nucleus $(Z, A - 1)$ will, as soon as it is produced by β-decay of the nucleus $(Z - 1, A - 1)$, capture another neutron to form nucleus (Z, A). Thus (where τ is a lifetime $= t_{1/2}/0.693$),

7.5.6 $$dN_{Z,A}/dt = N_{Z-1,A-1}/\tau_{Z-1} - N_{Z,A}/\tau_{Z}.$$

As with the s-process, when the flow into nucleus (Z, A) is essentially the same as that out of that nucleus, which occurs at some level of approximation even for the few seconds during which the r-process is thought to take place (the time to establish a steady state is the destruction timescale, which is τ_Z for nucleus Z and τ_{Z-1} for nucleus $Z - 1$), the rate of change of (Z, A) will be effectively zero, so we can write

7.5.7 $$N_{Z-1}/\tau_{Z-1} = N_Z/\tau_Z.$$

Thus those neutron closed shell nuclei with the largest half-lives will also tend to have the largest abundances. This is as would be expected because, as the r-process flows from lighter to heavier nuclides, abundance will tend to pile up at the nuclei with the longest "dwell times", that is, those for which the neutron-capture cross section is small and the half-life is long.

7.5.2 An r-Process Model: Neutrino Wind, Alpha-Rich Freezeout, and Neutron Captures

The neutron density required to drive the r-process path the 10–20 neutrons beyond the most neutron-rich stable nuclides to produce the abundance peaks at 130 and 195 u would certainly also satisfy the constraint on the r-process that it produce the $A > 209$ nuclei. However, the requirement that the r-process be primary is not so obviously satisfied, although two models currently under study do satisfy that constraint. It is important to emphasize, however, that the site of the r-process is not known at present; this is an active area of investigation by both theorists and observers. One model that does satisfy the condition that the r-process be primary is the neutron star merger model, to be discussed further below. Another is the recently developed scenario of the dynamical r-process that occurs in the neutrino wind of a supernova following α-rich freezeout (Meyer et al. 1992; Woosley et al. 1994; Takahashi, Witti, and Janka 1994; Hoffmann, Woosley, and Qian 1997). This is thought to occur in the hot neutrino wind that follows stellar collapse. In this scenario, the initial high temperatures, of order $T_9 = 10$, will reduce all preexisting nuclei to protons and neutrons by photonuclear processes. The nuclei up to roughly mass 80–100 u can then be reassembled within nuclear statistical equilibrium (see sec. 6.5) in the second or so required for the environment to cool down to below freezeout temperature, assuming the parameters of α-rich freezeout are chosen correctly. The remaining neutrons are then captured in the r-process, initially very rapidly, then more slowly, capturing the remaining neutrons and extending the neutron richness of the final nuclei as they shift toward stability, undergoing successively slower β-decays. However, the neutron density must be high; in order to promote $A \sim 80$ nuclei to $A \sim 200$ nuclei requires $\sim$120 neutrons per seed nucleus. In addition, the α-rich freezeout parameters, specifically the entropy, must be such as to produce a seed nucleus distribution up to 80–100 u, as abundances up to that mass range appear to be essential for seeding the ensuing r-process. That this scenario can indeed produce a successful r-process is shown in figure 7.9, produced in a study by Cowan et al. (1999).

The possible choices of the entropy, processing time, and electron fraction that can produce a successful r-process from its α-rich freezeout seeds have been studied by Hoffman, Woosley, and Qian (1997) and by Meyer and Brown (1997); the results of their work are indicated in figure 7.10 (Y_e is the number of leptons per baryon = proton density divided by the baryon density). There it can be seen that there are many possible choices of these parameters that will produce a successful r-process. However, they are not obviously realizable in a realistic supernova model. Indeed, it was concluded in that study that, in a spherical explosion model, the existing entropy is about a factor of 2 less than that needed to produce the required mass 80–100 u seed nuclei. The high entropy is required to regulate the reaction sequence

7.5.8 $$^4\mathrm{He} + {}^4\mathrm{H} \rightarrow {}^8\mathrm{Be} + \mathrm{n} \rightarrow {}^9\mathrm{Be} + \alpha \rightarrow {}^{12}\mathrm{C} + \mathrm{n}.$$

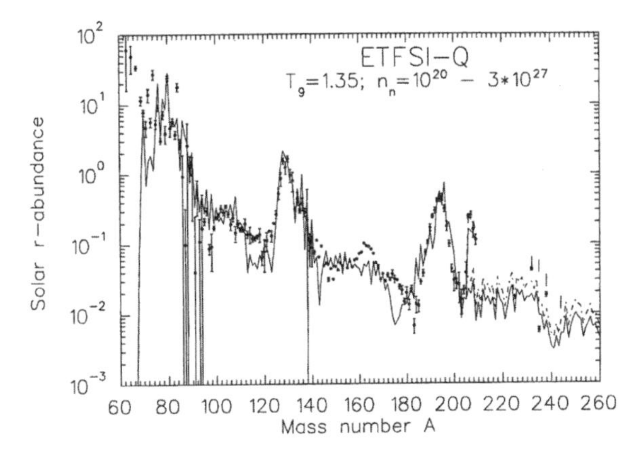

Fig. 7.9. Predicted r-process abundances compared with the solar abundances as a function of mass number. The dashed curve (at higher masses) has a slightly different weight for A $>$ 206. The predictions are seen to predict abundances that are too low in the $A = 110$–120 mass region and seem to be out of phase with the data in the 160–180 mass regions. However, they do well at the $A \sim 130$ and 195 mass peaks. From Cowan et al. (1999).

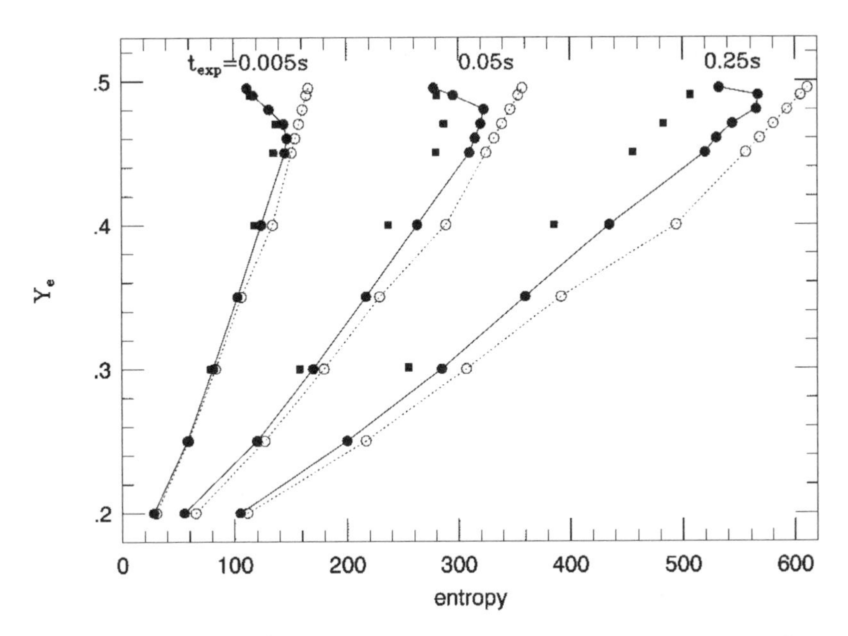

Fig. 7.10. Combinations of Y_e, entropy, and expansion time required for production of the $A = 195$ r-process peak. The circles connected by lines are for various fixed expansion times. Shown are the values derived in a numerical study (*filled circles*) and those from an analytic approximation (*open circles*). The filled squares represent results from the numerical survey that used an exact adiabatic equation of state. From Hoffmann, Woosley, and Qian (1997).

The nucleus ^{9}Be is not formed in NSE. It has an extremely low neutron binding energy, so it is easily destroyed by photonuclear reactions in the high-temperature α-rich freezeout environment. This, however, is crucial since low entropy translates into an insufficient number of high-energy photons to limit ^{9}Be production (see article by Hoffman and Timmes in Wallerstein 1997; see also Meyer, Krishnam, and Clayton 1996, 1998). Then too many of the neutrons and α-particles will be consumed by the many ^{12}C seeds to push the seed nuclei up to mass 80–100 u, and having the abundance peak produced from NSE at too low a mass will not adequately seed the r-process. This can be compensated for by having more neutrons, that is, by lowering the electron fraction, but the required combination seems to be unphysical, at least within the supernova models that have been studied thus far. Sasaqui et al. (2006) found an alternate reaction chain, $^4\mathrm{He}(t,\gamma)^7\mathrm{Li}(n,\gamma)^8\mathrm{Li}(\alpha,n)^{11}\mathrm{B}$, that can circumvent the mass 8 gap but, since this chain could produce more heavy nuclides, it required an even higher entropy than that necessitated by the reactions considered by Hoffman, Woosley, and Qian (1997).

Thompson, Burrows, and Meyer (2001) also studied the conditions under which the α-rich freezeout r-process could occur and found that the only plausible models, at least in terms of the entropy and other parameters of the r-process site, that could produce all the r-process peaks, especially that at mass 195 u, were those in which the r-process occurred on a very short timescale, of order 1 ms. However, even this was only achieved with some modest modification of the spectra of the neutrinos and other small modifications, which were necessary to shorten the timescale sufficiently for this r-process to occur.

As noted in chapter 6, recent work by Burrows et al. (2006) has studied a model, based on an 11 $M_\odot$ progenitor, in which the inner core of the collapsing star produced acoustic power with a period of 25–30 ms that began roughly 200 ms after the first core bounce. The pulses from the core become shock waves that appear ultimately to be capable of exploding the star. In addition, this scenario has another interesting feature, that is, it increases the entropy in the region in which the neutrino driven r-process would occur by a factor that may be large enough that the apparent problem of insufficient entropy to run the r-process may also be solved. Further work on these models will tell us if this is the solution to identifying the r-process site.

Despite its caveats, the naturalness of the α-rich freezeout model in producing the r-process is very appealing, and more theoretical effort is going on to try to invoke variations of the model assumed by Hoffman, Woosley, and Qian (1997) to modify the parameters in the direction needed for a successful r-process. The α-rich freezeout scenario is to be compared with the description of the r-process that has existed for much longer, that is, that which requires the superposition of several r-processes with varying neutron densities. It is not clear that such a scenario can satisfy the requirement that the r-process be primary, and the lack of a site for such a multistep r-process is also troublesome. However, it has been known for many years (Kratz et al. 1993) to give a good representation of the r-process abundances, albeit without a real model scenario in which it might occur, and it has allowed the study of the conditions that are required for the r-process to succeed, wherever it occurs.

Although the fits to the r-process abundances have been remarkably successful for the $A = 130$ to 200 r-process nuclides, it has also been generally recognized that the lighter nuclides, most notably the pure r-process nuclide ^{129}I, are not well represented. This led Wasserburg et al. (1996), and later Wasserburg and Qian (2000), to hypothesize that there are actually two r-process sites, one producing primarily the heavier nuclides, and the other producing primarily the lighter nuclides. One motivation for a scenario involving more than one r-process site can be seen in the poor representation of the light r-process nuclides shown in figure 7.9. However, the inference by Wasserburg et al. (1996) was based on a detailed analysis of the two short-lived nuclei of r-process origin, ^{129}I and ^{182}Hf, which were found to have been alive in early solar system condensates at relative ratios that are incompatible with a unique r-process distribution for stars of any mass.

It has also been suggested (Lattimer and Schramm 1976; Meyer 1989; Freiburghaus et al. 1999; Rosswog et al. 1999, 2000) that merging neutron stars might be responsible for the r-process. The appeal of this scenario is that the neutron stars could certainly produce the high neutron density required for the r-process. However, a subsequent study (Argast et al. 2004) appears to show that this scenario could not be solely responsible for r-process nucleosynthesis, as neutron star mergers simply do not occur with sufficient frequency to produce the observed galactic abundance.

Finally, more recent studies have been of the potential nucleosynthesis that could be done during gamma-ray bursts by hypernovae. Specifically, Nomoto et al. (2005) studied the synthesis that could be performed of nuclei up to the Fe-Ni peak. In the context of the r-process McLaughlin and Surman (2005) studied the possibility that the neutrino winds from accretion disks surrounding the black hole at the center of a hypernova might synthesize heavy nuclides. The issue with which one must deal is that of forming extremely neutron-rich matter in the disk; McLaughlin and Surman showed that in disks with high accretion rates, $\bar{\nu}_e$ capture can create the necessary conditions. Indeed, they

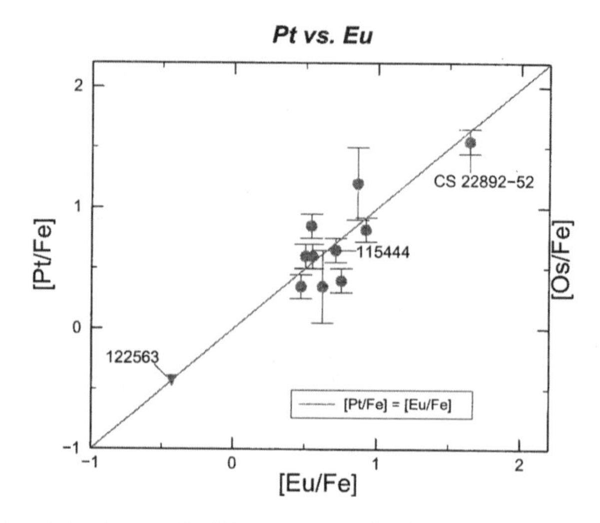

Fig. 7.11. Correlation between [Pt/Fe] and [Eu/Fe] abundances for metal-poor stars. The essentially perfect correlation between Pt and Eu shows that Pt is produced in the r-process. From Cowan and Sneden (2005).

showed that the r-process production appears to be quite promising. This scenario will certainly be the subject of further investigation.

7.5.3 r-Process Related Astronomical Data

In sorting out such possibilities the astronomical data are crucial. In some cases studies have been done to determine whether an element is produced in the r-process by observing how well its abundance, observed in very low-metallicity stars, is correlated with that of some element that must be produced in the r-process and, of course, is clearly observable in the low-metallicity stars. Eu is usually chosen as the r-process standard. Figure 7.11 shows the correlation between the Pt abundance and that of Eu, actually as [Pt/Fe] versus [Eu/Fe]. Since the Fe abundance builds up in time, that abundance can be used as a clock. The bracket notation used for the abundances, for example, [Eu/Fe], is defined as

7.5.9 $$[\mathrm{Eu/Fe}] = \log[(\mathrm{Eu/Eu_{\odot}})/(\mathrm{Fe/Fe_{\odot}})].$$

where, for example, $\mathrm{Fe}_{\odot}$ is the solar Fe abundance.

The range of the metallicities of the stars observed varies from about -4 (which corresponds to 10^{-4} of solar abundance) to about -1 (1/10 of solar). The correlation between Pt and Eu is essentially perfect; Pt is clearly produced in the r-process. A similar plot was seen for Os.

By contrast, two correlation plots were made for Ge; these are seen in figures 7.12 and 7.13. In the first plot, of [Ge/H] versus [Fe/H], it is suggested that Ge does not begin to be produced until metallicity about −2.5, well after Eu begins to be produced. In the second, plotted in a different way, it is seen that the Ge and Eu abundances are completely uncorrelated; this is a plot of [Ge/H] versus [Eu/Fe]. Thus Ge is not produced in the early Galaxy by the r-process.

Yet a third correlation plot is seen in figure 7.14, which shows [Ba/Eu] versus [Fe/H]. Ba is thought to be produced in both the r- and s-processes. This plot exhibits a remarkable separation of the correlations into two blobs, with the upper one ranging down to [Fe/H] = −2.5, and the other to as low as metallicities are observed. This separation is interpreted as indicating that the upper group of stars are those in which Ba is produced in the s-process, which apparently "turns on" at about [Fe/H] = −2.5, and the lower group has Ba produced via the r-process, which of course has created elements in the oldest stars. Plotting the data in this way produces a separation of the products of these two processes of nucleosynthesis at least for barium.

Another potentially important astronomical development is the determination of isotopic abundances in heavy nuclei. For elements that are produced purely by the r-process, this is not an issue, but for elements that can have mixed production processes, for example, Ba, this can be very important in determining the onset of the s-process. Figure 7.15 shows the current state

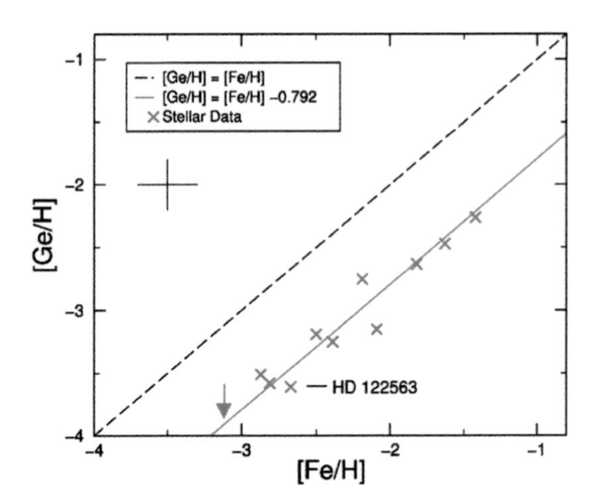

Fig. 7.12. [Ge/H] versus [Fe/H] in metal-poor stars. Note that Ge only begins to be produced at [Fe/H] = −2.5, suggesting that whatever process produces it turns on later than the early-Galactic r-process does. From Cowan et al. (2005).

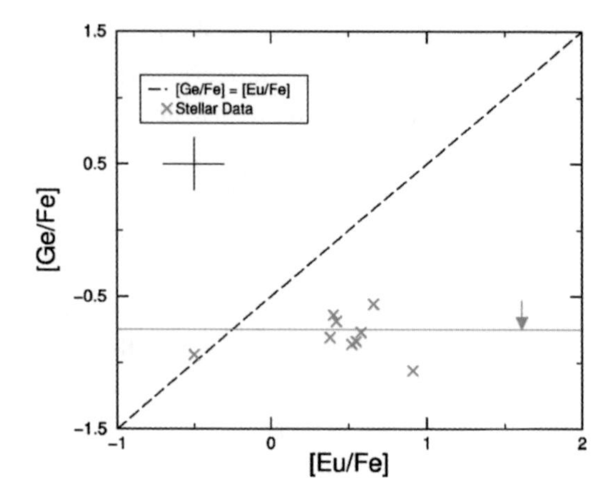

Fig. 7.13. [Ge/H] versus [Eu/Fe]. The complete lack of correlation is another way besides that shown in figure 7.12 of demonstrating that Ge is not produced in the early-Galactic r-process. From Cowan et al. (2005).

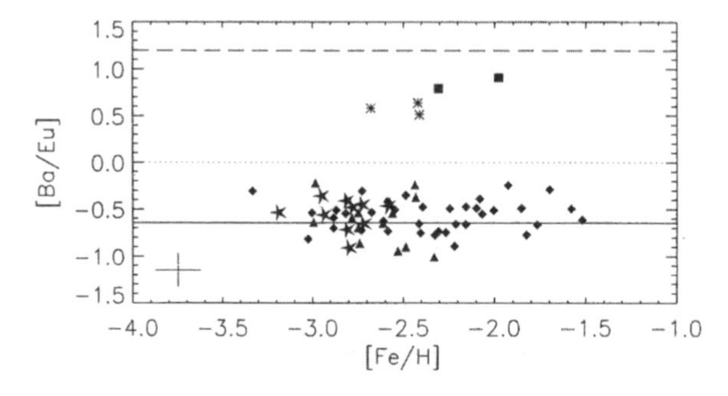

Fig. 7.14. [Ba/Eu] versus [Fe/H]. The separation of the data into two regions suggests that two mechanisms, that is, the r-process and the s-process, are producing Ba, as would be expected. From Barklem et al. (2005).

of the art in this context. Shown there is the spectrum through the region in which one would expect to observe lines of Eu. Eu has two stable isotopes, ^{151}Eu and ^{153}Eu, with solar abundances of 48% and 52%, respectively. Figure 7.15 shows the fit to the spectrum where the emission lines from these two isotopes are expected; it is seen that a mixture of about 50:50 of the two isotopes produces an excellent reproduction of the spectrum, whereas one predominately of either isotope produces a poor fit. This sort of analysis is in its infancy, although its difficulties are made clear from figure 7.15. The line broadening is not related to the telescope on which the observations were made; rather it is due to the thermal motion of the atoms in the surface of the star.

So are there two r-process sites? Or perhaps are there multiple r-process sites defined by different parameters? At the time this book is being written, this remains an open question; it is certainly one that is being pursued actively, both theoretically and observationally.

7.5.4 Nuclear Data Aspects of the r-Process

As indicated in equation 7.5.5, the half-lives of the neutron closed-shell nuclei far from stability are the most crucial data inputs for the r-process. As noted in chapter 3, remarkable progress has been made in measuring those relevant to the $A = 130$ ($N = 82$) r-process peak, mostly at ISOLDE at CERN. This

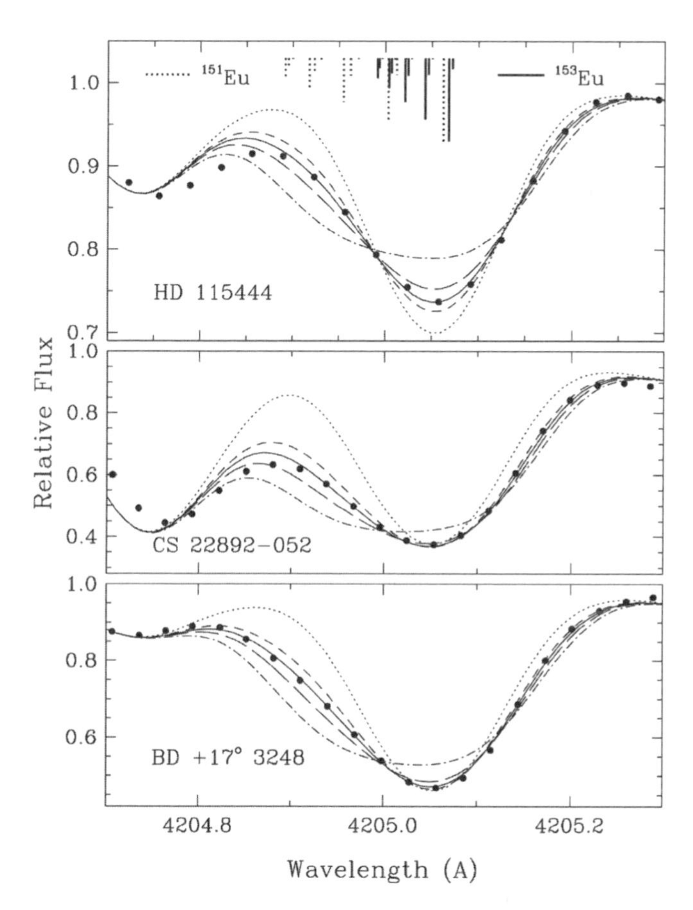

Fig. 7.15. Observed and synthetic spectra of the Eu II 4205.04 A line in three metal-poor stars. The observed spectra (*filled circles*) are compared with synthetic spectra with ^{151}Eu comprising 0.000 (*dotted line*); 0.350 (*short-dashed line*), 0.478 (*solid line*), 0.650 (*long-dashed line*), and 1.000 (*dot-dashed line*) of the total Eu yield. Since Eu has only two stable isotopes, the ^{153}Eu fraction is 1.0 − the ^{151}Eu fraction. In the top panel, vertical lines are added to indicate the wavelengths and relative strengths of the hyperfine components of the isotopes ^{151}Eu (*dotted lines*) and ^{153}Eu (*solid lines*). The absolute vertical line lengths are normalized by an arbitrary constant for display purposes. From Sneden et al. (2002).

facility utilizes a high-energy proton beam on an appropriate target to create the nuclides of interest. These are extracted rapidly in order to avoid large losses from decay of the short-lived nuclides created. Some chemical selection is also usually utilized to enhance the nuclides of interest in competition with other nuclides, often more strongly produced isobars. The beams of the short-lived nuclides are extracted from the primary target, mass analyzed, and transported to a secondary target where they are usually implanted into a detector. At this second station, the decay products of the nuclides are measured—betas, gammas, and possibly neutrons.

It was noted (Kratz et al. 1988, 1993; Rauscher, Thielemann, and Kratz 1997) that the neutron closed shells thought for decades to be responsible for the r-process abundance peaks may be "quenched" in the nuclei through which the r-process passes, that is, the same closed shells that exist for the nuclei near stability may not apply as strongly to the extremely neutron-rich nuclei. The existence of effects such as these will only be confirmed or rejected finally by experimental studies of these nuclei, and these require RNB facilities that do not exist at present. The interpretation of the basic features of the r-process hinges on the structure of these nuclei; it is of paramount importance that their details be obtained as rapidly as advances in technology allow. Recent work (Bernas et al. 1997) has provided a significant advance on the available information on very neutron-rich nuclei, with many new nuclei observed and numerous half-lives measured in the $A = 80$–110 mass region, but much remains to be done. Specifically, half-lives for all of the neutron closed-shell nuclei through which the r-process passes must be determined, as well as their decay modes, that is, whether they decay by β-emission or by β-delayed neutron emission, or by a more complicated mode. In addition, enough nuclear structure information must be gained to determine if the shell closures far from stability are as strong as those near stability. Finally, it would be desirable to measure some neutron radiative capture cross sections for some of the very neutron-rich nuclei. We will return to this below.

Much work (Kratz et al. 1986; Bjornstad et al. 1986; Omtvedt et al. 1995; Zhang et al. 1996; Walters et al. 2004 and references therein) has focused on the neutron-rich nuclides around the mass 130 u peak, with the experimental information now reaching well into the nuclides that are important to the r-process. Among the nuclei studied that are progenitors of nuclei in that r-process abundance peak are ^{128}Pd, ^{129}Ag, ^{130}Cd, ^{131}In, and ^{132}Sn (a doubly magic nucleus). Not only have the half-lives of these nuclei been determined, but other properties related to their nuclear structure have been observed by

Table 7.1 Properties of Neutron-rich Nuclides Relevant to the 130 u r-Process Peak

Nucleus	$T_{1/2}$ (in ms unless specified)	β-n Prob (percent)
^{127}Ag	79	0
^{128}Ag	58	0
^{129}Ag	46	0
^{130}Ag	50	0
^{130}Cd	162	3.5
^{131}Cd	68	3.5
^{132}Cd	97	60
^{131}In	280	<2.0
^{132}In	207	6.3
^{133}In	165	85
^{134}In	140	65
^{135}In	92	>0
^{130}Sn	3.7 m	0
^{131}Sn	56 s	0
^{132}Sn	40 s	0
^{133}Sn	1.45 s	0.08
^{134}Sn	1.05 s	17
^{135}Sn	530	21
^{136}Sn	250	30
^{137}Sn	190	58
^{138}Sn	200	50
^{139}Sn	100	75
^{140}Sn	100	50
^{136}Sb	923	16.3
^{137}Sb	330	49
^{138}Sb	250	50
^{139}Sb	150	90
^{137}Te	2.49 s	2.7
^{138}Te	1.4 s	6.3
^{139}Te	400	12

Note From Walters et al. (2004) and http://www.nndc.bnl.gov/nudat2/sigma_searchi.jsp.

producing isobars of the nucleus of interest and then observing the subsequent β- and γ-decays. In addition, if they decay by β-delayed neutron emission, that will shift the mass to which that progenitor decays from that of the progenitor. Some of the nuclei that could contribute to the mass 130 ± 3 r-process peak, along with some of their observed properties, are indicated in table 7.1.

While cross sections tend to be less important to the r-process than half-lives and masses, certainly in determining the r-process path, they are important in determining the neutron closed shell nuclides at which neutron capture can begin to break out from that particular group of closed-shell nuclei. Of course, performing neutron-capture measurements on short-lived nuclides is not possible at this time, but one can determine much of the needed information by invoking a surrogate reaction, (d,p). This neutron transfer reaction can map out possible resonant states as well as measure total and partial widths. Unfortunately, what is needed is (d,p), or one neutron transfer, reaction studies on the same nuclei that have extremely small neutron-capture cross sections, so these experiments will be difficult. However, even poor statistics data would help greatly in enhancing our understanding of the r-process.

Measurement of half-lives of the progenitors of the nuclides at the 195 u abundance peak is not possible at present, although next-generation RNB facilities may make some of these measurements possible. Specifically, one would like to know the half-lives of nuclei such as ^{192}Dy, ^{193}Ho, ^{194}Er, ^{195}Tm, ^{196}Yb, ^{197}Lu, and ^{198}Hf. Note that these are simply the $N = 126$ nuclides that are close to the 195-u peak provided no additional shifting of the peak occurs. However, it would not be surprising if some β-delayed neutron emission, possibly even of two or three neutrons, would occur and would affect the location of the peak. For example, ^{196}Yb, although located slightly to the high-mass side of the mass 195 r-process peak, might well be expected to emit one or more neutrons in its decay chain, shifting its production location to slightly below the mass 195 peak. Working in the opposite direction might be late-time neutron captures near the end of the r-process that would increase the neutron abundance of the nuclides being produced.

Even the next-generation RNB facilities will not be able to generate all of these nuclides as they are too far from stability to have an adequate production yield for even the most basic studies. However, even a few of them can, assuming their identification is well established, provide a good enough half-life measurement to be of great help in checking the validity of theoretical predictions, and hence, in understanding the r-process.

At what mass does the r-process terminate? In principle, if the neutron density were high enough and the r-process lasted long enough it could push to higher and higher mass. However, there are nuclear physics reasons why this is not the case. As nuclei become more and more massive, the repulsive (between protons) Coulomb force, which is long range, begins to dominate over the stronger, but shorter range, (attractive) nuclear force. When nuclei reach roughly mass 270 u, they become unstable to fission, which breaks the nuclei into two comparable halves and usually emits some neutrons as well. When this occurs, the nuclides involved return to masses around 120 u and begin capturing more neutrons so as to work their way to higher and higher masses. This "fission recycling" is incorporated in the r-process codes, as it must be to produce accurate characterizations of this process of nucleosynthesis. However, much remains to be discovered about fission of the heaviest nuclides, especially those neutron-rich ones that would be made in the r-process. The deformation and the height of the fission barrier will determine the fission products, and these will be critical to determining the ultimate r-process abundances.

Given their extreme neutron richness, might it be possible for these nuclei along the r-process path to undergo the transition to strange quark matter (see chap. 6)? If such matter is truly the ground state of nuclear matter, this should be possible in principle. However, for that to occur there would have to be many nearly simultaneous strangeness changing weak decays of both neutrons and protons to strange quarks, a very unlikely possibility given that the timescale for weak decays is generally very slow—a fairly large fraction of a second. Thus it would seem that the transition to strange quark matter could only occur in the environment at the center of a neutron star, if it can even there.

7.5.5 Role of Neutrinos in Nucleosynthesis

THE r-PROCESS. What role do neutrinos play in the r-process? Although their precise effects are complex, it is clear that they can be important. They can both help and hinder the r-process. The former effect results from the reaction

7.5.10 $$\nu_e + Z_A \rightarrow e^- + (Z+1)_A,$$

that is, from the perspective of the nuclei, they produce the same result as β-decay. This can be most helpful in pushing the nuclei produced by many neutron captures past the neutron closed shells, where half-lives can be the order of a second. This can obviously be a problem for a process that may only last for roughly that period of time, as it must first produce flow past the $A = 130$ $(N = 82)$ neutron closed shell and then past the $A = 195$ $(N = 126)$ neutron closed shell. A high-flux neutrino wind could greatly expedite this flow to the heaviest nuclei that the r-process must synthesize.

However, these same neutrinos can also kill the r-process, as the reaction

7.5.11 $$\nu_e + n \rightarrow e^- + p$$

will convert neutrons to protons, which are quickly captured into ^4He nuclei, thereby reducing the neutron density below the value required to run the r-process. The problems presented by this possibility were discussed in detail by Meyer, McLaughlin, and Fuller (1998), who noted that the process represented in equation 7.5.11 will generally dominate over those of equation 7.5.10 because of the limited number of final states for capture inside a nucleus.

A possible solution to this problem has been offered by Schirato and Fuller (2002). They suggested that the ν_es might oscillate as they pass through the dense matter near the core of the nascent neutron star so as to depopulate the ν_es. However, this scenario would require a sterile neutrino species, so there would be a state into which the ν_es could disappear and not be regenerated by some other flavor oscillating back into the ν_es. Thus this solution must await confirmation of the existence of sterile neutrinos before it will achieve general acceptance.

Of course the problem is complicated by the fact that the different neutrino flavors, and even antineutrino flavors in the case of ν_e and $\bar{\nu}_e$, may have different energy spectra. This was discussed in chapter 6; it results from the fact that the ν_es and to a lesser extent $\bar{\nu}_e$s can interact via both neutral-current and charged-current interactions, whereas the ν_μ, $\bar{\nu}_\mu$, ν_τ, and $\bar{\nu}_\tau$ will only interact via the neutral-current interactions. This will produce neutrinospheres of the different flavors at different radii, hence different temperatures at the neutrinos' surfaces of last scatter, thereby producing the different distributions.

THE ν-PROCESS. The ν-process of nucleosynthesis (Woosley et al. 1990; Wallerstein et al. 1997; Boyd 1999) has been shown to describe the synthesis of some nuclides that had previously evaded theoretical understanding: ^{180}Ta, ^{138}La, ^{19}F, ^{11}B, and ^{7}Li. It does this, at least to some extent, via neutrino-induced reactions on nearby abundant nuclides in the hot neutrino wind from a supernova. The observed abundances of these nuclides do agree reasonably well with those predicted by ν-process production, although a detailed comparison will require a better defined knowledge of the energy distribution of the neutrinos.

In another context, Haxton et al. (1997) and Qian et al. (1997) have suggested that neutrino processing of the nuclei in the r-process abundance peaks could be responsible for the abundances of the nuclei just below those peaks. While this does not really constitute a confirmation of the α-rich freezeout plus r-process model, it is at least a consistency check of the conditions within the r-process site. The nuclei they studied were those from mass 124 to 126 u, just below the 130 u peak, and those from mass 183 to 187 u, just below

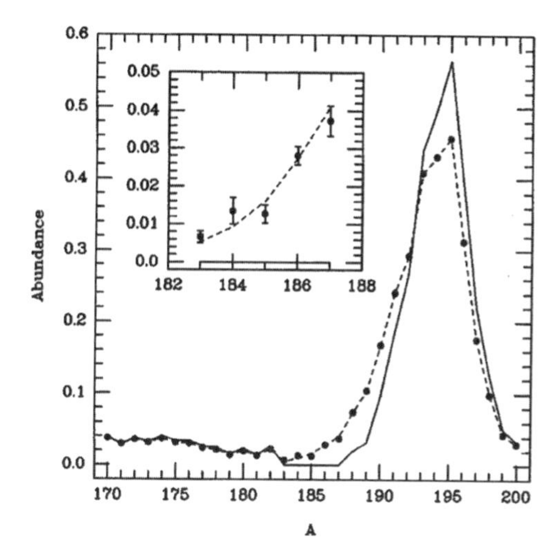

Fig. 7.16. The effects of neutrino fluence on producing the nuclides just below the main r-process abundance peak at 195 u. The abundances before and after processing are given by the solid and dashed curves, respectively. The filled circles (some with error bars) give the solar r-process abundances. Reprinted with permission from Qian et al. (1997) Copyright 1997 by the American Physical Society.

the 195 u peak. Their result assumed that the amount of neutrino-induced knockout of neutrons would be a function of the neutrino fluence, essentially the time integral of the neutrino flux. It was found that the fluences required to produce the lower mass nuclei were the same for the three nuclei and that those for the higher mass nuclei were different from those for the lower mass nuclei but the same for those five higher mass nuclei. The study concluded that realistic neutrino fluences could have produced those eight nuclei by neutrino postprocessing on their nearby abundant r-process neighbors and that the fact that the required fluences were the same within each group supported this interpretation. Furthermore, the fact that the required fluence for the upper mass group was lower than that for the lower mass group is consistent with the r-process scenario, in which the mass 130 u peak nuclides would have been synthesized first and so would have experienced the neutrino processing for a longer time than would the mass 195 u peak nuclides. Some of the results of this study are shown in figure 7.16.

Additional work by Surman et al. (1997) has suggested an explanation of the broad r-process abundance peak at 160 u. That study assumed modern nuclear structure descriptions of the nuclei in the mass region along the r-process path, which suggest that such nuclei will be highly deformed and that this will produce a shell closure–like effect on the movement of the r-process through the progenitor nuclei of those in that mass region. The result does indeed produce a good representation of that peak, although the authors note that the usual r-process calculation, that is, one that terminates the calculation at the point at which the neutrons have been used up, is insufficient for this mass region. Continuation of the calculation to a temperature at which (γ, n) processes cease to be important appears to be required in order to produce the peak. Thus this result is a freezeout phenomenon, so it involves many detailed rates for capture and decay, confirming the need for as much experimental information as possible about nuclear structure for the r-process nuclei.

It has been suggested (Hillebrandt 1984) that an 8–10 $M_\odot$ star might burn to a degenerate ONeMg core (unlike more massive stars, which burn Ne and O in nondegenerate cores), which would then collapse by undergoing electron capture. This might produce an r-process (Wheeler, Cowan, and Hillebrandt 1998) and might also produce a collapsing iron core that is smaller than that from more massive stars. This in turn would more readily allow ejection of the newly synthesized material into the ISM than has been found to be the case for more massive stars. The resulting r-process would satisfy the requirement that it be primary. However, it would also satisfy an additional constraint, pointed out by Mathews, Bazan, and Cowan (1992), that the iron abundance in the Galaxy may have begun to increase from the material synthesized in the big bang before that of r-process nuclides such as europium. Since iron is produced in massive stars and since they finish their stages of stellar evolution sooner than less massive stars, this would be consistent with the $\sim 10\ M_\odot$ stars providing the r-process nuclei (Wheeler et al. 1998).

7.6 Nucleocosmochronology

Our understanding of the r-process has allowed its application to nucleocosmochronology, the use of the long-lived heavy nuclides to determine the age of the universe. In principle this consists of comparing the present abundance of some radionuclide, with a half-life comparable to the age of the universe, with the abundance it would have had at the time the Milky Way Galaxy was formed and then using the difference and the decay lifetimes to set a lower limit (because the time required for Galaxy and star formation would still need to be accounted for) on the age of the universe. The basic scheme is indicated in figure 7.17. As indicated in the figure, the big bang occurred at time zero, the Galaxy formed somewhat later in time, at T_G, the solar system formed after that, at time T_S, and an additional 4.7 Gy have elapsed since T_S to get us to the present, at T^*.

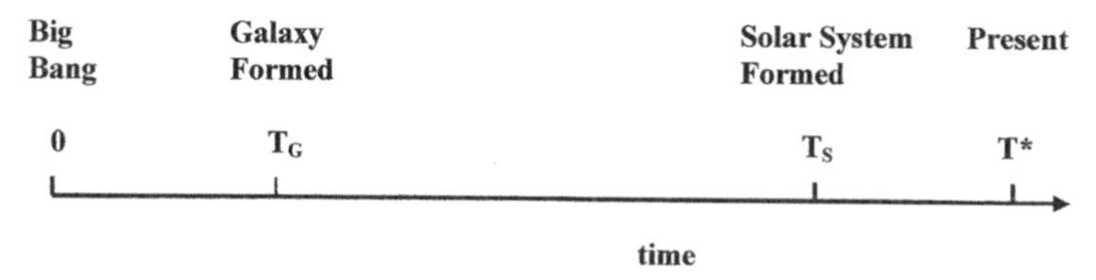

Fig. 7.17. Relationship between times of various events in the evolution of a galaxy.

The simplest model of nucleocosmochronology observes abundances of cosmochronometers in very metal-poor stars, that is, stars that are comprised of material that was synthesized shortly after the Galaxy was formed. This technique was adopted by Cowan et al. (1997) to use the observation of thorium in CS 22892-052 to infer the age of the Galaxy. In order to relate the observed thorium abundance in CS 22892-052 to the age of the Galaxy, one needs to predict accurately the amount of thorium that is synthesized in r-process nucleosynthesis compared with other r-process nuclides to determine how much thorium has decayed in the intervening eons. For this star it can be assumed that there was a single production event, the core-collapse supernova that produced the observed r-process abundances, assumed to have occurred essentially at T_G, and that the thorium has been decaying ever since.

To develop the model of galactic chemical evolution (GCE) that applies to this situation, it is assumed that the r-process nuclei considered are produced in core-collapse supernovae, and that the relative abundances of the r-process nuclei synthesized, N_r (or N_{Th} for thorium), are constant. Thus, for this simple case in which the thorium atoms are synthesized by the single supernovae that occured at τ_G and destroyed only by their decay, that is,

7.6.1 $$dN_{Th}(t)/dt = N_{Th}(\tau_G)\delta(t - \tau_G) - N_{Th}(t)/\tau_{Th}.$$

The solution to this differential equation can readily be seen to be, for $t > T_G$

7.6.2 $$N_{Th}(t) = N_{Th}(\tau_G)\exp[-(t - \tau_G)/\tau_{Th}],$$

where τ_{Th} is the lifetime of Th, 20.27 Gy. In this equation, N_{Th} can be either the density of thorium atoms in the Galaxy (assuming the Galaxy to be well mixed, or for this case that all the r-process nuclides from the original supernova were accumulated in the observed star in the ratios in which they were produced) or it could also refer to the mass of thorium in the Galactic gas. In what follows it will be useful to define it as the latter.

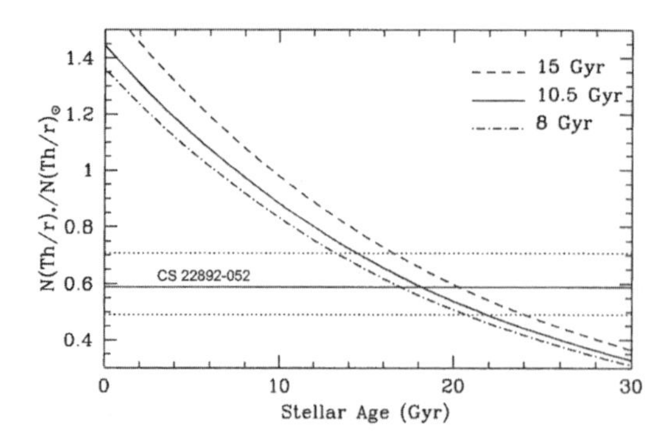

Fig. 7.18. Age dependence of the observed thorium/*r* ratio (in units of the observed solar system value), based on a simple model of chemical evolution and three different assumed ages for the Galactic disk: 8, 10.5, and 15 Gyr. The horizontal lines represent the observed thorium/*r* ratio in CS 22892-052 with 1 sigma uncertainty; the best-fit age is 18 Gyr, with an acceptable range from 14 to 22 Gyr. From Cowan et al. (1997).

The evolution of thorium is commonly expressed as a ratio to some other r-process nuclide that is stable; Eu is often the choice, since it is produced almost entirely by the r-process. Since Eu is stable, the ratio $N_{\mathrm{Th}}(t)/N_{\mathrm{Eu}}(t)$ will be steadily decreasing in time, that is, with the age of the Galactic disk, as the thorium decays. This is indicated in figure 7.18; the Galactic age determined from this star is 14–22 Gyr, in reasonable agreement with the age of the universe determined from the Hubble constant or from the anisotropy measurements of cosmic microwave background of about 14 Gyr (see chap. 9).

Although consideration of thorium in an old star would seem to provide the ideal chronometer, determining its abundance can be difficult, and systematic uncertainties can produce large errors in the result. Thus the ideal chronometer is actually thorium/uranium, since consideration of the ratio cancels out many of the systematic uncertainties. Unfortunately, both are difficult to observe in old stars (see, e.g., Cayrel et al. 2001; Hill et al. 2002), but interpretation of the ratio requires only minimal GCE (see below), and the two nuclides are near in mass to each other. However, use of this ratio does require an accurate description of the r-process, but that is a necessity for any of the chronometers.

The above model does not deal with a situation in which the chronometers were not produced as a single production event but possibly in a single event at some random Galactic time or in multiple randomly spaced production events over the age of the Galaxy. To accommodate these possibilities, at least formally, we need to return to our discussion of GCE in chapter 4 but to adapt it to include the cosmochronometers. In the following, we will follow fairly closely the discussion of Symbalisty and Schramm (1981). A similar description has been given by Cowan, Thielemann, and Truran (1991).

The equation that describes the rate of change of the density of nuclear species i having total Galactic mass N_i, where τ_i is the lifetime of nuclide i, is

7.6.3 $$dN_i(r,t)/dt = -N_i(r,g)/\tau_i + B(t,r,N_i), \text{ for } 0 < t < T$$

throughout the age of the Galaxy, where $B(t, r, N_i)$ is the generalized production and destruction function (excepting radioactive decay) and T is the total duration of nucleosynthesis. The function B may depend on time t, Galactic location r, and the abundance N_i. However, for this equation to be useful, one needs to develop the features that determine $B(t, r, N_i)$. To do so, consider the developments of GCE discussed in chapter 4:

7.6.4 $$dM/dt = F - E,$$

where, as in chapter 4, F is the Galactic infall rate, E is the rate at which gas is expelled in a Galactic wind, and M is the total mass of the Galaxy. The total mass of gas, g, is converted into stars at rate Ψ. Since we are interested in r-process chronometers, and only rather massive stars produce the r-process, we can safely assume instantaneous recycling. If we denote the fraction of the gas that goes into stars that is returned to the ISM as R, then

7.6.5 $$dg/dt = -\Psi(1 - R) + F - E.$$

To describe the evolution of a single nuclide, we use the mass fraction Z_i to indicate the abundance of nuclide i by $m_i = gZ_i$. Then we can write

7.6.6 $$dm_i/dt = -m_i/\tau_i - Z_i\psi(1 - R) + Y_{iS}\psi(1 - R) + Z_{iF}F - Z_{iE}E,$$

where the first term on the right-hand side has been added in to account for the fact that the nuclide being described may be radioactive, a fact that has not been included in the general description of the GCE of nuclides until now. In principle, nuclide i could be produced by decay of some other nuclides, but we will ignore that possibility for the moment. Y_{iS} is the mass of nuclide i produced and ejected by stars per unit mass retained by stars. Note that the Z_{iF} of the infalling material may well be very different from the Z_{iE} of the Galactic wind.

It is convenient to rewrite m_i as

7.6.7 $$m_i = Am_H N_i,$$

where A is the atomic weight of the radionuclide of interest and m_H is the mass of the hydrogen atom. Then equation 7.5.6 becomes

$$\textbf{7.6.8} \qquad dN_i/dt = -N_i/\tau_i - Z_i\psi(1 - R)/Am_H + Y_{iS}\psi(1 - R)/Am_H + (Z_{iF}F - Z_{iE}E)/Am_H.$$

Note that quantities such as Z_i, Y_{iS}, and Z_{iE} may be functions of time. Recognizing that $Am_H = gZ_i/N_i$, equation 7.5.8 can be rewritten as

$$\textbf{7.6.9} \qquad dN_i/dt = -N_i/\tau_i + Y_{iS}\psi(1 - R)/Am_H - N_i[Z_i\psi(1 - R)/gZ_i + (Z_{iF}F - Z_{iE}E)/gZ_i].$$

The first term to the right of the equal sign, of course, is the term that represents the decay of nuclide i. The second term represents stellar "production" of nuclide i, while the last term is the "production/depletion parameter" times N_i. This allows us to write

$$\textbf{7.6.10} \qquad dN_i/dt = -N_i/\tau_i + P_i\psi + \omega_i N_i.$$

For completeness, a general solution for this differential equation can be written as

$$\textbf{7.6.11} \qquad N_i = \exp[-(t/\tau_i + \nu_i)] \int_0^t \exp[t'/\tau_i + \nu_i(t')] P_i\psi(t')dt',$$

where

$$\textbf{7.6.12} \qquad \nu_i = \int_0^t \omega_i(t')dt'.$$

Specification of the various processes represented in the GCE model would, in principle, predict the GCE of the cosmochronometers. However, in lieu of that, with this discussion of GCE to guide us, several scenarios can be assumed in which the factors that can affect the abundances of the cosmochronometers can be specified. Formally, the nucleosynthesis occurs through the Y_{iS} term in equations 7.6.6 and 7.6.8, but this depends on the star formation rate, specifically that for massive stars, which synthesize the cosmochronometers through the r-process. Within any section of a galaxy the temporal distribution of such supernovae may not be uniform; indeed one model might be that a single event may have produced the chronometers.

The Galaxy is thought to have formed in roughly 1 Gyr after the big bang, the solar system is thought to have formed 4.7 Gyr ago, and the universe is thought to be about 14 Gyr old. That leaves roughly 8 Gyr over which a production event could have occurred, the actual time of which will obviously matter greatly to the abundance of a radionuclide with a half-life the order of 10 Gyr or less. Of course, there may have been many production events, which might lead to the other extreme, a fairly uniform series of production events. Indeed, this would appear to be a very plausible scenario, given that the r-process abundances shown in figure 7.8, presumably from a single r-process production event, are found to be about 1/40 of those observed of the solar r-process abundances. Although a number of factors could contribute to low r-process abundance values, that from a single star does suggest that a fairly large number of r-process events was necessary to synthesize the present-day r-process nuclides. The production-depletion term in equation 7.6.9 could also be important; not only can the cosmochronometers be swallowed by stars, but they can be diluted by galactic infall, which presumably is essentially primordial in abundance and so has essentially no metals, and the term describing the gas that is being expelled may well be very rich in cosmochronometers, since it is often associated with core-collapse supernovae (see, e.g., Galactic fountains, as discussed by Corbelli and Salpeter 1988).

In this context, a considerably more complicated case than that in which the abundance of a cosmochronometer in a single metal-poor star is used is that provided by the $^{238}U/^{235}U$ chronometers. Chemical selectivity can produce anomalous effects when trying to compare daughter and parent nuclides, but comparing ratios of isotopes of the same element can circumvent this complication. The relevant half-lives are ^{238}U, 4.47 Gyr; and ^{235}U, 0.70 Gyr. Since one can only guess at the nucleosynthesis history, one can make different assumptions as to what that is and then see how much that affects the result. What is better known, however, is that the r-process not only synthesizes these nuclides but also nuclides that decay to these two nuclides. For example, ^{238}U is made directly by the r-process, but so are ^{242}Pu (half-life $= 3.87 \times 10^5$ years, which α-decays to ^{238}U, and ^{246}Cm (half-life = 4800 years), which α-decays to ^{242}Pu and then to ^{238}U. The nucleus ^{250}Cm has a 10 % decay branch to ^{246}Cm, which ultimately decays to ^{238}U. The remainder of the time it fissions, producing much lighter nuclides. In the case of ^{235}U, ^{255}No (half-life = 3.1 minutes), ^{251}Fm (half-life = 5.3 hours), ^{247}Cf (half-life = 3.1 hours), ^{243}Cm (half-life = 28.5 y), and ^{239}Pu (half-life = 24,119 years), all decay ultimately to ^{235}U.

In order to determine the time that the two isotopes have existed since they were made, one needs to calculate the r-process production. One such estimate was made by Thielemann et al. (1983); it is $[^{235}U]/[^{238}U] = 1.24$–$1.42$ in detailed r-process calculations. For the cosmochronometer pair ^{232}Th and ^{238}U, which we will also consider, Thielemann et al. (1983) obtained an r-process ratio of $[^{232}Th]/[^{238}U] = 1.39$–$1.80$.

Tailoring equation 7.6.10 to the specific case, we can write for either isotope

7.6.13
$$dN_i/dt = -N_i/\tau_i + P_i\psi + \omega_i N_i + \sum_j N_j/\tau_j,$$

where the last term represents the decays from heavier isotopes discussed above. Since for both isotopes the heavier nuclides all decay with much shorter lifetimes than either ^{235}U or ^{238}U, the abundances of the heavier nuclides can simply be subsumed into those of either ^{235}U or ^{238}U.

Because we are taking ratios, many of the details that went into the above general discussion of GCE will cancel to a large extent. Furthermore, since we are taking a ratio of isotopes of the same element, chemical selectivities would be expected to be small. Thus, the terms $P_i\psi + \omega_i N_i$ are rewritten as $f(t, t_n, \lambda, T_R)$, where t_n refers to the time of the onset of nucleosynthesis, λ is the (time-dependent) rate of production of N_i, and T_R is the characteristic time of nucleosynthesis of N_i. In analyses such as these, time is often taken to run backward, so that the present time is 0 and the time T_S, the time of formation of the solar system, is 4.7 Gyr (see fig. 7.17). Then the relationship between the abundances of ^{235}U and ^{238}U at the two times is

7.6.14
$$[N(^{235}\mathrm{U})/N(^{238}\mathrm{U})]_{\mathrm{TS}} = [N(^{235}\mathrm{U})/N(^{238}\mathrm{U})]_{\mathrm{now}} \exp(T_S/\tau_{235} - T_S/\tau_{238}),$$

which gives a value of that ratio at T_S of 0.33, compared with its present-day value of 0.00725.

At this point the specific production scenario becomes important. In considering such, we will ignore the Galactic infall and outflow terms, assuming they will cancel to a large extent when we deal with the ratios.

Single Production Event Scenario

In this scenario, a single r-process event occurs at $t_n = T_S + \Delta$, so that the differential equation to be solved for nuclide i is

7.6.15
$$dN_i/dt = -N_i(t)/\tau_i + \Lambda_i\delta(t_n - t),$$

where Λ_i specifies the production history of nucleus i. The constraint on Λ_i is given, again assuming time is increasing toward the big bang, by

7.6.16
$$\Lambda_{235}/\Lambda_{238} = [N(^{235}\mathrm{U})/N(^{238}\mathrm{U})]_{\mathrm{TS}} \exp(\Delta/\tau_{235} - \Delta/\tau_{238}).$$

Of course, this expression is nothing more than a manifestation of the usual radioactive decay law. This case corresponds to the one discussed above for a single cosmochronometer ^{232}Th, except in this current situation one would be taking a ratio of cosmochronometer abundances.

When the ratio intersects the calculated abundance ratio at T_S, this gives the age of the Galaxy, subject to the caveat that the single production event is the correct description of the production history. This is not likely to be the correct picture, of course, unless one is looking at a single metal-poor star, as was done for CS 22892-052; r-process production is an ongoing Galactic process. If, however, one does assume this to be the correct picture, one achieves an age of the Galaxy of $T_G = 6.2$ Gyr. Although this is an interesting number, and it is certainly the correct order of magnitude, it is not likely to be correct. Note that the same procedure could have been followed for the two cosmochronometers ^{232}Th and ^{238}U (τ_{Th} is 20.27 Gyr). In this case the Galactic age determined would be roughly between 8 and 10 Gyr.

Uniform Production Scenario

In this scenario, the production is assumed to be uniform, beginning at t_n and continuing until the present time. Of course, only the production that occurs prior to the formation of the solar system is relevant to the radionuclides we would measure terrestrially, so the production is assumed to be uniform from t_n to T_S. Now the differential equation to be solved is

7.6.17
$$dN_i/dt = -N_i(t)/\tau_i + \Lambda_i.$$

Defining $t_n - T_S = \Delta =$ the time during which production will have occurred, the solution for Λ_i will be

7.6.18
$$\Lambda_i = N_i(T_S)/\{\tau_i[1 - \exp(-\Delta/\tau_i)]\}.$$

Thus the resulting equation for the ratio of the Λ_is is

7.6.19
$$(\Lambda_{235}/\Lambda_{238}) = [N(^{235}\mathrm{U})/N(^{238}\mathrm{U})]_{TS}(\tau_{238}/\tau_{235})[1 - \exp(-\Delta/\tau_{238})]/[1 - \exp(-\Delta/\tau_{235})].$$

Note that this expression applies at T_S; from T_S until the present the ratio of the abundances will decay according to the exponential decay law, since production for the solar system abundances will have ceased. Thus the abundance ratio will appear as in figure 7.19.

The increase in the ^{235}U/^{238}U abundance ratio as time decreases back to the big bang (or as the "historical time" increases back to the big bang) reflects the fact that the half-life of ^{238}U is larger than that for ^{235}U. The solution for the age of the Galaxy T_G in this scenario occurs when the calculated abundance ratio crosses the cross-hatched region, that is, when the abundance ratio is equal

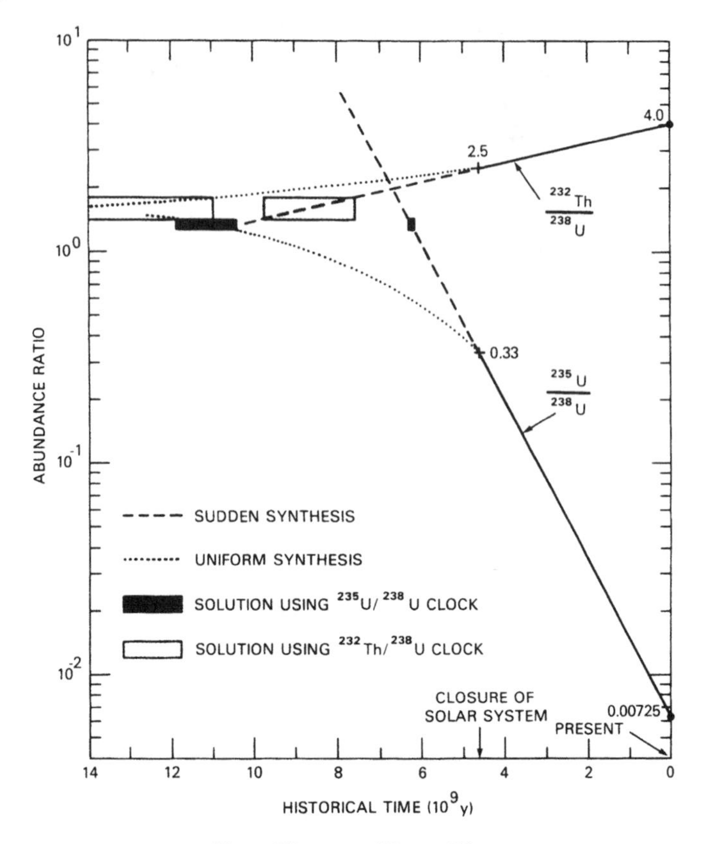

Fig. 7.19. Abundance ratio of ^{235}U to ^{238}U and of ^{232}Th to ^{238}U as a function of historical time using both a sudden synthesis scenario and a uniform synthesis scenario. The time prior to the closure of the solar system (T_S in the above discussions) is given by the solution to the differential equation 7.6.17, with the Λ_i given by equation 7.6.18. At that time, the abundance ratios must match those determined by extrapolating the current abundance ratios back to the beginning of the solar system. After that time the abundances decay according to the simple radioactive decay law. The block and open boxes to the left indicate the region estimated to be the r-process production ratio for the two nuclides; where those regions intersect the predicted r-process abundance ratio represents a possible solution to the age of the Galaxy. Those ages overlap for the two chronometers between 11 and 12 Gyr. From Rolfs and Rodney (1988). Courtesy of C. Rolfs.

to that predicted by r-process calculations. That age is around 10.5–12.0 Gyr. Given that this production scenario is considerably more realistic than the single production event scenario, except when one is dealing with a very old star, this is expected to be a more realistic value for the age of the Galaxy and a lower limit on the age of the universe. A similar approach could also be done for the chronometers ^{232}Th and ^{238}U; these are also shown in figure 7.19. The value obtained for this abundance ratio for the age of the Galaxy is seen to be >11

Gyr. Thus the two chronometers are seen to produce consistent values for the age of the Galaxy in the region of about 11–12 Gyr. Of course, other production scenarios could be, and have been, assumed, but they would not be expected to give results that would differ greatly from the uniform production scenario.

Other chronometers have also been used; for example, the pair of nuclides ^{187}Re and ^{187}Os. However, one of the major uncertainties affecting one of the nuclei often used as a chronometer, ^{187}Re, has been the possibility that it could undergo bound-state β-decay (as discussed in sec. 3.5). Atomic ^{187}Re β-decays to the ground state of ^{187}Os with the cosmologically interesting half-life of 4.35×10^{10} years. However, in bound-state β-decay, the emitted electron can go directly into one of the inner atomic shells rather than being expelled into the continuum. Since the electron then does not need to have enough energy to escape the attractive electrostatic potential well of the nucleus, it can have much less energy than would be required for decay from the atom. This effectively increases the Q-value of the transition, so it can have a huge effect on half-lives of nuclei that barely have enough energy to undergo β-decay or, in some cases, are even stable. Normally bound-state β-decay is forbidden by the Pauli principle, as the inner atomic shells are filled with the atomic electrons. However, when the nucleus is fully stripped of its electrons, as is the case in the high-temperature environments of stellar cores, bound-state β-decay becomes possible. In the case of ^{187}Re, the effective Q-value increase allows a transition to the first excited state of ^{187}Os, a much more favorable transition than that to the ^{187}Os ground state but one that is energetically forbidden for atomic ^{187}Re. The effect on the half-lives of a number of nuclei of this process was estimated theoretically (Takahashi et al. 1987), but only recently has it become possible to measure the effects of bound-state β-decay on the half-lives. This has been done by storing them in the heavy-ion storage ring at *GSI* (Germany) and then observing the decay with time of the magnitude of their circulating beam to determine the half-life. It was found to change from 4.35×10^{10} years for the atomic case to about 33 years (Bosch et al. 1996; Klepper 1997) for fully stripped ^{187}Re. This might well have a large effect on the determination of the age of the universe using ^{187}Re as a chronometer; even though the time the nuclei spent in the cores of stars was small, the change in the half-life is enormous.

CHAPTER 7 PROBLEMS

1. An r-process code is available on the Internet at http://nucleo.ces.clemson.edu/home/online_tools/three_phase_ism/. Run an r-process simulation using as initial parameters some of those typical of an r-process: $T_9 = 10$ and density and

entropy from the values in figure 7.10. Try several plausible values of the neutrino flux to observe its effects on the final distributions. Observe the effects of parameter variations on your results.

2. Show that equations 7.3.18–7.3.21 satisfy the differential equation given by equations 7.3.10–7.3.12.
3. Show that equation 7.6.14 satisfies the differential equation in equation 7.6.13.
4. Show that equation 7.6.16 satisfies the differential equation in equation 7.6.15.
5. Show that equation 7.6.11 is the solution of equation 7.6.10.

8

NUCLEOSYNTHESIS ON THE PROTON-RICH SIDE OF STABILITY, X-RAY BURSTS, AND MAGNETARS

8.1 Introduction

In contrast to the processes of nucleosynthesis that occur by neutron capture and operate on the neutron-rich side of stability, there are several processes that are lumped under the title of the p-process that synthesize nuclides on the proton-rich side of stability. Some of these processes occur by successive proton radiative captures and occasional accompanying positron decays or electron captures in a high-temperature hydrogen-rich environment. As such, they do not occur in the cores of stars, as the core temperatures are too low until later phases of stellar burning, prior to which the hydrogen has been consumed. However, they can occur in accretion onto a white dwarf ($T_9 = 0.1$–0.5) or neutron star ($T_9 = 1$–2), or in some of the burning scenarios that can occur in type I or II supernovae. These processes are interesting in their own right. For example, the rapid proton capture process, or rp-process, is one such mode of nucleosynthesis; it operates by successive proton captures and occasional β^+ decays or e^- captures. However, the Coulomb barrier is too high for the rp-process to operate on the heavier nuclei, but the γ-process, at temperatures of several billion K, will drive the heavy-seed nuclei toward the proton drip line via photonuclear reactions initiated with the high-energy photons in such an environment, such as (γ, α) and (γ, n), to also synthesize the proton-rich nuclides. Thus the rp-process generally occurs together with the γ-process. The γ-process is apparently responsible for synthesizing most of the heavy p-process nuclei, or p-nuclei, those nuclei that are blocked from production by the r- or s-processes by the presence of stable nuclides that prevent β-decay to them, while the rp-process appears to be responsible for synthesizing some of the lighter p-nuclei.

Both of these processes are among the ways in which p-nuclei are synthesized. The p-nuclei definition given above immediately suggests the major task of any p-process description: to overcome the inability of β-decays to convert the usually more abundant r- and s-process nuclides into the p-nuclides. The rp- and γ-processes do exactly that, and so they are intimately tied in with the synthesis of the p-nuclei. Two recent reviews have been written on the p-nuclides, one by Lambert (1992) that emphasizes observational aspects, and one by Boyd in Wallerstein (1997) that emphasizes more nuclear features. We will return to the characteristics of the rp- and γ-processes below.

However, there are some problematic p-nuclei that do not seem to be produced by either of these processes; this has led to development of other possible processes. The α-rich freezeout model in type II supernovae has suggested a new mechanism for synthesis of light p-nuclides that would occur in nuclear statistical equilibrium in the neutrino wind resulting from supernova collapse. Other p-nuclides may be synthesized by reactions involving neutrinos in either the ν-process or the νp-process. These processes will also be discussed in this chapter.

The modern formulation of the rp- and γ-processes was proposed by Wallace and Woosley (1981) primarily as a description of the nucleosynthesis that would occur in a fairly high temperature hydrogen-rich environment. However, recent work by Strohmayer and Bildsten (2005), Schatz et al. (1998, 2001), Arnould and Goriely (2003), and Woosley et al. (2004) has greatly elucidated these processes, notably by providing the essential details about the sites in which they must occur. In any version of these processes, the seed nuclei contained in the rp- and γ-process environment are driven close to the proton drip line by successive proton radiative captures or photo dissociations, from which they β-decay (positron decay or electron capture) after the high temperatures have subsided. In the Wallace-Woosley description, temperatures as high as $T_9 \approx 2$ were considered. It was found that, even at temperatures appreciably less than T_9 of 1, significant abundances of ^{74}Se and ^{78}Kr could be produced. At lower temperatures, a wide variety of lower mass nuclei can be synthesized, depending on the thermodynamic and seed conditions.

The high temperatures being considered in the rp- and γ-processes change even the basic structure of the processes of nucleosynthesis. For example, the CN and CNO cycles discussed in chapter 5 are changed profoundly; they become the hot CNO cycle. This, and the low-temperature rp-process, could result from accretion from a star onto a companion white dwarf; this would produce temperatures of the infalling matter of up to as high as $T_9 \approx 0.5$. If the accretion occurred onto a neutron star the temperature would be expected to reach $T_9 = 1$ or 2, a result of the much higher gravitational energy associated with the (much smaller) neutron star. If that sort of temperature lasted for tens of seconds, it could drive the higher mass rp-process up to about 100 u. This situation would not be expected to occur from accretion onto a white dwarf because the charged particles would not have sufficient energy at the temperatures typical of those for matter accreted onto a white dwarf to overcome the Coulomb barriers presented by medium-mass nuclei. But they would have sufficient energy from accretion onto a neutron star. In either case, the accreted material will also contain the "catalyst" nuclei necessary to run, for example, CNO cycles. Much of the basic astrophysics we have already discussed will be relevant to these processes. However, the peculiarities of each environment change the details of each process.

It should be emphasized that all of the information obtained about the abundances of the p-nuclei, and hence about many of the details of the processes that produce them, result from terrestrial sources; for the most part, astronomical observations cannot resolve lines from different isotopes. Additional details on abundance determinations can be found in the review article on the p-nuclei by Lambert (1992). In particular, he notes that most of the information on the p-nuclear abundances has been derived from studies of meteoritic samples.

The generally accepted list of p-nuclides is given in table 8.1. With few exceptions, most notably the Mo and Ru p-nuclides, their common characteristic is their rarity with respect to the other isotopes of their respective element. As indicated in table 8.1, some of them can also be made by r- or s-processes, but those processes are not thought to contribute greatly to any particular p-nuclide's abundance.

8.2 White Dwarfs and Novae

Subramanyan Chandrasekhar predicted that there was a limiting mass for white dwarf stars (see chap. 4): no white dwarf could be stable against gravitational collapse if it exceeded this mass, which is about 1.4–1.5 solar masses, depending on its detailed composition. Although this idea was not well received at the time it was proposed, it was later accepted within the astrophysics community, and Chandrasekhar eventually won a Nobel Prize for his theoretical contributions to astrophysics. A white dwarf may have an extremely high surface temperature. However, despite the high temperature, white dwarfs are dying stars; they have used all their nuclear fuel and can no longer produce energy by thermonuclear processes.

However, if the white dwarf has a companion star that is sufficiently close that the white dwarf can accrete matter from it, the result can be a nova. In

Table 8.1 The p-Process Nuclides

Element	Z	A	N_A [a]	$N_A/\Sigma N_A$	Comments
Se	34	84	0.55	0.88	
Kr	36	78	0.15	0.34	
Sr	38	84	0.13	0.56	
Nb	41	92	0.0	0.0	β-decays to ^{92}Mo; $T^{b}_{1/2} = 3.5 \times 10^{7}$ years
Mo	42	92	0.38	14.84	
		94	0.24	9.25	Also possibly produced by s-process
Ru	44	96	0.10	5.52	
		98	0.035	1.88	
Pd	46	102	0.014	1.02	
Cd	48	106	0.020	1.25	
		108	0.014	0.89	
In	49	113	7.9×10^{-3}	4.3	Also produced by r- and s-processes
Sn	50	112	0.037	0.97	
		114	0.025	0.66	Also produced by s-process
Te	52	120	4.3×10^{-3}	0.09	
Xe	54	124	5.7×10^{-3}	0.12	
		126	5.1×10^{-3}	0.11	
Ba	56	130	4.8×10^{-3}	0.11	
		132	4.5×10^{-3}	0.10	
La	57	138	4.1×10^{-3}	0.09	β-decays to ^{138}Ce; $T^{b}_{1/2} = 1.5 \times 10^{11}$ years
Ce	58	136	2.2×10^{-3}	0.19	
		138	2.8×10^{-3}	0.25	
Sm	62	144	8.0×10^{-3}	3.10	
		146	0.0	0.0	α-decays to ^{142}Nd; $T^{b}_{1/2} = 1.03 \times 10^{8}$ years
Dy	66	156	2.2×10^{-4}	0.06	
		158	3.8×10^{-4}	0.10	
Er	68	162	3.5×10^{-4}	0.14	
Yb	70	168	3.2×10^{-4}	0.13	
Hf	72	174	2.5×10^{-4}	0.16	
Ta	73	180	2.5×10^{-6}	0.01	Actually ^{180}Ta^{m}; $T^{b}_{1/2} = 1.2 \times 10^{15}$ years. β^{-}-decays to ^{180}W or e^{-}-captures to ^{180}Hf. Also produced by s-process.
W	74	180	1.7×10^{-4}	0.13	Also possibly produced by s-process
Os	76	184	1.2×10^{-4}	0.02	
Pt	78	190	1.7×10^{-4}	0.01	
Hg	80	196	5.2×10^{-4}	0.15	

[a] From Anders and Grevesse (1989).

[b] From Tuli (2000).

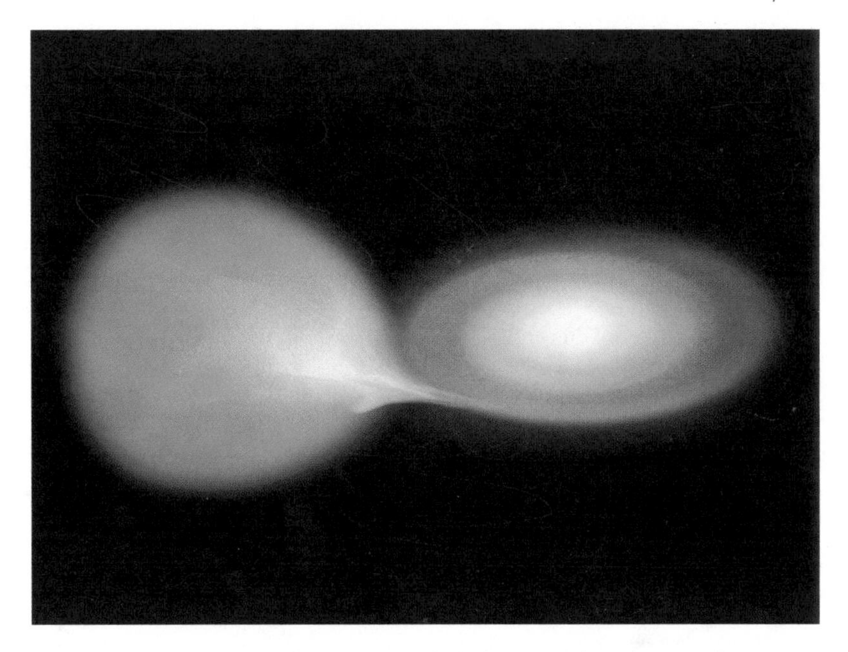

Fig. 8.1. Artist's conception of the accretion of a hydrogen-rich layer from a red giant companion star onto the surface of a white dwarf, producing an accretion disk (or ring) around the central star. Reprinted courtesy of NASA/CXC/SAO.

this situation, matter from the outer portion of the star accretes in a thin layer onto the surface of the white dwarf and eventually ignites in a thermonuclear explosion possibly, though not necessarily, under (electron) degenerate conditions.

The result of this accretion process is an explosion that blows a thin surface layer off into space, causing a sudden large rise in light output from the system. The accretion process is illustrated in figure 8.1.

With a neutron star, the accretion process again forms a thin layer on the surface of the star. Thermonuclear runaway occurs, and an explosion results. However, the higher temperature involved increases the energetics of all aspects of the process, and the result is an X-ray burst. Much effort has gone into understanding these events; we will return to some of the details of X-ray bursts at the end of this chapter.

8.3 The Hot CNO Cycles

One result of the elevated temperatures that result from accretion onto a white dwarf is that the normal CNO cycle, discussed in chapter 5, is modified by

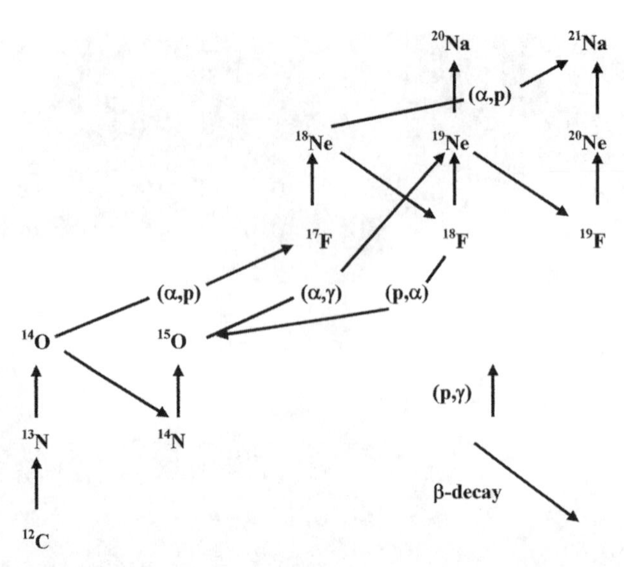

Fig. 8.2. Possible pathways of the hot CNO cycles, along with the breakout paths via $^{15}O(\alpha, \gamma)^{19}F$ and $^{18}Ne(\alpha, p)^{21}Na$. Most of the nuclei indicated are radioactive; only ^{12}C, ^{14}N, ^{19}F, and ^{20}Ne are stable.

the high temperature, producing the hot CNO, or HCNO, cycle. The details of this were first suggested by Hoyle and Fowler (1960) and Fowler and Hoyle (1964), extended considerably by Wallace and Woosley (1981), and reexamined by Champagne and Wiescher (1992); they are shown schematically in figure 8.2. There it is seen that the assumption made for normal hydrogen burning, that the β-decays always occur on a timescale that is short compared with the timescale for proton captures, no longer applies. The cycle proceeds directly from ^{12}C to ^{14}O without waiting for the β-decay of ^{13}N, even at temperatures as low as $T_9 = 0.2$. Since ^{15}F is proton unbound, the HCNO cycle must then either await the β-decay of ^{14}O ($t_{1/2} = 70.6$ s) or find some other way around the impasse. One possibility is the $^{14}O(\alpha, p)^{17}F$ reaction, by which nuclei can break out of the light HCNO cycle; this is thought (Champagne and Wiescher 1992) to occur at temperatures around $T_9 = 0.4$ or higher. This could be followed by $^{17}F(p, \gamma)^{18}Ne$, which could then either undergo an (α, p) reaction to ^{21}Na; or a β-decay to ^{18}F, which would most likely be followed by an $^{18}F(p, \alpha)^{15}O$ reaction. Another possibility is β-decay of ^{14}O followed by proton capture to ^{15}O, then an (α, γ) reaction to ^{19}Ne, a branch thought to occur at even higher temperatures, around $T_9 = 0.5$. These pathways are indicated in figure 8.2.

However, the detailed understanding of these processes will require a great deal of experimental effort. These reaction pathways are considerably less well established than those for lower temperature hydrogen burning are. In all cases of interest, the nuclei that would normally be the targets in experiments have very short half-lives, so they become the beam in an RNB experiment. Measurement of these reactions has become an industry in recent years, although the studies are still difficult and the data are rarely of the quality to which nuclear astrophysicists are accustomed. Beam intensities are the obvious reason for the deficiency, although it should also be noted that the energies that are relevant to the HCNO cycle reactions are considerably higher than those that are relevant to normal hydrogen burning, so the Coulomb barrier presents less of a problem than it would for study of reactions of interest to lower temperature burning scenarios.

One HCNO cycle reaction that has been studied in detail is $^{13}N(p, \gamma)^{14}O$. This reaction is dominated by a resonance at 541 keV, which makes its measurement considerably easier than it might otherwise be. The strength of the resonance has been measured three times, once by a direct measurement, using the ^{13}N beam from the Louvain-la-Neuve RNB facility (Decrock et al. 1991), and twice by the Coulomb breakup technique. One of those measurements was performed by Motobayashi et al. (1991), at the Institute for Physical and Chemical Research, RIKEN, in Japan, and the other by Kiener et al. (1993), at Grand Accelerateur National d Ions Lourds, GANIL, in France. The latter approach enjoys some enormous advantages over direct measurement. As discussed in chapter 3, it determines the cross section for the reaction of interest by studying the inverse reaction. It does so by bombarding a target of a heavy nucleus, for example, lead, with a beam of the nucleus of the product of the reaction of interest, ^{14}O in this case. The photon field resulting from the interaction of the two nuclei serves to break up the beam nuclei, producing as reaction products the two particles that would be in the incident channel of the astrophysical reaction of interest. The yield is large for the Coulomb breakup process, so use of RNBs is not necessarily a detriment to obtaining good data. It should be noted that in cases where excited states of the product nucleus of the astrophysical reaction can contribute to the yield, the Coulomb breakup approach will not provide the entire answer desired, as it is not sensitive to those states. Furthermore, it may excite states in what would be the target nucleus for the astrophysical reaction, so good resolution may be required to determine the desired information. However, for the $^{13}N(p, \gamma)^{14}O$ reaction, this is not a problem; ^{14}O has only one state, the ground state, that is proton bound. Therefore, this technique should produce results that should be applicable to this astrophysical reaction. Thus two studies were performed with ^{14}O RNBs: one using the facilities of RIKEN, and the other using those at GANIL.

The three experiments gave the following results for the resonance strength. That of DeCrock et al. (1991), which was the direct measurement, gave 3.1 ± 1.2 eV. The two Coulomb breakup measurements gave 3.1 ± 0.6 eV (Motobayashi et al. 1991) and 2.4 ± 0.9 eV (Kiener et al. 1993). The agreement among the three results is excellent at least at this level of accuracy.

There have been several studies aimed at understanding the $^{14}O(\alpha, p)^{17}F$ reaction, primarily because of its importance as a breakout reaction from the HCNO cycle. Two of these (Harss et al. 1999; Blackmon et al. 2001) were studies of the $^{17}F(p, \alpha)^{14}O$ inverse reaction. These are interesting experiments in themselves, as ^{17}F is an RNB. Furthermore, use of the inverse reaction is further complicated in this case by the fact that the $J^{\pi} = 1/2^{+}$ first excited state in ^{17}F is expected to contribute to the $^{14}O(a, p)^{17}F$ reaction, so its contribution needs to be taken into account in inferring the rate of that reaction. However, inelastic scattering measurements in $^{17}F(p, p')$ studies (Harss et al. 2002; Blackmon et al. 2003) allowed an estimate of its effect on the overall reaction rate. Finally, a direct measurement of the $^{14}O(\alpha, p)^{17}F$ cross section was performed (Notani et al. 2002) using an ^{14}O beam and a cryogenic helium gas target. This utilized the *CRIB* facility at RIKEN (see chap. 2) to produce the ^{14}O beam from a $^{1}H(^{14}N,^{14}O)n$ reaction and separate it from the elastically scattered beam and then direct it to a ^{4}He target where the desired reaction products are observed. Unfortunately, the results of all these experiments do not yield the agreement that might be hoped for, so more work needs to be done to resolve the discrepancies.

Should the $^{14}O(\alpha, p)^{17}F$ reaction provide a breakout reaction from the HCNO cycle, the next reaction in the chain, $^{17}F(p, \gamma)^{18}Ne$, will also be critical in moving the initial nuclei to higher mass. This reaction appears to depend critically on the possible existence of a $J^{\pi} = 3^{+}$ state, which is expected from the ^{18}O mirror nucleus. This state was first identified in experiments performed at the Oak Ridge National Laboratory using their ^{17}F beam (Bardayan et al. 1999, 2000); data from that work, along with the result of the analysis, are shown in figure 8.3. Although this work has determined the parameters of this state, more work will be needed to pin it down with sufficient accuracy to define the rp-process through this mass region.

A vast amount of work has been done using nuclear reactions on stable nuclei to infer the information needed to determine indirectly the reaction rates for the reactions important to the HCNO cycle and the low-mass rp-process. This work was reviewed by Champagne and Wiescher (1992) and will not be repeated extensively here. However, some comments on this work are relevant to this treatise in that they can illustrate the considerations necessary

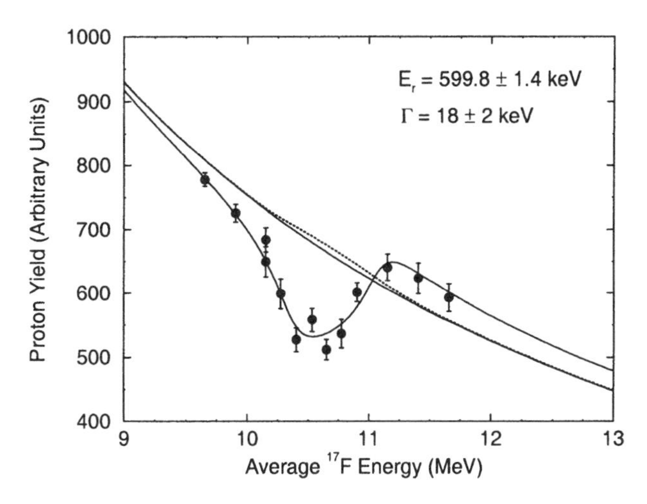

Fig. 8.3. Excitation function of ^{1}H (^{17}F, ^{17}F) elastic scattering, along with the fit to the data, in the region of the ^{18}Ne(3^+) state. Reprinted with permission from Bardayan et al. (2001). Copyright 1999 by the American Physical Society.

for understanding the reactions that are important to nuclear astrophysics. Specifically, nuclear structure information on ^{20}Na is crucial for understanding possible breakout of the HCNO cycles via ^{19}Ne(p, γ)^{20}Na. Ideally, one would like to infer the cross section for such a reaction by studying a proton-stripping reaction, such as (d, n) or (^{3}He, d), as these must be sensitive to the same terms in the wave function as would participate in a (p, γ) reaction. However, this is very difficult for this particular reaction. Thus, it has been studied by the ^{20}Ne(^{3}He, ^{3}H)^{20}Na reaction (Kubono et al. 1988). The ^{19}Ne(p, γ) threshold is at an excitation energy of 2.20 MeV in ^{20}Na, and the Gamow window is at 130 keV (244 keV) at $T_6 = 150$ ($T_6 = 400$). Thus levels from 100 to 300 keV above the threshold or 2.3 to 2.5 MeV in the center of mass will be important to this reaction at the temperatures at which the low-temperature rp-process operates, especially if they can be populated by s-wave proton capture. A level was found at 2.64 MeV in ^{20}Na which, given the width of the Gamow window (244 keV at $T_6 = 400$), might well influence the reaction network for the higher end of the temperature range associated with novae resulting from accretion onto a white dwarf. The ^{19}Ne(p, γ)^{20}Na reaction has also been studied (Michotte et al. 1996) using a ^{19}Ne beam from the Louvain la Neuve RNB facility. While the yield was too small to observe gamma rays, the experiment was able to establish an upper limit on any resonant strength for that reaction.

Another reaction, which might follow ^{19}Ne(p, γ)^{20}Na, is ^{20}Na(p, γ)^{21}Mg. This has a threshold at an excitation energy of 3.22 MeV in ^{21}Mg, and a Gamow window at 170 keV (325 keV) at $T_6 = 150$ ($T_6 = 400$). Thus, levels at an excitation energy of roughly 3.39–3.55 MeV may have an important influence on this reaction at rp-process temperatures. In this case, the $^{24}Mg(^3He,^6He)^{21}Mg$ reaction indicated (Kubono et al. 1992) a level at 3.216 MeV, which, given the width of the Gamow window (144 keV at $T_9 = 0.15$ and 325 keV at $T_9 = 0.4$), could exert significant influence on this reaction over the entire temperature range of the rp-process as it applies to novae.

Additional work with RNBs has been directed at reactions on ^{18}F, with two experiments studying the $^{18}F(p, \alpha)$ reaction. Rehm et al. (1996) did direct measurements of the cross section for that reaction through a resonance at about 640 keV. While the experiment was difficult, given the intensity of the beam, it did establish the existence of the resonance and gave some indication of its width. Unfortunately, the energy broadening intrinsic to the experiment was large and had to be removed from the observed width of the resonance to infer its natural width. The resulting value, 13.6 ± 4.6 keV, was considerably less than either the observed width or the corrections to it. A second experiment (Coszach et al. 1995) utilized an ^{18}F beam from the RNB facility at Louvain la Neuve but attempted to determine the width of the resonance by elastically scattering the ^{18}F beam from a hydrogen target. The energy of the resonance they observed is within error bars of that seen by Rehm et al. (1996), but they find a considerably larger width, 37 ± 5 keV, for the resonance. That larger value is also consistent with widths for the corresponding state inferred from other nuclear reaction measurements. Finally, the Louvain la Neuve group (Graulich et al. 1997) also studied a lower energy region, 265–535 keV. They found a resonance at 324 ± 7 keV, which is consistent with a state seen in the $^{20}Ne(^3He, \alpha)^{19}Ne$ reaction. Each of these experiments was difficult; all were just at the margin of what can be done with existing technology. The RNB facilities of the future, which should have considerably higher intensity beams, should readily resolve any outstanding issues on this and other reactions.

A particularly interesting reaction study is that of the $^{21}Na(p, \gamma)^{22}Mg$ reaction, performed (D'Auria et al. 2004; Azuma et al. 2003) using the Isotope Separation and Acceleration facility (see chap. 2) RNB facility at TRIUMF, along with the complex reaction product analysis system Detector of Recoils and Gammas of Nuclear Reactions, or DRAGON. The latter system was designed to analyze the forward-going reaction products while eliminating the incident beam with very high efficiency. This experiment utilized a ^{21}Na RNB to perform this reaction study, which provided an excellent mapping of the resonance shown in figure 8.4 but which is also a precursor of experiments to be done in the future.

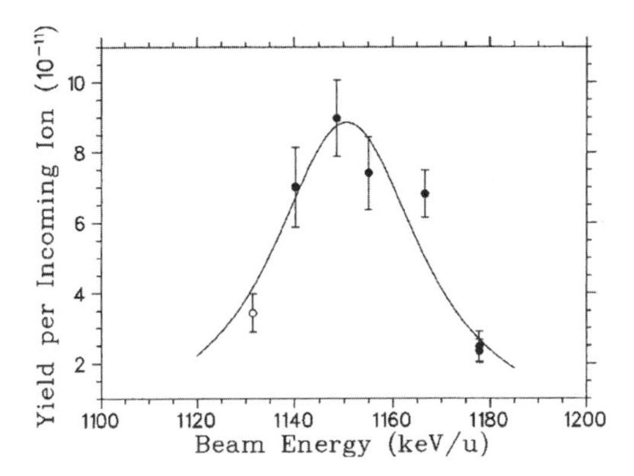

Fig. 8.4. The experimental yield curve, which is proportional to the cross section of the 1101-keV resonance in the $^{21}Na(p, \gamma)^{22}Mg$ reaction as a function of the beam energy entering the gas target. The measurement was made using the DRAGON and *ISAC* facilities at TRIUMF. The solid curve is the resulting least squares fit of the yield function to the data. Reprinted with permission from D'Auria et al. (2004). Copyright 2004 by the American Physical Society.)

More recent work (Schatz et al. 1998) has measured the properties of excited states of ^{24}Si, of importance to a possible pathway, involving two-proton captures on ^{22}Mg, which might reduce the yield of ^{22}Na in novae if the conditions were conducive to two-proton captures. This latter experiment would have implications to the observation of signatures of ^{22}Na from gamma-ray astronomy. Another situation involving two-proton capture is discussed in section 8.6.

Some spectroscopic information might be obtained even for reactions that cannot be studied via single proton stripping by comparing the final-state nuclei with their analog nuclei, which have often been studied via nuclear reactions. In the case of the rp-process, this usually means comparing levels in unbound nuclei with those that are bound in the analog nuclei. This necessitates use of an energy shift to identify the analog states, the Thomas-Ehrman shift, the calculation of which (Thomas 1952; Wiescher, Gorres, and Thielemann 1988) is straightforward, albeit nontrivial (see discussion in sec. 3.8).

8.4 The Lower Temperature rp-Process

The rp-process follows in mass the HCNO cycles. Its details have been studied recently by Champagne and Wiescher (1992) and by Thielemann et al. (1994), and for the higher mass parts (to which we shall return below) by Schatz et al.

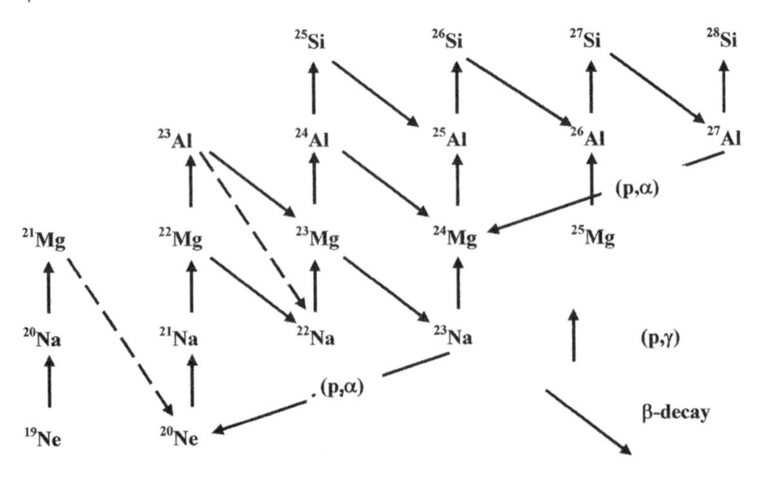

Fig. 8.5. Possible reaction pathways through the mass 19- to 28-u nuclei. For this figure it was assumed that β-decays with half-lives greater than 4 s would not have time to occur, that is, that those nuclei would undergo a reaction. Note, however, that the half-lives for ^{23}Mg and ^{27}Si are 11.32 and 4.16 s, respectively; their decays were indicated to close the NeNaMg and MgAlSi cycles. Note also that ^{26}Al has an isomeric state, with a half-life of 6.3 s that decays to ^{26}Mg; this represents a potential nucleus through which reactions can pass that is not indicated in the figure. Finally, ^{21}Mg and ^{23}Al can decay by β-delayed proton emission, which changes the baryon number by 1 unit and the charge by 2 units.

(1998). As shown by Thielemann et al. (1994), the rp-process occurs in many instances through cycles involving three successive proton radiative captures, two β-decays, and, to complete the cycle, a (p, α) reaction. However, myriads of branches also exist, so that the details of the path of this process are very complex and could require many reaction studies for satisfactory elucidation. While many of these are planned for the near future using RNB facilities, the only studies to date have been of the nuclei themselves through which the rp-process path proceeds.

Of specific interest is the mass region from about 20 to 30 u, since it impacts isotopic anomaly data taken from meteoritic sample studies of the Ne and Mg isotopes and also data from gamma-ray astronomy. Some of the nuclei and reactions that might be involved in a network for this mass region are shown in figure 8.5. There it can be seen that, if the temperature is not too large (see, e.g., Champagne and Wiescher 1992), NeNaMg and MgAlSi cycles can operate much like the CN cycle, with four protons being converted to an α-particle in each cycle. The number of reaction paths is sufficient that, at different temperatures, the nucleosynthesis from those cycles could be quite different (see, e.g., Buchmann et al. 1984).

Isotopic anomaly data (Lee, Wasserburg, and Papanastassiou 1976, 1977), shown for ^{26}Mg in figure 8.6, confirm that nucleosynthesis is actively going on in the Galaxy. Because the ^{26}Mg abundance seen in the meteoritic samples so clearly track those of ^{27}Al, some Al was originally included in the sample as ^{26}Al and then β-decayed to ^{26}Mg to produce the abundance anomaly. The rather short half-life of ^{26}Al, 7.4×10^5 years, confirms that active nucleosynthesis processes have been going on fairly close to our solar neighborhood.

This is also consistent with the observation of 1.809-MeV gamma-ray lines observed in several gamma-ray satellites. These are produced by the β-decay of ^{26}Al, which proceeds as follows:

8.4.1 $$^{26}\mathrm{Al}\ (T_{1/2} = 7.2 \times 10^5 \text{ years}) - \beta \rightarrow {}^{26}\mathrm{Mg}^*.$$

Most of the decays to ^{26}Mg go to its first excited state, indicated as $^{26}\mathrm{Mg}^*$. It immediately decays by gamma-ray emission, which can be detected. The β-decay of ^{26}Al to the first excited state of ^{26}Mg instead of the ground state can be easily understood in terms of the rules for β-decay transitions given in chapter 3. The ground state of ^{26}Al has a spin/parity of 5^+, so in order for the decay to go to the 0^+ ^{26}Mg ground state, $5\hbar$ of angular momentum would have

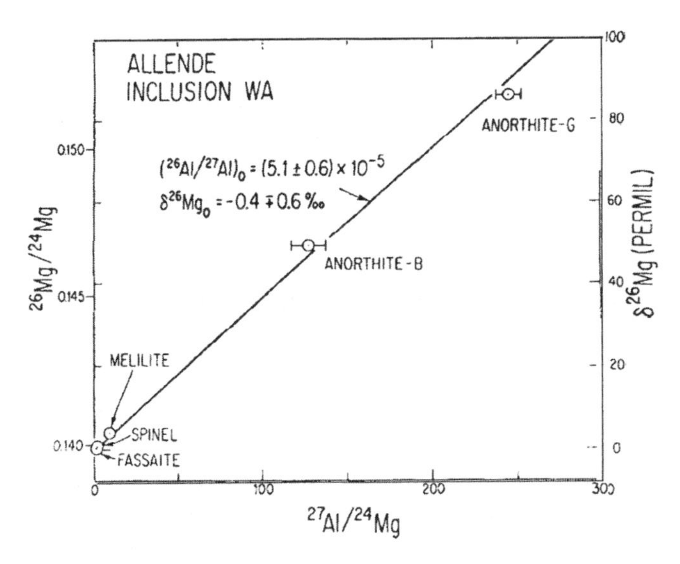

Fig. 8.6. The correlation of the ^{26}Mg abundances with the Al abundance. These data came from different minerals from the same (Allende meteorite) inclusion. The linear dependence suggests that the ^{26}Mg excess must originally have been included in the meteorite as ^{26}Al and decayed subsequent to its formation. From Lee, Wasserburg, and Papanastassiou (1976, 1977).

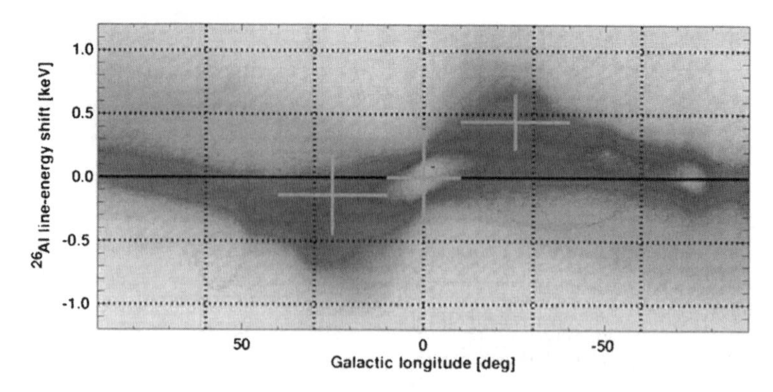

Fig. 8.7. Energy shift of the 1.809-MeV line, which results from the decay of ^{26}Al, distributed over the Galactic coordinates. The red- and blueshifts are consistent with the Galactic rotation at the 94 % confidence level. The error bars indicated are standard deviations. Reprinted by permission from Macmillan Publishers Ltd., copyright 2006.

to be transferred; this would be a greatly inhibited transition. Although the decay to the 2^+ ^{26}Mg first excited state is also inhibited, it is certainly much less inhibited than that to the ground state. Indeed, the required angular momentum transfer to the first excited state of $3\hbar$ is the reason that ^{26}Al has as long a half-life as it does.

The gamma rays from the decay of ^{26}Mg* were first seen by the *HEAO* satellite (Mahoney et al. 1984), studied considerably more extensively with the *Compton Gamma Ray Observatory* (Diehl et al. 1995; see fig. 2.14), and most recently with *INTEGRAL* (see sec. 2.6.5 and Diehl et al. 2005). In the context of ^{26}Al, there has been a longstanding question as to whether the source of ^{26}Al is novae or supernovae or some combination thereof. Diehl et al. (1995) mapped the distribution of the galactic sources of ^{26}Al to attempt to answer this question. If novae were the source, the ^{26}Al would be produced in small amounts by many sources, whereas if supernovae were the source, each would be expected to produce a large amount but the number of sources would be small. The distribution was found to be clumpy, as would be more consistent with massive stellar sources such as core-collapse supernovae than with novae. It was also found that there was less ^{26}Al in the galaxy than had been previously suspected (Clayton 1984; Clayton and Leising 1987), as a considerable amount of the observed flux seemed to originate with the Vela supernova remnant, which is relatively close to Earth. *INTEGRAL* has achieved a level of precision in its gamma-ray observations that it can now observe the slight line shifts in gamma rays that indicate the rotation of the Galaxy; these are shown in figure 8.7. In any event, with these measurements, gamma-ray astronomy has unquestionably proved its worth as an astrophysical diagnostic tool.

It should also be noted in the reaction network for the MgAlSi cycle, shown in figure 8.6, that ^{26}Al has an isomeric state, at 228 keV excitation, that could have a large influence on the reaction processes in that mass region. The isomeric state has a much shorter half-life, 6.3 s, than does the ground state, so excitations to that state in high-temperature environments in which the ^{26}Al is undoubtedly produced could dramatically change the effective half-life. In addition, nuclear reactions through that state could have a large impact on the nucleosynthesis of ^{26}Al. However, since even the ground state is radioactive, any observation of reactions on ^{26}Al is difficult. Nonetheless, data for proton capture on the ground state do exist (Buchmann et al. 1984).

The "ashes" of the rp-process will be dominated by the nuclear "waiting points," which can occur for several reasons. These are the nuclei on which researchers have focused their experimental efforts. One such point occurs around ^{32}Cl. Proton capture on it produces ^{33}Ar, which cannot capture another proton, as ^{34}K is proton unbound. ^{33}Ar has two states that could produce resonant proton capture, which would greatly affect the rate at which the $p+{}^{32}Cl$ reaction will proceed. An experiment (Clement et al. 2004) in which a hydrogen target was bombarded with an ^{34}Ar beam (half-life = 845 ms) was conducted at the Michigan State University NSCL (see chap. 2), with the deuterons and ^{33}Ar produced in some of the reactions being detected, along with gamma rays from the decay of ^{33}Ar excited states. This produced information, most notably the excitation energies of ^{33}Ar levels, that reduced the uncertainty on the reaction rate for the $^{32}Cl(p, \gamma)$ reaction from a factor of 10^4 to a factor of 3. This, of course, greatly refines the predictive capability of rp-process models in this mass region.

Several other nuclei near the proton drip line through which the rp-process passes can delay its progression to higher masses because of the low binding energies of the nuclei produced in subsequent proton captures, less than 1 MeV in some cases. These nuclei will produce waiting points; at the relatively high temperatures of the rp-process, photonuclear reactions will quickly destroy nuclei with low binding energies, preventing successive proton captures. This feature, discussed in detail by Champagne and Wiescher (1992), will undoubtedly focus the efforts of experimentalists using RNB facilities to determine the extent to which such nuclei inhibit the progression of the rp-process. Specifically, the proton capture reactions on ^{22}Mg, ^{26}Si, ^{30}S, ^{34}Ar, ^{42}Ti, and others produce very loosely bound nuclei, so these nuclei probably have to undergo β-decay before the rp-process can proceed past them.

In addition, as also noted by Champagne and Wiescher (1992), and discussed for the specific case of ^{33}Ar, some bottleneck reactions resulting from proton capture on nuclei can inhibit the rp-process flow for another reason. This is because they have very low reaction Q-values, so they will involve the low-excitation region of the final state nuclei. This would apply to proton captures on nuclei such as ^{23}Mg, ^{27}Si, ^{31}S, and ^{35}Ar. Nuclei in this mass region usually have a low level density, so one would expect few resonances to enhance capture reaction cross sections to allow these reactions to compete strongly with β-decay, or even (p, α) back to lower masses. Thus these reactions would also be expected to slow the progress of the rp-process.

However, understanding the rp-process in totality requires understanding all the thermonuclear reaction rates that might occur on the nuclei through which the rp-process path runs. Although much information has been obtained, clearly much work will be necessary before this important process is well understood.

8.5 Neutron Stars and X-Ray Bursts

An astronomical development that has occurred in the past decade that has resulted in a rich bounty of observational phenomena, a concomitant theoretical understanding, and numerous opportunities for nuclear astrophysicists is that of X-ray bursts. The recent developments have been driven by the advent of X-ray satellites: specifically the *Rossi X-ray Timing Explorer* (*RXTE*) (see sec. 2.6.2) and the *BeppoSAX* mission (see sec. 2.6.3), but also other satellites that were sensitive to X-rays. The general category discussed here will be low-mass X-ray binaries (LMXBs). These are well documented to be the result of matter being accreted by a neutron star from a companion star into an accretion disk (see fig. 8.1, but with the white dwarf replaced by a neutron star) with orbital periods ranging from 11 minutes to many days. In so doing, the matter falls into the neutron star's very deep gravitational well, releasing roughly 200 MeV per nucleon. The surface of the neutron star is at a temperature of 10^7 K, which is consistent with the gravitational energy that is released. X-ray bursts often result from these conditions; they are indicative of thermonuclear burning and have the following properties (Strohmayer and Bildsten 2005):

- They generally produce of order 10^{39} ergs of energy in some tens of seconds. This is considerably less than, for example, a supernova, which produces $\sim 10^{53}$ ergs of neutrinos in about 10 seconds, but, as will be discussed below, the X-ray bursts repeat every few hours.

- Burst rise times are generally shorter than their decay times; roughly 2 s compared with the tens of seconds.
- Burst profiles are shorter at higher energies, a result of the cooling of the neutron star surface with time.
- Burst profiles are fairly smooth on long timescales, decaying with close to an exponential profile. However, they can exhibit strong oscillations on ms timescales.

The spectra are observed (Swank et al. 1977; Hoffman, Lewin, and Doty 1977) to be roughly blackbody, at least when viewed with coarsely binned energy. If the total luminosity is L, then the radius of the object can be determined (assuming the distance d to the object is known, which is the case in some globular clusters; Kuulkers et al. 2003) from $R = [L/(4\pi\ \sigma T^4)]^{1/2}$, where σ is the Stefan-Boltzmann constant. The sizes of the source objects have thus been found to be about 10 km, consistent with their being a neutron star.

If the burst is bright, the local X-ray luminosity may reach the Eddington limit, that is, the point at which the gravitational acceleration is equal to the radiation pressure gradient. This depends on the mass, radius, and atmospheric opacity of the neutron star, M, R, and κ. Then the luminosity is given by (Strohmayer and Bildsten 2005):

8.5.1 $$L_{\rm Edd} = (4\pi cGM/\kappa)(1 - 2GM/c^2 R)^{-1/2} = 4\pi R^2 \sigma T_{\rm eff}^4.$$

In this situation, the photospheric layers can be lifted off the neutron star by radiation pressure. What is observed in these bursts is that the blackbody temperature decreases while the inferred blackbody radius simultaneously increases, and the total X-ray flux stays roughly constant or declines if the $T_{\rm eff}$ goes into the ultraviolet. These (brighter bursts) are called photospheric radius expansion (PRE) bursts. The point at which the photosphere falls back to the neutron star surface is called "touchdown." As discussed in Strohmayer and Bildsten (2005), the PRE bursts could be understood if the atmospheres of the PRE bursts were hydrogen poor and those of the fainter bursts were hydrogen rich. Ebisuzaki and Nakamura (1988) suggested that the hydrogen-rich envelope was ejected during bright PRE bursts, resulting in a pure helium atmosphere.

The amount of photospheric uplift can vary greatly from burst to burst in single X-ray bursters and between different X-ray bursters. However, the Eddington limit should impose an upper limit to burst fluxes (van Paradijs 1978). This produces a critical luminosity of about 3.8×10^{38} ergs/s for hydrogen-poor matter from a neutron star, which was found (Kuulkers et al.

2003) to be realized, and constant to a level of about 15%. Since it is not expected that very many LMXBs accrete pure helium from their companion, this relative constancy suggests that these LMXBs are blowing off their surface hydrogen in a wind.

Regarding the recurrence rates of X-ray Bursts, Strohmayer and Bildsten (2005) summarize the recurrence times in terms of the mass accretion rate. They quote van Paradijs, Penninx, and Lewin (1998) as summarizing the change of recurrence times as dM/dt increases as

- the recurrence time increases from 2–4 hours to 10 hours;
- the bursts burn less of the accumulated fuel; and
- the duration of the bursts decreases from $\sim$30 s to $\sim$5 s.

However, these facts are not necessarily easy to understand; indeed, not all X-ray bursters follow the same patterns. Furthermore, even the ones that follow these trends do not have dM/dt values that correspond to those obtained from theory. However, some of the puzzle disappears if one assumes that only part of the surface of the neutron star is covered by the accreted fuel before it ignites (Bildsten 2000), and there is evidence that this is indeed occurring, as is discussed in the next section.

One important feature of X-ray bursts is the energetics: as noted above, the gravitational energy released per accreted baryon is $GMm_p/R \approx 200$ MeV (M is the mass of the neutron star, R is its radius, and G is the gravitational constant), whereas the typical energy produced per baryon by nuclear fusion is $\approx$5 MeV when baryons are burned to heavy elements, as noted by Strohmayer and Bildsten (2005). In order for the fusion yield to have any impact, the accreted matter must be stored for some time and then burned quickly in the burst of energy seen in X-rays.

The nuclear processes that occur do so at a sufficiently high temperature, $T_9 \sim 1$, that both hydrogen and helium burning can occur at the same time in the same place, an event that does not normally occur. This burning is thought to occur in a thin shell that surrounds the neutron star (Hansen and Van Horn 1975). This same thin-shell instability was indicated in section 8.2 as also resulting from accretion onto a white dwarf, although, because the temperatures reached would not be as high, the burning is slower. Both instabilities are the result of the extreme sensitivity of the nuclear burning rates to temperature (see sec. 3.9); the result is a classical nova if the accreting star is a white dwarf or an X-ray burst if it is a neutron star.

The high temperature is achieved by a balance in pressure between the gravitational attraction of the shell, given by $g_{NS}y$, where g_{NS} is the gravitational acceleration at the neutron star's surface and y is the areal density of the shell, and the radiation pressure, given by $\sigma T^4/3c$, where σ is Boltzmann's constant. This relationship results from the fact that the shell is so thin that the pressure is essentially constant throughout.

A large advantage of studying this instability in neutron star systems is that the timescales are "observationally accessible" (Strohmayer and Bildsten 2005). Bursts can repeat themselves on timescales of hours, resulting in many events to be observed from the same system. This is not the case in novae and, indeed, is the case for very few time-varying astronomical objects.

As material is accreted onto the neutron star surface, hydrostatic compression increases the density and temperature until they are sufficient to trigger the burst. Thus the burst frequency will depend on the accretion rate dM/dt.

Prior to burst ignition, the HCNO cycle commences; it is shown in figure 8.2. The energy generation rate clearly depends on the amount of the CNO cycle catalyst, carbon, that is contained in the accreted matter. This burning phase is thermally stable; this type of burning occurs in the "accumulation phase" of the accretion process. This will occur when (Strohmayer and Bildsten 2005)

8.5.2 $$dM/Adt > 900\ \mathrm{g\ cm^{-2}s^{-1}}(Z_{\mathrm{CNO}}/0.01)^{1/2},$$

where A is the surface area of the neutron star.

The slow hydrogen burning during accumulation allows for a more accelerated burning mode, H/He burning, at high mass transfer rates. This occurs when (Bildsten 1998; Cumming and Bildsten 2000)

8.5.3 $$dM/Adt > 2 \times 10^3\mathrm{g\ cm^{-2}s^{-1}}(Z_{\mathrm{CNO}}/0.01)^{13/18}.$$

The strong temperature dependence of the helium-burning rates leads to the thin-shell instability that produces the X-ray burst. Note that the mass transfer rate is crucial; stable burning sets in at higher dM/Adt values, as the helium-burning temperature sensitivity becomes weaker than the cooling rate sensitivity (Ayasli and Joss 1982; Taam, Woosley, and Lamb 1996). This picture is supported by the absence of bursts from high-field X-ray pulsars, where the accreting material is concentrated onto a polar cap, increasing dM/Adt and stabilizing the burning (Joss and Li 1980; Bildsten and Brown 1997).

To summarize then, the mass transfer rates, assuming $Z_{\mathrm{CNO}} \approx 0.01$, (Strohmayer and Bildsten 2005), are

1. Mixed hydrogen and helium burning triggered by thermally unstable hydrogen ignition for $dM/Adt < 900\ \mathrm{g\ cm^{-2}s^{-1}}$;

2. Pure helium shell ignition for $900 \text{ g cm}^{-2}\text{s}^{-1} < dM/Adt < 2 \times 10^3 \text{ g cm}^{-2}\text{s}^{-1}$ following completion of hydrogen burning; and
3. Mixed hydrogen and helium burning triggered by thermally unstable helium ignition for $dM/Adt > 2 \times 10^3 \text{g cm}^{-2}\text{s}^{-1}$ (or $dM/dt > 4.4 \times 10^{-10}$ $M_\odot$ year^{-1}).

If Z_{CNO} is less than that assumed, the transition accretion rates are reduced, and the range for pure helium ignition is narrowed.

It is also important that the energy from this burning is supplemented by pycnonuclear burning (see sec. 6.10) of the material below the surface, thereby greatly influencing the energy of the observed burst. The current estimate is that between 10 % and 100 % of the heat thus produced emerges through the crust (Brown, Bildsten, and Rutledge 1998; Brown 2000; Colpi et al. 2001).

8.6 The High-Temperature rp-Process

The nucleosynthesis that occurs in these high-temperature hydrogen and helium mixed environments that appear to characterize X-ray bursts and rp- and γ-process burning has been described in detail in several studies. Beyond iron, the rp-process was found to be capable of populating some of the progenitors of the p-nuclides. The details of the burning were worked out by van Wormer et al. (1994), who assumed temperatures ranging from T_9 of 0.15–1.5, densities around 10^6 g cm^{-3}, and processing times of up to 1000 s for the lowest temperatures. Typical rp-process pathways are shown in figure 8.8 (from Schatz et al. 1999). Very few of the cross sections relevant to the higher temperature rp-process have been measured; Hauser-Feshbach (statistical model; see sec. 3.11) cross sections are used instead. Van Wormer et al. (1994) attempted to extend the time-honored approach (Woosley et al. 1978; Thielemann, Arnould, and Truran 1987) to calculating these cross sections by introducing several improvements, while a more recent effort by Rauscher, Thielemann, and Kratz (1997) applied a "back-shifted Fermi gas formalism" and a global parameterization of nuclear level densities to provide a very general prescription. The higher temperature network calculations (van Wormer et al. 1994) become quite complex since, in addition to a myriad of radiative capture and particle transfer reactions, one has also to include the photonuclear reactions of the γ-process because of the high abundance of high-energy photons that accompany the high temperatures of this process. Some nuclei with low photoproton Q-values, for example, ^{57}Cu, can become important waiting points for the cycle. In addition, long-lived nuclei that are populated by reactions with low or negative Q-values might slow the rp-process progression to higher masses. In this context, Blank et al. (1995) found that ^{69}Br, which would be formed by ^{68}Se(p, γ)^{69}Br, is proton unbound. Since the half-life of ^{68}Se is 35.5 s (Tuli 2000), appreciably longer than, or at least comparable to, the time over which the high-temperature conditions of the rp-process are usually thought to persist, the rp-process would be expected to be slowed appreciably at $A = 68$. A more recent measurement of the ^{68}Se mass (Clark et al. 2005) confirmed the previously observed fact that ^{68}Se is a waiting point in the rp-process, and similar results have been obtained for ^{64}Ge (Clark et al. 2004) and ^{72}Kr (Rodriguez et al. 2004).

Despite these hindrances, it was found (van Wormer et al. 1994) that, at the upper end of the temperature range studied, significant abundances of ^{74}Se and ^{78}Kr could be produced. Of course, some leakage past ^{68}Se will occur from the β-decays that do occur, especially if the high-temperature conditions last for a time comparable to the ^{68}Se half-life. An extension of the van Wormer et al. (1994) work was performed by Schatz et al. (1998), who studied the extremes of the rp-process and who included two-proton captures in their networks. These might operate as sequential proton captures through a resonant state that is sufficiently long-lived that a second proton capture can occur before the state proton decayed. Schatz et al. (1998) note that these may also be capable of circumventing some of the termination points in particularly hot dense environments.

Unquestionably, the masses of the nuclei all the way to the proton drip line are crucial to predictions of the rp-process, and considerable progress has been made in determining these. These are summarized on the Atomic Mass Evaluations, so they will not be repeated here. Half-lives are also important, as rapid β-decays can compete with the reaction processes, producing parallel paths. In addition, they determine the nucleosynthesis that can occur, as well as the timescale of the process. The level of accuracy of the masses required for accurate reaction rate determination is roughly 10 keV (Schatz and Rehm 2006), but few masses along the high-temperature rp-process path have been determined yet to that level. Some of the techniques used in the mass measurements were described in chapter 3; they include the ESR storage ring at GSI but usually in the isochronous mode for the short-lived nuclei along the rp-process path; endpoint measurements (Wohr et al. 2004); and time of flight techniques (Chartier et al. 1998). However, the use of Penning traps, especially on the short-lived nuclei, has revolutionized this field, as this permits mass measurements to well below the required level (Stolzenberg et al. 1990; Savard et al. 1997; Kolhinen et al. 2004; Schwarz et al. 2003). Use of this

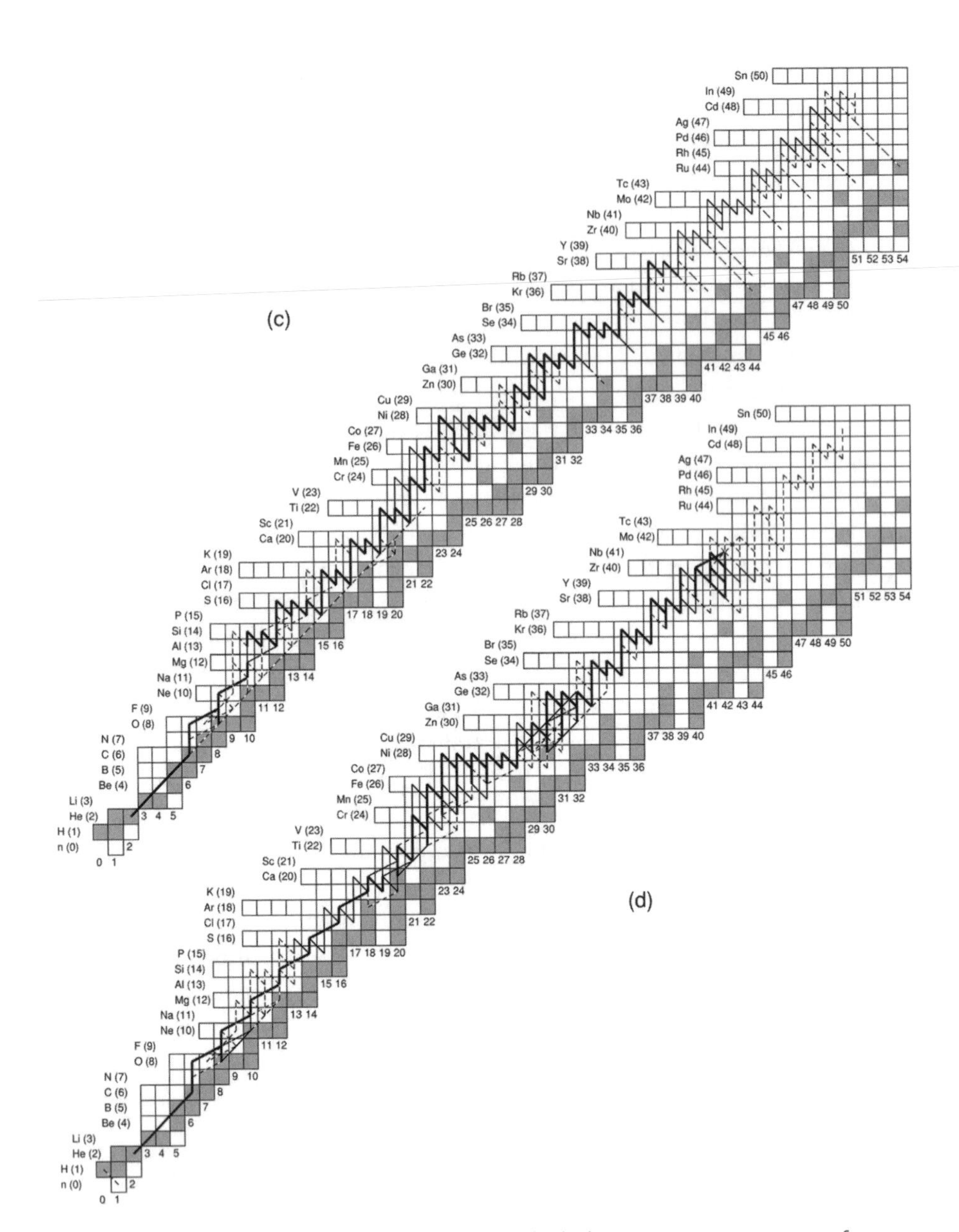

Fig. 8.8. Paths of the rp-process as it occurs in the high-temperature environment of an accreting neutron star for two different accretion rates, with the lower right pathway (*d*) being 2.5 times as large as that for the upper left one (*c*). The thickness of the line indicates the strength of the reaction flow. Each square represents a proton-stable nucleus, and filled squares are for stable nuclei. The extent of the rp-process is seen to increase in mass with increasing mass accretion rates. In this particular model, the accretion rate was sufficiently high that the rp-process terminated around Sn due to (p, α) reactions that operate on the proton-rich nuclides produced. The same would be true for all higher mass accretion rates that produced an X-ray burst. From Schatz et al. (1999).

technique requires only that the half-life exceed 100 ms, which is the case for nearly all nuclei along the rp-process path.

Many of the longer half-lives have now been measured with RNB facilities. However (Schatz and Rehm 2006), the β-decay strength functions for many of the nuclei along the rp-process path are not known, so they have to be inferred from theory. This can also be important to calculations of rp-process burning, as it will determine the neutrino losses from the β-decays.

With respect to the crucial Mo and Ru p-nuclides, Schatz et al. (1998) found that, at temperatures well in excess of $T_9 = 1$ and processing times of tens of seconds, large abundances of these nuclides could be achieved. However, they also found that this scenario tended to produce about the same overproduction factors, the ratio of the abundances of each nuclide divided by the corresponding solar abundance, for some non-p-nuclei such as ^{80}Kr. This might have been a problem for this model, as that isotope is produced by other processes, so its total predicted abundance becomes much larger than nature can accommodate. However, some nuclear physics measurements have played an important role in resolving this apparent problem. The half-life of ^{80}Zr was measured (Ressler et al. 2000), and the result was $4.1 + 0.8/-0.6$ s. This is significantly less than the value used in the theoretical simulations of the rp-process, 6.855 s. Changing this half-life reduced the ^{80}Zr abundance during the cooling phase of the rp-process by a factor of 2. In the experiment it was also found that the ^{80}Zr daughter, ^{80}Y, had a previously unknown isomeric state that resulted in an effective reduction of the ^{80}Y half-life, thereby also reducing its abundance during the cooling phase. Then fast proton captures on ^{80}Zr and ^{80}Y during the freezeout phase further reduced the mass 80 abundance, providing a reduction in that abundance of about an order of magnitude from that resulting from use of the larger ^{80}Zr and ^{80}Y half-lives. This certainly alleviates the ^{80}Kr overabundance problem. However, that does not solve all the potential difficulties with this model; the very high temperatures that seem to be required to produce high abundances of the Mo and Ru p-nuclei might be difficult to sustain in any plausible cosmic environment. More data than just these are needed to understand the high-temperature rp-process; half-life and decay mode data and data on the possible existence of isomers are needed for all the proton drip line nuclei in this mass region. Among the experiments performed to obtain such data are those of Hencheck et al. (1994), Hellstrom et al. (1996), Gorska et al. (1997), and especially Faestermann et al. (2002).

In the study of Schatz et al. (1999), the hydrodynamics of the environments in which the high-temperature rp-process might occur was given special

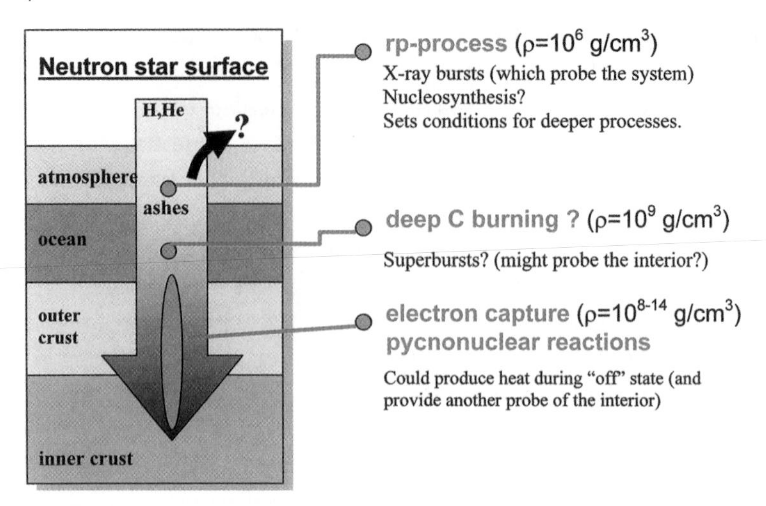

Fig. 8.9. Schematic diagram of the different layers on a neutron star. Reprinted with permission from H. Schatz.

attention. The scenario depicted in their analysis is indicated in figure 8.9. The physics associated with the bursts involved both the means by which the energy is generated and the nucleosynthesis that would be performed in the resulting environments. Although this sort of burning was shown (Schatz et al. 2001) to produce a wide variety of medium-mass nuclei, including the Mo and Ru p-nuclei, the model that describes this sort of burning strongly suggests that the nuclei produced would never become a part of the interstellar medium. In order to do so, the nuclei would have to surmount the 200 MeV per nucleon gravitational barrier. In the one-dimensional models studied, this appears to be impossible. Furthermore, these nuclides would not even be able to participate in subsequent burning phases, as they would sink below the neutron star's surface.

However, Paczynski and Proszynski (1986) and, recently, Weinberg, Bildsten, and Schatz (2006) have shown that in situations in which the output of the thermonuclear reactions exceeds the Eddington limit, the resulting radiation-driven wind can mix a small fraction of the rp-process nuclides with the other material that achieves a sufficiently large radius in the PRE phase of the X-ray burst that it can be expelled from the neutron star. Furthermore, some of these same nuclides could also remain on the surface of the neutron star, so they could be involved in the next burning phase of the star. The study also notes the importance of (α, p) reactions in pure helium-burning layers, as the protons emitted from those reactions will be captured on ^{12}C, and the resulting

^{13}N will undergo a ^{13}N(α, p)^{16}O reaction, greatly expediting the production of ^{16}O from ^{12}C and producing an important energy source for the burst.

This might be a testable conclusion; observation of spectral features would provide the information needed. Although searches for spectral features have provided very little information to date, enough has been found theoretically to at least raise the question about emission of newly synthesized nuclei by observation (Cottam, Paerels, and Mendez 2002) of highly redshifted (from the gravitational well of the neutron star) iron absorption lines in X-ray bursts from one source. The level of iron implied is well in excess (Bildsten, Chang, and Paerels 2003) of what could possibly be produced and expelled from an X-ray-emitting region. However, subsequent work (Chang, Bildsten, and Wasserman 2005) suggested that the source of the iron is the accreted material from the companion, not the synthesized material at the surface of the neutron star.

Another interesting feature is the possible termination point of the rp-process via nuclear systematics, instead of the limitations of time and temperature during the rp-process. It was noted by Schatz et al. (2001) that the *Q*-values of the nuclei around Sn, Sb, and Te would produce a natural cycle much like the CN cycle, so that no matter how long the burst lasted or how high the temperature was (if it was not high enough to photodisintegrate all

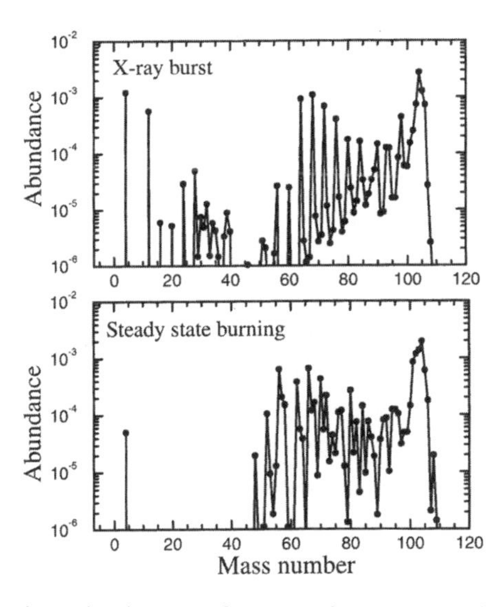

Fig. 8.10. Final abundance distribution as functions of mass number for an X-ray burst and for steady state burning at a particular mass accretion rate. Reprinted with permission from Schatz et al. (2001). Copyright 2001 by the American Physical Society.

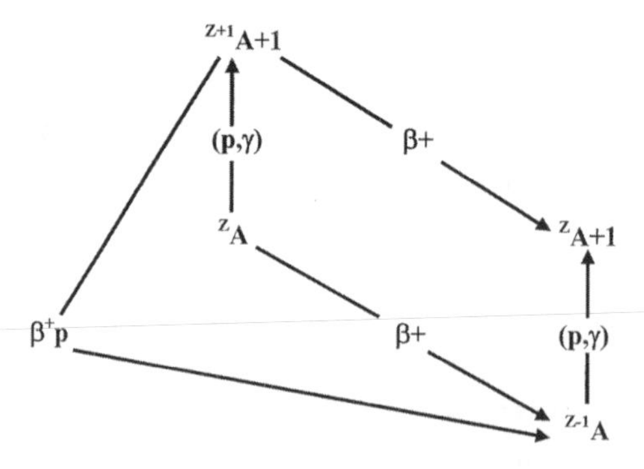

Fig. 8.11. This shows the various alternatives that could occur if one of the nuclei produced in the rp-process, in this case, nucleus $^{Z+1}$A, can undergo β-delayed proton emission.

the heavy nuclei), one would expect the highest masses involved in the rp-process to be those Sn, Sb, and Te nuclei close to $N = Z$, which would decay to stable nuclei no more massive than the p-process Cd isotopes. The nuclides produced in two scenarios, one involving an X-ray burst and the other involving steady state burning, which is thought to occur at higher accretion rates, are indicated in figure 8.10.

Of course, the proof of this rp-process endpoint would be in observing X-ray lines from Cd in a long X-ray burst. Unfortunately, it has proved difficult to observe such spectroscopic identifications in bursts; even those from iron (Cottam, Paerels, and Mendez 2002) required heroic effort.

8.7 β-Delayed Proton Emission

One of the curiosities that might affect the abundances produced in the rp-process is β-delayed proton emission. Some of the very proton-rich nuclides through which the rp-process passes are so far from stability that they can decay by positron emission followed immediately by proton emission, thereby possibly shifting the abundance peaks from those of the progenitors to a slightly lower mass. Some examples of nuclei that can undergo this mode of decay are given in table 8.2.

What effect might β-delayed proton emission have on the nucleosynthesis produced in the rp-process? Figure 8.11 indicates the scenarios that exist in this situation (Boyd 2000). Assume that the seed nuclei have already undergone several proton captures, driving them close to the proton drip line, to nucleus ^{Z}A. Assume further that nucleus ^{Z}A can undergo another proton radiative capture to take it to nucleus $^{Z+1}(A+1)$. If nucleus $^{Z+1}(A+1)$ undergoes

Table 8.2 Some Nuclei That Might Undergo β-Delayed Proton Emission

Nucleus	Half-life	Percent of β-p emission
^{21}Mg	122 ms	32.6
^{23}Al	0.47 s	1.1
^{24}Si	140 ms	38
^{52}Ni	38 ms	17
^{84}Nb	9.5 s	0
^{93}Pd	1.35 s	1.5
^{94m}Ag	0.47 s	0
^{96}Ag	4.4/6.9 s[a]	8.1/18.0
^{100}In	5.9 s	1.6

SOURCE: http://www.nndc.bnl.gov/nudat2/sigma_searchi.jsp.
[a]Two different degenerate states.

simple β-decay, it will produce nucleus $^{Z}(A+1)$. However, if it can undergo β-delayed proton emission and its half-life is short compared with the duration of the rp-process, the result of that decay would be nucleus ^{Z-1}A, producing a mass number change of one unit but a change of proton number of two units. In the high-temperature proton-rich environment of the rp-process, this nucleus would be expected to capture another proton quickly, producing nucleus $^{Z}(A+1)$, giving the same net result as if the original nucleus had captured a proton and undergone a simple β-decay.

However, now suppose that the half-life of nucleus $^{Z+1}(A+1)$ is long—longer than or comparable to the duration of the rp-process. In this case, the β-delayed proton emission that produces nucleus $^{Z-1}$A will not be followed by another proton radiative capture but rather will be followed by successive β-decays until stability is reached. In this case, the β-delayed proton emission will have resulted in a shift of the mass peak downward by one unit; this will affect the nucleosynthesis resulting from the rp-process and must be taken into account in the rp-process simulations.

8.8 Details of Some "Other" Processes of Nucleosynthesis: The γ-Process, the ν-Process, the νp-Process, and α-Rich Freezeout

8.8.1 The γ-Process

Since the γ-process is so intimately connected with some sites of the rp-process, certainly the higher mass rp-process, it is appropriate to discuss it at this time. The current description of the γ-process was formulated by Woosley and Howard (1978), who included (p, γ), (n, γ), (α, γ), (p, n), (α, p), and (α, n) reactions and their inverses in their network to study the synthesis resulting

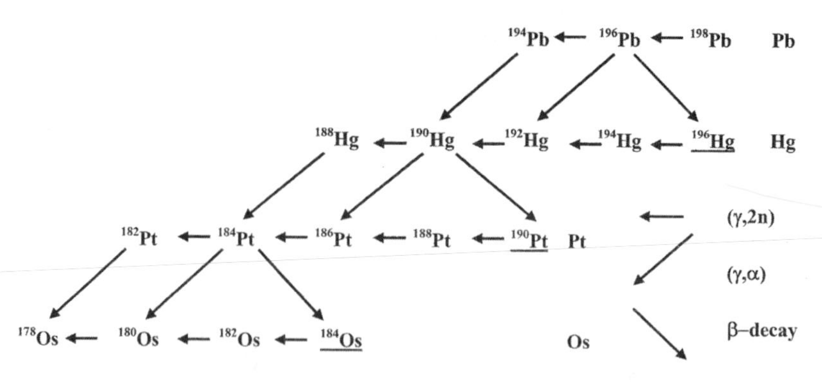

Fig. 8.12. Possible pathways of the γ-process through a section of the periodic table. The actual photonuclear reaction that occurs at each stage will depend on the *Q*-values for the competing reactions. The seeds for the γ-process in this mass region are the stable nuclides of each element, which are indicated just as Pb, Hg, Pt, and Os on the figure. The p-process nuclides produced by the γ-process are underlined. From Woosley and Howard 1978.

from such a process. They also noted that a short timescale, of the order of a second, would be necessary as the duration of this process so as not to require inordinate amounts of energy. They found that, with a few notable exceptions, the predicted ratios of the abundances of all the heavy p-nuclides corresponded reasonably well with those observed.

One of the exceptions, ^{146}Sm, was subsequently explained by Woosley and Howard (1990) by a more careful handling of the nuclear physics details. The other two, ^{180}Ta and ^{138}La, are extremely rare and are thought to be produced at least partially in the ν-process, as discussed below. Both solar and s-process enhanced seeds were tried. The latter, of course, enhanced the p-nuclide abundances but did little redistribution of them. However, the light p-nuclides were all underproduced in the original γ-process scenario, with the Mo and Ru p-nuclides having especially low productions.

In the γ-process, photons from a high-temperature ($T_9 \approx 2$–3) bath containing previously synthesized heavy nuclides initiate successive (γ, p), (γ, n), and (γ, α) reactions on those nuclides to synthesize the somewhat lighter p-nuclides. Typical processing times must be less than a second, as longer times would, at these high temperatures, destroy all the heavy nuclides by photonuclear processes. The path of this process is indicated in figure 8.12. As can be seen, any sequence of photonuclear reactions in which 14 neutrons and four protons are removed from (the abundant) ^{208}Pb would produce the p-nuclide ^{190}Pt. Initially the (γ, n) reactions would dominate, positioning the resulting nuclides to the proton-rich side of stability, but (γ, p) and (γ, α) reactions become more probable as the photonuclear reactions proceed. In some cases proton-rich unstable nuclides are formed that β-decay back to the stable p-nuclides after the γ-process conditions have subsided. Of course, the *Q*-values and, for charged particles, Coulomb barriers will largely determine the probability of each reaction from any intermediate nucleus.

In its original form, the γ-process was assumed to occur in an inner shell of an exploding supernova so would not necessarily have any hydrogen with which to drive (p, γ) reactions and so to synthesize sufficient abundances of the light p-nuclides. Thus the underproduction of the light p-nuclides in the original scenario was not surprising. Since other sites exist for producing the light p-nuclides, this is not necessarily a problem for the model. The primary issue for the γ-process appeared to be whether or not a site for it could be found that would generate the requisite conditions for producing the heavy p-nuclei. Rayet, Prantzos, and Arnould (1990) and Prantzos et al. (1990) all made important extensions of the original Woosley and Howard (1978) model, both in the context of the astrophysical site and in the handling of the thermodynamics of the environment. The model used by Prantzos et al. (1990) utilized the thermodynamics of a type II supernova that had been shown by Nomoto and Hashimoto (1988) to give a good representation of SN 1987a. The results of both studies were p-nuclide overproduction factors similar to those achieved by Woosley and Howard (1990). Notably, the Mo and Ru p-nuclides were still badly underproduced, and the overproduction factors of the lighter p-nuclides, ^{74}Se, ^{78}Kr, and ^{84}Sr, were roughly the same as those for the more massive nuclides. Finally, Rayet et al. (1995) considered a range of supernovae from 13 to 25 $M_{\odot}$ and found that, although the variations in nucleosynthesis details were sometimes large, the above qualitative conclusions were essentially unchanged.

An extension of the p-process models, by Howard, Meyer, and Woosley (1991), studied one potential site for the γ-process that also produced most of the light p-nuclei. The requisite thermodynamic and seed conditions occurred when a carbon-oxygen white dwarf exploded, either as type Ia or a subclass of a type II supernova. The resulting high temperatures, $T_9 \approx 2$–3, produced the high-energy photons necessary to photodisintegrate the heavy nuclear seeds to yield the heavy p-nuclides, and the carbon-burning reactions of the white dwarf produced enough protons to fuel the (p, γ) reactions necessary to synthesize the lighter p-nuclides. Virtually all of the p-nuclei heavier than ^{92}Mo achieved overproduction factors of order 10^4, which were within an order of magnitude of each other (excepting ^{180}Ta, ^{146}Sm, and ^{138}La). However, the light p-nuclides were overproduced appreciably relative to their more massive counterparts: the ^{74}Se, ^{78}Kr, and ^{84}Sr overproduction factors were roughly a factor of 2 above those of the heavier nuclides. Unfortunately, other processes, such as the rp-process and α-rich freeze-out, are also thought to be possible producers of those lighter p-nuclides. The resulting total overproduction could thus present a serious conflict with the observed abundances. Finally, as usual, the Mo and Ru p-nuclides were underproduced.

Jordan and Meyer (2004) generalized these considerations by studying the nucleosynthesis produced in fast expansions of high-entropy proton-rich matter. They discovered that this scenario can produce a wide variety of abundance patterns, resulting from the fact that, in the rapid expansion, the nucleons are not necessarily in nuclear statistical equilibrium with ^{4}He, resulting in a wide range of densities of the nucleons that can be captured on the heavier nuclei for different expansion timescales. If the reassembly of ^{4}He is efficient, the depletion of neutrons drives the products of the nucleosynthesis down toward the Fe-Ni region. If it is inefficient, the high density of nucleons can produce a variety of heavier nuclides. Although a site for these conditions was not identified, it was found that some scenarios could produce high abundances of the Mo and Ru nuclides, while others resulted in relatively high abundances of the usually rare nuclides such as ^{180}Ta.

Essential inputs to any γ-process description are the properties of the very proton-rich lighter nuclides that are produced primarily by rapid proton captures. In particular, the masses are crucial, as the high temperature environment that characterizes the γ-process will certainly produce (γ, p) and (γ, α) reactions to compete with the (p, γ) reactions, and the probabilities of the photonuclear processes depend critically on the reaction *Q*-values. Those nuclides with long half-lives will tend to inhibit the flow of the γ-process and will build up in abundance. The decay modes may also be important as, for example, β^+ decay and β^+p-decay will produce different final nuclides from the γ-process. Thus significant effort has gone into observing these properties of the proton-rich nuclides from stability to the γ-process progenitors (Boyd 1994; Winger et al. 1993; Blank et al. 1995), although at present most of these studies have been conducted only for relatively light γ-process nuclides. Finally, the photonuclear cross sections are also important for the γ-process. However, virtually none of those relevant to the γ-process has been measured; they are instead generally calculated from a Hauser-Feshbach (statistical model) formalism. It is anticipated that, as more cross section data are acquired, the statistical model calculations should improve greatly in sophistication and accuracy. Such measurements are presently possible in special cases by inverse reaction techniques, so some have been initiated recently (see, e.g., Fulop

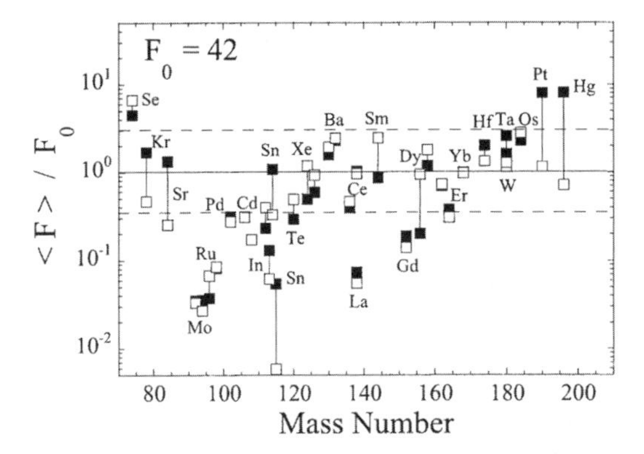

Fig. 8.13. The p-nuclide overproduction factors obtained for a solar abundance 25 solar mass model star with MOST (Goriely 2001 and http://www-astro.ulb.ac.be) rates and in the absence of neutrino nucleosynthesis (*open squares*). For comparison the results of Rayet et al. (1995), which were obtained with a different set of nuclear reaction rates, are shown (*filled squares*). The upper and lower limits correspond to values obtained with different nuclear inputs to the Hauser-Feshbach model (see sec. 3.11). The dashed horizontal lines define a factor of 3 around the mean value (*solid line*); most of the values, except the abovementioned Mo-Ru region, fall within this factor of 3. From Goriely et al. (2001).

et al. 1996; Somorjai et al. 1998; Chloupek et al. 1999; Ozkan et al. 2001, 2002; Gyurky et al. 2003; Harissopulos et al. 2005).

A particularly interesting nucleus is ^{138}La, the focus of the paper by Goriely et al. (2001) (see fig. 8.13). It is one of the lowest abundance nuclides in the periodic table. When the abundance of a particular nuclide is extremely small, several processes may contribute to its production, complicating its understanding. This appears to be the case for ^{138}La, as two processes seem to be possible contributors to its abundance: the ν-process (see the next section and Kaeppeler et al. 2004) and the γ-process. The former could produce a significant abundance for ^{138}La, although that calculation involves a delicate balance between the ^{137}La(n, γ) and ^{138}La(n, γ) reactions. However, Goriely et al. (2001) studied the possible effects on the ^{138}La abundance of the uncertainties in these two cross sections and found that even extreme changes in those (unmeasured) cross sections could not boost its abundance up to a value close to the level predicted for the other p-nuclides when just the γ-process was assumed. However, when the ν-process was included by Goriely et al. (2001), it was found that the ^{138}La abundance increased by roughly an order of magnitude. Although the parameterization of the ν-process has a large number of uncertainties (Goriely et al. 2001 assumed two different sets of parameters of the neutrino distributions, both reasonable, which produced about a factor of 3 difference in the predicted ^{138}La abundances), this analysis does seem to suggest that the ν-process must be a major contributor to the ^{138}La abundance. The ν-process is described in further detail in the next section.

8.8.2 The ν-Process

The ν-process, originally proposed by Woosley et al. (1990), and updated by Heger et al. (2005), has been shown to be capable of producing some of the lowest abundance p-nuclides. The process is thought to occur in the neutrino wind generated by stellar collapse in supernovae. The nuclides synthesized clearly depend on the shell in which the ν-process occurs. For example, ^{11}B and ^{19}F would be expected to be made in shells in which the dominant constituents were ^{12}C and ^{20}Ne, respectively, both by processes in which a neutrino would excite the target nucleus via the neutral-current interaction (which therefore allows interactions with all neutrino flavors) to an excited state from which the nucleus could emit either a proton or a neutron.

The ν-process could also make two of the rarest stable nuclides in the periodic table: ^{138}La and ^{180}Ta. The latter would be made by the $^{181}Ta(\nu, n)^{180}Ta$ (neutral current) reaction and possibly the $^{180}Hf(\nu_e, e^-)^{180}Ta$ (charged-current) reaction, which appear to produce an abundance consistent with that observed. Similarly, the $^{139}La(\nu, n)^{138}La$ (neutral current) reaction, together with the $^{138}Ba(\nu_e, e^-)^{138}La$ (charged current) reaction, appear capable of synthesizing roughly the observed ^{138}La abundance. In principle, the $^{138}Ce(\bar{\nu}_e, e^+)^{138}La$ reaction could also make ^{138}La, but the abundance of ^{138}Ce is so small in any scenario of nucleosynthesis that this would not be expected to contribute appreciably. Thus, the ν-process seems to provide a natural mechanism for synthesis of ^{138}La and ^{180}Ta, the abundances of which have evaded description for several decades, as well as some other nuclides. However, it should be noted that its results are somewhat uncertain due to questions about the neutrino spectrum resulting from a type II supernova (Myra and Burrows 1990). Thus more work needs to be done on this process to understand fully its capabilities and properties.

8.8.3 α-Rich Freezeout

In the past several years, a new process, α-rich freezeout, has emerged as the possible underpinning of some of the processes associated with synthesis of the p-nuclides, as well as of the r- and ν-processes. The α-rich freezeout model is thought to occur in neutrino-driven winds within about 1 s after core bounce in supernovae (Woosley and Hoffman 1992; Takahashi, Witti, and Janka 1994; Witti, Janka, and Takahashi 1994; Woosley et al. 1994; Hoffman, Woosley, and Qian 1997) in the high-entropy high-temperature (initially $T_9 \approx 10$) bubble near the core.

In this process, discussed previously in sections 6.5 and 7.5.2, the photons in the high-temperature environment reduce all the preexisting nuclei down to essentially protons and neutrons, then into α-particles, and subsequently into many other nuclei as the environment cools. Critical parameters in characterizing the environment include the electron fraction $Y_e = p/(n + p)$, where n (p) is the neutron (proton) density, the entropy, and the processing time. These work together to determine the ability of this site to produce the r-process nuclides. As the bubble cools and the α-particles are formed, then two α-particles and a neutron can combine to form ^{9}Be. Following formation of ^{9}Be, two-body reactions determine the abundances of the remaining nuclei synthesized in nuclear statistical equilibrium (in which the strong and electromagnetic interactions are in equilibrium, but the weak interaction is not; see sec. 6.5) until the nuclear reactions initiated with charged particles freeze out. Then the remaining neutrons synthesize all the r-process nuclides. However, there are an enormous number of nuclear reactions that are important to the resulting nucleosynthesis, most of which have not been studied. For example, a systematic study of the significance of specific reactions by Jin et al. (1997) suggested that the ^{45}V(p, γ)^{46}Cr reaction presented the largest uncertainty in the calculations being performed, at least up to masses around the iron peak nuclei. Measurement of the cross section for this reaction presents a formidable challenge for present users of RNB facilities. However, its reaction rate has been estimated theoretically (Horoi et al. 2002).

The α-rich freezeout environment has been shown by Hoffman et al. (1996) to be capable of creating the light p-nuclides through ^{92}Mo, both in the correct relative abundances and with an appropriate absolute magnitude, again, if the parameters are chosen appropriately. Some of these are actually synthesized from nearby r-process products through charged-current interactions with the electron neutrinos. For example, Fuller and Meyer (1995) showed that neutrino capture will dominate over antineutrino capture, so that processes such as ^{92}Zr(ν, e$^-$)^{92}Nb(ν, e$^-$)^{92}Mo, possibly as much a part of the ν-process as of α-rich freezeout, become possible and could make a great deal of ^{92}Mo. Since such processes can proceed through stable nuclei, they do allow production of p-nuclides from neutron-rich seeds. However, the extent to which such processing can occur is limited, as too many neutrino-initiated interactions would smear out and shift the well established r-process abundance peaks resulting from the neutron closed shells. As noted in section 7.5.5, a study (Haxton et al. 1997; Qian et al. 1997) of r-process yields just below the r-process peaks suggests that those abundances are produced primarily from neutrino postprocessing of the nuclides in the r-process progenitor peaks. However, the fact that those nuclides are only a few nucleons below the peaks suggests that as a limit to the extent to which such ν-processing can occur; shifts of only a few nucleons are permitted, consistent with the limit from the sharpness of the r-process peaks.

The α-rich freezeout model does have some deficiencies that have persisted since its inception. Those associated with the r-process are discussed in the section on that subject. From the perspective of the p-process, however, although this process does make ^{92}Mo with enough abundance to provide all that is observed in nature, it is unable to produce even close to the abundances of the next several p-nuclides. In addition, the results of the model calculations exhibit an uncomfortable sensitivity to the neutron excess η. Only future work will tell if these problems can be solved and the α-rich freezeout model can provide the description of the nucleosynthesis of the light p-nuclides.

8.8.4 The νp-Process

The possibility of a new (as of 2006) process of nucleosynthesis that would occur in a core-collapse supernova environment was proposed by Frohlich et al. (2006) and independently by Pruet et al. (2006). This process was shown in both studies to be capable of synthesizing nuclei heavier than iron, but, unlike the r-process, it can synthesize nuclides to the neutron-poor side of stability and even produce some s-process nuclides. In the presence of a large proton abundance, fusion between protons and heavy nuclei can occur, leading to the nucleosynthesis path of the rp-process. As discussed above, this process can be delayed, certainly for far longer times than those thought to characterize the neutrino burst from a core-collapse supernova, by waiting point nuclei such as ^{64}Ge ($T_{1/2} = 63.7$ s). The new process was dubbed the νp-process, as it proceeds similarly to the rp-process, but with the benefit of the intense blast of electron antineutrinos from the supernova. The $\bar{\nu}_e$s would interact with the inner-shell ejecta from the supernova, converting the protons there to neutrons. The neutrons would interact essentially instantly, via (n, p) reactions, which produces the same net effect as β^+ decay. This moves the proton-rich nuclei away from their rp-process waiting points toward stability on a very short timescale, greatly expediting this rp-like process. Because the neutrons interact so quickly, the ν_es that would also be present would have very little opportunity to convert the neutrons back to protons.

In this scenario, as soon as the heating and expansion of the inner star regions has lifted the electron degeneracy, the reactions

8.8.1 $$\nu_e + n \leftrightarrow p + e^-$$

and

8.8.2 $$\bar{\nu}_e + p \leftrightarrow n + e^+$$

drive the composition toward proton richness because of the smaller mass of the proton (as in big bang nucleosynthesis; see chap. 9). Alpha-rich freezeout then forms nuclei up to at least ^{56}Ni in NSE, with considerable amounts also of ^{4}He and protons. With no further neutrino interactions, these will be the primary constituents of this region.

However, the $\bar{\nu}_e$s, as in equation 8.8.2, will create neutrons, which can greatly change this scenario. Pruet et al. (2006) studied the sensitivity of the νp-process to several of the possible parameters that could influence its outcome, including the neutron to proton ratio, temperature, and entropy, among others. What they found was that the νp-process occurs in three, or possibly four, stages. In the first stage, the neutrons all combine with protons to form α-particles. When the temperature has cooled below 5×10^9 K, a small fraction of the protons and α-particles combine to form nuclei around iron: ^{56}Ni, ^{60}Zn, and ^{64}Ge. In the second stage, the temperature drops to around 3×10^9 K, and nucleosynthesis proceeds along the usual rp-process path, essentially the $N = Z$ nuclei, albeit with modifications at the waiting points from (n, p) reactions.

In the third stage, however, the temperature is now around 1.5×10^9 K, and charged-particle reactions are greatly inhibited. Neutrons are still being produced, so they will initiate (very rapid) (n, p) reactions to drive the path of the νp-process away from the $N = Z$ nuclides along the rp-process path toward the stable nuclides. The atomic mass number A does not change much, since the dominant reactions, (n, p), only change the Z of the nuclei in which it operates. In this stage, some nuclei generally thought to be synthesized only by the s-process are made. Note that this is truly striking: s-process nuclides could be produced in a core-collapse supernova, a primary site, by the νp-process. This could certainly complicate the usual separation between the r- and s-processes, although it might also solve some troublesome problems in nucleosynthesis.

If the neutrino flux is still large, a fourth stage may occur. In this situation, the neutron density is sufficiently high that the nuclei along the path now are to the neutron-rich side of stability, as in the r-process. Remarkably, this seems to work best when the proton density is high but the temperature is too low for proton captures; the protons only act as a source of neutrons.

The νp-process has two especially interesting results. One is that it seems capable of synthesizing the neutron-poor Mo and Ru nuclides, possibly solving a longstanding puzzle. The second is that, since the νp-process occurs in core-collapse supernovae, it is a primary process, in the same way that the r-process is primary. The νp-process has been found to be able to synthesize ^{92}Nb, which has been argued (Dauphas et al. 2003) to be a nucleus that is produced in a primary process. Thus the νp-process would also solve this problem, since it could produce ^{92}Nb in first-generation stars.

8.9 Special Observations from X-ray Bursts

Oscillations in the Signals from X-Ray Bursts

The level of detail to which studies of X-ray bursts have evolved in a very short time is remarkable and is closely linked to the information that can be obtained about the site of the burst. Specifically, millisecond oscillations have been observed in the several-second-long X-ray bursts from these objects. Some of the relevant data are shown in figure 8.14. Although the time resolution and statistics of the signal are not sufficient to map out the oscillations, a Fourier power analysis of the data does produce a well defined frequency, as shown in the insert. Although the data appear to have stochastic variations, these are really just the result of the burst oscillations as can be seen with the time resolution of *RXTE*.

Strohmayer and Bildsten (2005) have described in detail the theoretical analyses that have been performed to understand these oscillations. As they note, the burning times associated with the thermonuclear runaway are much shorter than the time required to accrete a critical mass of fuel, so it is unlikely that the ignition will occur simultaneously over the entire surface of the neutron star. They believe that the ignition is local, that is, initially a "hot spot," and the burning front then spreads, with speeds of up to 5×10^6 cm/s, over the rest of the surface of the star that has been fuel loaded. The time required to engulf the entire star is long compared with the rotation period in many instances, and (Bildsten 1995), if the burning front is not strongly convective, the distribution of the nuclear fuel may be "patchy." Thus, as noted in Strohmayer and Bildsten (2005), "the rotation of the neutron star can modulate the inhomogeneous or localized burning regions, perhaps allowing for direct observation of the spin of the neutron star." Indeed, they state, "Although many detailed questions remain [regarding the burst oscillations], there is now little doubt that spin modulation of the X-ray burst is the basic mechanism responsible for these oscillations."

The burst oscillations are not observed in all X-ray bursts, however, so they must be related to the detailed conditions that produce the burst. This might suggest that the magnitude of the oscillations would depend on the mass accretion rate. There is also evidence that the burst oscillations are related to the existence of PRE (Smith, Morgan, and Bradt 1997; Strohmayer, Zhang, and Swank 1997). Although the basic features of the oscillations seem to be understood in terms of simple models, the understanding may be somewhat illusory. For example, the frequency drift of the millisecond oscillations and even the existence of the burst tails when the burning should be spread out are not understood. Even the origin of the oscillations is not entirely clear: it may be some oscillatory mode of the surface or a spinning hot spot. However, recent work (Piro and Bildsten 2004) suggests that the oscillations have their origin at the discontinuity that occurs at the interface between the neutron star crust and the ocean below (a similar phenomenon could also occur in novae; Piro and Bildsten 2005a). The oscillation produces a radial displacement due to flexing of the crust of the neutron star. However, this would only be expected to produce the oscillations observed as the burst luminosity is increasing.

Oscillations are observed both in the rising portion of the burst time profile and in the decaying portion. However, the latter oscillations become problematic for the hot-spot scenario, as it is difficult to understand how the rotation of the neutron star could modulate the signal after the entire star has been engulfed with the thermonuclear burst. A number of suggestions have been put forth (see Strohmayer and Bildsten 2005), but none has the simplicity of the explanation of the modulation of the hot spots. The oscillations in the rising portion can have very large amplitudes, as can be seen in figure 8.15. In coarse time resolution one sees only a somewhat ragged increasing burst profile, but in finer resolution one can see clear evidence of large oscillations, having amplitudes almost as large as the signal itself. As noted by Strohmayer and Bildsten (2005), the amplitude drops as the flux increases toward its maximum, which is consistent with the expectation from spin modulation of an initially localized X-ray hot spot that expands in the order of 1 s to engulf the neutron star. This simply reflects the fact that the amplitude would be expected to be largest, that is, the spin modulation would be expected to be maximum, when the hot spot is smallest and then decreases as the hot spot engulfs the entire star. However, Piro and Bildsten (2005b) offer another possible explanation; they studied the case of X-ray bursts from a nonmagnetic neutron star. They found (theoretically) that a surface wave in the shallow

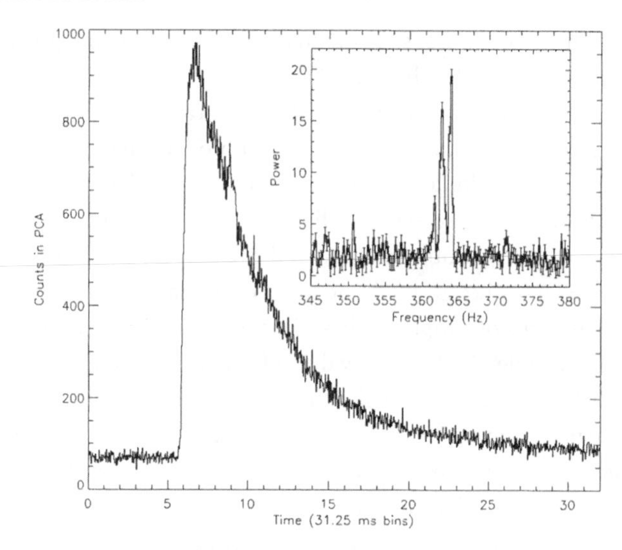

Fig. 8.14. An X-ray burst from 4U 1728-34 observed with the Proportional Counter Array (PCA), one of the instruments onboard *RXTE*. The main panel shows the X-ray counts observed by the PCA in (1/32) s bins. The inset panel shows the power spectrum in the vicinity of 363 Hz. From Strohmayer et al. (1996).

burning layer changed to a crustal interface wave as the envelope cooled and that the surface modulations decreased dramatically as the mode switched. This could explain why burst oscillations often disappear even before the burst cooling ceases. Perhaps more importantly to nuclear astrophysicists, they also compared the results of their model with the observed drifts of the frequency of the oscillations seen in the X-ray bursts and found that neutron stars with a higher average accretion rate showed smaller drifts, both in their model and in the data. Furthermore, the drifts depended on the composition of the crusts; those observed were consistent with iron crusts, as would be expected.

Spitkovsky, Levin, and Ushomirsky (2002) studied the effects of rotation of the neutron star on the development of the X-ray burst. They found that in this scenario the equatorial belt would be ignited at the beginning of the burst and that the burning front would then evolve from the equator to the poles. They also found that inhomogeneous cooling, also beginning at the equator and moving to the poles, could produce vortices, which the authors conjectured could produce the modulation observed in the tails of some X-ray bursts.

Superbursts

Another phenomenon that has been observed also involves X-ray bursts from neutron stars but is qualitatively different from those described above because

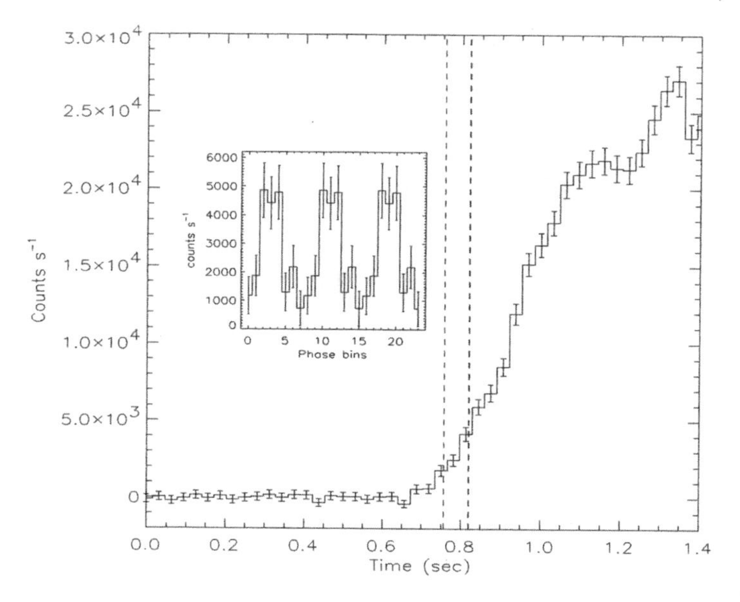

Fig. 8.15. X-ray timing evidence indicating a spreading hotspot at the onset of thermonuclear bursts. The main panel shows a burst from 4U 1636–53 with large-amplitude, 581-Hz oscillations on the rising edge of the profile. The inset shows the pulse profile, with much higher time resolution phase bins, during the interval marked by the vertical dashed lines. These data were folded at intervals of $5P_{spin}$, where $P_{spin} = 1.725$ ms. The preburst count rate intensity of about 35 counts/s has been subtracted. Adapted from Strohmayer et al. (1998).

of the duration of the burst. While the bursts described above lasted for tens of seconds or perhaps minutes, the superbursts continue for hours, scaling their total energy output accordingly. The "hardness" of the spectra, that is, the comparison of the higher energy X-rays to the lower energy X-rays, does decrease with time, arguing for a response indicative of cooling of the surface of the neutron star. In addition, these superbursts are observed in systems that also produce X-ray bursts, that is, they are also accreting binary systems. However, the possibility that this very different energy output and timescale could be fueled by the same nuclear burning mechanism as that responsible for the shorter time bursts is very improbable (Fryxell and Woosley 1982; Zingale et al. 2001). This argues for a different fuel source, that is, one located at depths below the column density where the helium flashes are triggered (Strohmayer and Bildsten 2005). It has been argued (Strohmayer and Brown 2002; Cumming and Bildsten 2001; Schatz, Bildsten, and Cumming 2003) that this involves burning the carbon in the neutron star's liquid ocean (see fig. 8.9). Recent theoretical work on superbursts appears to confirm the

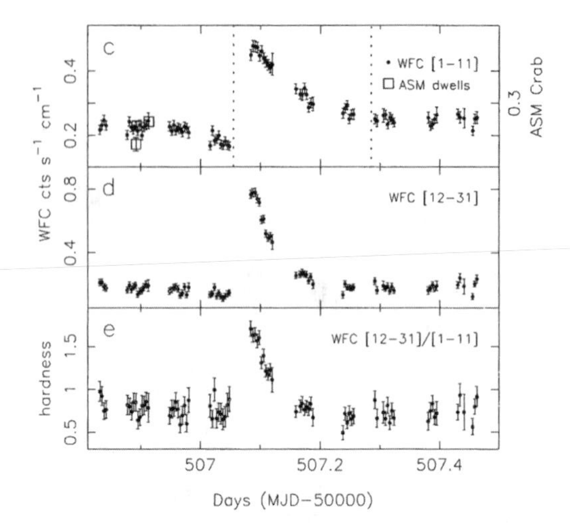

Fig. 8.16. The superburst from Ser X-1 observed with the Wide Field Camera on *BeppoSAX*. Note especially the timescale associated with the decay and the evolution of the hardness. From Cornelisse et al. (2002).

hypothesis that such bursts are the result of burning the carbon deep in the ocean of the neutron star. Particularly noteworthy is the fact that the cooling curve is consistent with this supposition (Cumming and Macbeth 2004). Another study, by Brown (2004), examined the properties of the crust as inferred by the superburst recurrence times and energetics.

An example of a superburst output, observed with the Wide Field Camera of *BeppoSAX* (Cornelisse et al. 2002), is shown in figure 8.16. The things to note are the timescales for the decay of the emission and the hardness. Another important feature, not shown in that figure, is the superburst recurrence timescale; it can be the order of 1 year (Kuulkers et al. 2004).

8.10 Magnetars

A phenomenon that also exhibits oscillations, but apparently with a completely different origin from those discussed above, is that of magnetars. These are objects known as Soft Gamma Repeaters, SGRs, which produce outbursts of energy fairly frequently. However, these occasionally also produce prodigious bursts of energy, apparently from "starquakes" that can occur on the crusty surface of a rotating neutron star. The magnetic field of the neutron star constrains the particles emitted from the fracture in the neutron star's surface into directed emissions of material that emerge from the surface of the neutron

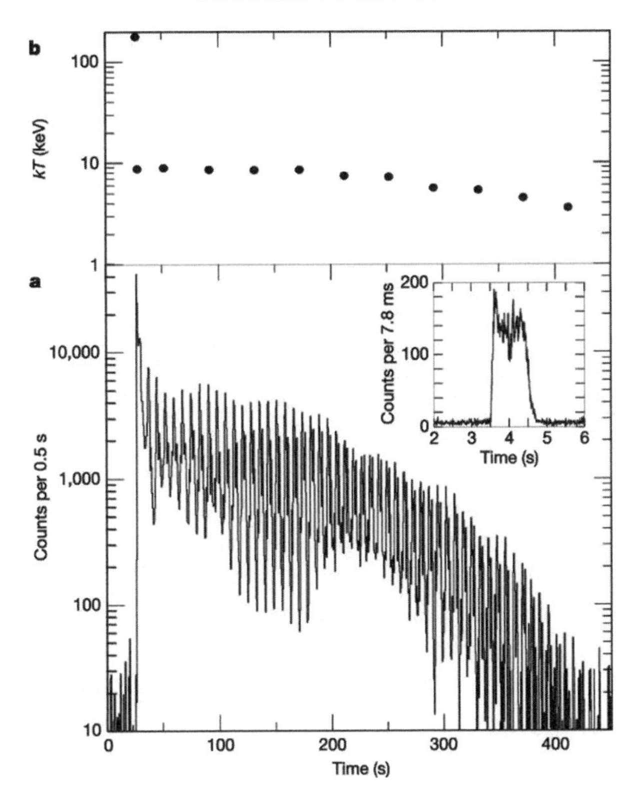

Fig. 8.17. Temperature (above) and time (below) profiles of the giant flare of December 27, 2004, from the magnetar SGR 1806–20, as seen by the NASA RHESSI satellite. An accurate brightness measurement of the initial 0.2s spike, which accounts for more than 99% of the flare's energy and corresponds to the energy output of the Sun in about 300,000 years, could not be obtained because it saturated the detectors. The 6-minute tail exhibits a 7.52-s periodicity that corresponds to the rotation period of the neutron star on which the giant flare occurred. The amplitude variations, recorded as all X-rays with energies above 3 keV, are unrelated to instrumental effects; they are real. The formal uncertainties in the oscillatory phase are smaller than the data points, so they are not shown. The inset shows the flux from the precursor per rotation cycle. Figure and comments are adapted from Hurley et al. (2005). Copyright 2005 by *Nature.*

star. This includes the photons that are a part of the energy release from this object; they are trapped as a result of the large scattering cross section the photons have from the electrons that are constrained by the magnetic field, thus resulting in the photons moving along with the particles. The signals from these events are extraordinary both in the amount of energy they emit (estimated by Eichler 2002 to be as much as 10^{46} ergs) and the oscillations they exhibit. One such magnetar's output, that from SGR 1806-20, is seen in figure 8.17, a result observed by the Reuven Ramaty High-Energy Solar Spectroscopic Imager satellite. The spectrum of the radiation is blackbody (Hurley et al. 2005), with a slowly falling temperature (upper portion of fig. 8.17). The periodicity observed from the magnetar has been interpreted as resulting from the hot spot on the neutron star's surface coming into and out of view as the star rotates; it is fully consistent with the known rotation frequency of this object (SGR 1806-20 was a well-known and well studied object before it produced the burst that was responsible for the data shown in fig. 8.17).

The magnetar's intense magnetic field is the result of its rapid rotation, which is produced as an initially spinning star collapses to a neutron star. Conservation of angular momentum then imposes the extremely rapid rotation, with the concomitant magnetic field, estimated (see, e.g., Duncan and Thompson 1992) to be in excess of 10^{14} G. Following this prediction, Kouveliotou et al. (1998) actually "measured" the magnetic field of a magnetar, SGR1806-20. They were able to measure not only the period of the oscillations seen but the derivative of the period. This allowed them to infer the spindown pulsar age to be 1500 years. Then, using an expression from Michel (1991): $B = 3.2 \times 10^{19} [PdP/dt]^{1/2}$, they were able to determine that the magnetic field was 8×10^{14} G, a value that is billions of times stronger than the strongest magnets that exist on Earth.

The 6-minute oscillating tail observed in figure 8.17 is thought to emanate from a localized hot plasma of electron-positron pairs, in equilibrium with the gamma rays that are trapped within the plasma (Schwarzschild 2005). However, the longer lasting emissions are thought to result from the interactions of the jets of particles with the ambient material that resides outside the neutron star (Gaensler et al. 2005), possibly from the nebula born from the same supernova that produced the neutron star.

A very nice discussion of many aspects of magnetars can be found on the Web page of R. Duncan, http://solomon.as.utexas.edu/~duncan/magnetar.html.

CHAPTER 8 PROBLEMS

1. You can get some feeling for how the rp-process operates by checking out the rp-process movie at the Web site, http://www.jinaweb.org/html/movies.html. You can also see three simulations of energy generation in an X-ray burst at the same Web site. These latter movies show three scenarios from a fiducial simulation of a 3-keV accretion disk interacting with the magnetic field of a 1.4 solar mass neutron star.

2. Perform your own rp-process calculation, using the code found at http://nucleo.ces.clemson.edu/home/online_tools/three_phase_ism/. Try variations in the parameters, for example, density and temperature, to see what effect this has on your calculation.
3. The thermonuclear reaction rate for the $^{45}V(p, \gamma)$ reaction has been identified as that with the largest uncertainty and largest effect on the results in rp-process calculations (Jin et al. 1997). Run a calculation with the above rp-process code, varying that reaction rate by a factor of 2, then by a factor of 10, to see what effect it has on the abundances of some of the light p-process nuclides.
4. Calculate the energy that a ^{112}Sn nucleus would have to have in order to escape the gravitational field of a neutron star.
5. The nucleus ^{26}Al in its ground state ($J^{\pi} = 5^{+}$) has a half-life, 7.4×10^{5} years, that is long compared with the processing times of the rp-process. However, its first excited (isomeric) state ($J^{\pi} = 0^{+}$) has a half-life, 6.3 s, which is short compared with those processing times. Both states β-decay. At the elevated temperatures at which the rp-process occurs, both states are populated in thermal equilibrium. Calculate the composite half-life of ^{26}Al at temperatures of $T_9 = 0.5$ and $T_9 = 1.5$.
6. Equate the effects of the gravitational force and the radiation pressure on an element in an accretion ring around a neutron star, and around a white dwarf, to derive an expression for the temperature in the ring in the two sites.

9

THE BEGINNING OF THE UNIVERSE

If an idea does not appear absurd at first then there is no hope for it.—*Albert Einstein*

Joy in looking and comprehending is nature's most beautiful gift.—*Albert Einstein*

9.1 Introduction

Although major technological advances have been made in virtually every aspect of astrophysics in the past decade, it would be difficult for any subfield to compete in that context with cosmology. The cornerstones of the big bang remain as they have been for decades, but the precision measurements and observations that have contributed to our present understanding of our universe's birth event during the past decade have been truly remarkable. This chapter will present those aspects of cosmology that relate to nuclear astrophysics but will also describe in some detail the scientific advances that have led to our current understanding of cosmology.

Modern cosmology has established convincingly that our universe began roughly 14 billion years ago as a hot dense fireball that encompassed all of space. The lynch pins upon which this conclusion is based are (1) the universal expansion, discovered by Edwin Hubble in 1929, (2) the 2.7 K microwave background radiation, and (3) big bang nucleosynthesis. The universal expansion has been assumed for many decades to be based on Hubble's law, which states that the velocity of recession of distant objects, v, is proportional to their distance from us, R, that is,

9.1.1 $$v = \mathrm{H}_0 R,$$

where H_0 is the Hubble constant. Thus, the most distant objects we can see are moving away from us at speeds approaching the speed of light. Only recently, however, has the Hubble constant, the inverse of which gives the age of the universe, been defined to better than a factor of 2. And even more recently, the accuracy of the measurements of the Hubble "constant" have shown it to be variable and to be considerably more complicated than would be implied by the word "constant." We shall return to this point subsequently. The background radiation, produced from the photons in the primordial fireball, was found, using the *Cosmic Background Explorer* (*COBE*) satellite, to be described by a black body radiation curve to incredible precision (Smoot et al. 1991, 1992). Recent work, which will also be discussed in some detail below, has not only refined those data but has now provided details of the fluctuations in the primordial soup that have determined the parameters of the big bang to unprecedented precision and to which theories of galaxy formation must now be compared. Big bang nucleosynthesis, which synthesized light nuclei from the protons and neutrons produced in the big bang, is at least a qualitative success story, although there are some interesting associated caveats. It will also be described in detail below.

The expansion of the universe is generally described in the context of the Friedman-Robertson-Walker metric. Models of this type were first studied by Alexandr Friedman in 1922 to 1924 and independently by Georges Lemaitre in 1927. It was then shown by Walker and Robertson in 1935 that the so-called Friedman-Robertson-Walker metric is the only metric consistent with a homogeneous isotropic universe, which we believe applies to the universe in which we live. If general relativity is correct, the evolution of the universe is given by the Friedman equation (see Leahy 2003; Schmidt et al. 1998; Peacock 1999):

9.1.2
$$(dR/Rdt)^2 = 8\pi G\rho/3 - kc^2/R^2,$$

where ρ is the energy density of the universe, including mass energy, G is the gravitational constant, k is the curvature constant $= 1, 0$, or -1, corresponding to a closed, flat, or open universe, respectively, and R is the scale factor. In an expanding universe, R changes with time, and thus with apparent distance. Thus the radius R of an imaginary sphere, of radius R_o, will change as the universe expands. Similarly, the wavelength of light moving through the universe will expand (redshift). The ratio R/R_o is thus related to the redshift:

9.1.3
$$R(z) = R_0/(1 + z).$$

Recall that $\lambda(z) = \lambda_o(1 + z)$ is the redshift relation for light (see chap. 1) of true wavelength λ_o emitted at redshift z and observed subsequently as having an increased wavelength $\lambda(z)$.

For small redshifts, the Doppler shift is approximately $z = v/c$, while the Hubble velocity is approximately $v = Hd$. Thus, the distance-redshift relation becomes to a good approximation for small enough scales

9.1.4
$$d = cz/H,$$

since it just involves the local expansion rate. (For additional information on this, see http://www.aoc.nrao.edu/~smyers/courses/astro12/L25.html.)

The left-hand side of equation 9.1.2 is the square of the Hubble constant

9.1.5 $$H_o = (1/R)(dR/dt).$$

Defining a critical energy density ρ_c as

9.1.6 $$\rho_c = 3H_o^2/8\pi G.$$

allows definition of the density parameters (to which we will return below) in terms of the critical density,

9.1.7 $$\Omega = \rho/\rho_C.$$

This allows the Friedman equation to be rewritten as

9.1.8 $$kc^2 = H_o^2 R^2(\Omega - 1).$$

Thus, if the density is greater than the critical density, the right-hand side is positive, $k = +1$, and a closed universe will result, that is, it will ultimately halt its current expansion and contract back to a "big crunch." If $\Omega = 1$, $k = 0$; this is referred to as a flat universe. For smaller densities, the universe will be open, that is, ever expanding. If $\Omega = 1$, then $k = 0$, in which case R is not determined absolutely. However, it can be chosen at some reference time and the Friedman equation then describes its evolution. Note, however, that these conclusions are based on the assumption that the cosmological constant, to be discussed below, is zero. This is now known not to be the case, and this changes the conclusions about the fate of the universe discussed above.

The very early universe is often said to have been radiation dominated, as opposed to matter dominated. This is because, as the universe expands, the matter density will decrease as

9.1.9 $$\rho_m \propto R^{-3},$$

But the radiation density will decrease faster than that because, in addition to the density of photons decreasing as R^{-3}, the energies of the individual photons expand with the scale factor of the universe, so their energy density decreases more rapidly than R^{-3}. Thus, if matter and radiation were the only considerations, the universe would now be matter dominated. We will therefore omit radiation energy density from the subsequent equations.

However, as noted above, we do need to worry about the cosmological constant Λ, on which much recent attention has been focused. It was originally put into the equations of general relativity by Albert Einstein in 1917 to accommodate a static universe (he referred to it as his "greatest blunder") and subsequently discarded by him. It is ironic that nearly a century later cosmologists have found that the cosmological constant is indeed essential in the general equation that describes the expansion of the universe and that Einstein should not have been so hasty in discarding it! It corresponds to an effective density that is independent of R and can be either positive or negative. As the universe expands, it will become increasingly important and eventually will dominate; indeed, it has been found to dominate at present. A positive cosmological constant corresponds to an effective negative pressure, equal and opposite to its energy density. Work done by expansion against this pressure creates the extra energy that allows the energy density to be constant even as the volume of each element of the universe is increasing (Leahy 2003). In general relativity, the true source of gravity is not just mass density but also includes the pressure P as

9.1.10
$$\rho + 3P/c^2 = \rho_m + \rho_\Lambda + 3(-\rho_\Lambda c^2)/c^2 = \rho_m - 2\rho_\Lambda.$$

It is customary to refer to each of the densities in terms of the critical density, that is,

9.1.11
$$\Omega_m = \rho_m/\rho_C,\ \Omega_\Lambda = \rho_\Lambda/\rho_C,\ \Omega_{\text{tot}} = \Omega_m + \Omega_\Lambda.$$

The terms associated with the cosmological constant Λ are currently the subject of several major research efforts; the cause of Λ is currently referred to as dark energy. So what is the dark energy? We do not know at present; it may be associated with supersymmetry or with extra dimensions. Perhaps it is a property of space-time itself (Carroll 2005). The studies that have been proposed, or that are being constructed, to search for the dark energy are directed more at the details of what it does than what it actually is. The understanding of it will certainly occupy the activities of astronomers and physicists for many years into the future.

History of the Early Universe

The evolution of the early universe is characterized is follows. As the big bang fireball expanded (see, e.g., Walker et al. 1991; Yang et al. 1984), possibly very rapidly in the initial instants, the inflationary era (for a very nice discussion of the means by which inflation produces a isotropic and homogeneous universe, in concordance with that observed, as well as other cosmological issues, see Liddle 1999), it cooled. After roughly 10^{-5} s had elapsed, the quark-gluon plasma that had existed up to that point had cooled enough to allow a phase transition in which the quarks combined into baryons. The primordial plasma also contained between 10^9 and 10^{10} times as many photons as baryons and numerous leptons, especially neutrinos, and their corresponding antiparticles.

At a somewhat later time, around $t = 10^{-2}$ s, the temperature would still have been high enough, $T = 10^{11}$ K (or $k_B T = 8.6$ MeV, an energy that is decidedly in the realm of nuclear physics!), that the protons and neutrons would be comparable in number and would be maintained in equilibrium by the interactions with the (much more numerous) leptons that also existed in the primordial soup (Walker et al. 1991; Yang et al. 1984). Indeed, the leptons were kept in thermal equilibrium by the neutral-current weak interaction as

9.1.12
$$e^+ + e^- \leftrightarrow \nu_i + \bar{\nu}_i .$$

The interactions that kept the baryons in equilibrium with the leptons are charged-current weak interactions:

9.1.13
$$p + e^- \leftrightarrow n + \nu_e ,$$

9.1.14
$$n + e^+ \leftrightarrow p + \bar{\nu}_e ,$$

9.1.15
$$n \leftrightarrow p + e^- + \bar{\nu}_e .$$

However, the slight mass-energy difference between the proton and the neutron (the neutron is more massive by $\Delta Mc^2 = 1.293$ MeV) gradually made it easier to convert neutrons into protons than the reverse, and the proton density [p] increasingly exceeded that of the neutrons [n], that is,

9.1.16
$$[n]/[p] = \exp(-\Delta Mc^2/k_B T).$$

At a slightly lower temperature, the equilibrium condition of equations 9.1.12–9.1.15 could no longer be maintained, as the universal expansion rate exceeded the rate at which neutrinos could interact. This occurred for $T = 4 \times 10^{10}$ K ($k_B T = 3.5$ MeV) for μ- and τ-neutrinos and for about $T = 2 \times 10^{10}$ ($k_B T = 1.7$ MeV) for e-neutrinos (because both neutral-current weak interactions and charged-current weak interactions occur for electron neutrinos).

9.2 Big Bang Nucleosynthesis (BBN)

As the temperature dropped slightly more, the neutrinos decoupled, and the reactions given in equations 9.1.13 and 9.1.14 proceeded only to the right. This temperature, however, was still sufficient to produce copious numbers of electron-positron pairs. However, at a slightly later time, about 100 s, and a temperature of about 10^9 K ($k_B T =$ just under 100 keV), the number of high-energy photons became insufficient to sustain pair production, and the positrons annihilated with electrons. The neutron-proton ratio at this freezeout temperature, T_F, was given by

9.2.1
$$[n]_F/[p]_F = \exp(-\Delta Mc^2/k_B T_F),$$

giving about 13 % neutrons and 87 % protons in the universe. Of course, a free neutron is an unstable particle; it decays to a proton with a lifetime of 885.7 ± 0.8 s (Particle Data Group 2004); this does produce a small but significant effect in the calculation of abundances from BBN.

As the universe expanded and cooled, the protons and neutrons interacted, and could, for an instant, form deuterons. However, the temperature was high enough that each resulting deuteron was immediately broken up into its constituent proton and neutron by high-energy photons until the temperature dropped to slightly less than 10^9 K. At that point, at about 100 s, BBN began and continued for roughly 20 minutes. The following reactions then resulted in the nuclides produced in the big bang:

9.2.2
$$\mathrm{p} + \mathrm{n} \rightarrow \mathrm{d} + \gamma,$$

9.2.3
$$\mathrm{d} + \mathrm{p} \rightarrow {}^3\mathrm{He} + \gamma,$$

9.2.4
$${}^3\mathrm{He} + \mathrm{n} \rightarrow {}^4\mathrm{He} + \gamma \text{ or } {}^3\mathrm{He} + \mathrm{d} \rightarrow {}^4\mathrm{He} + \mathrm{p},$$

9.2.5
$${}^3\mathrm{He} + \mathrm{n} \leftrightarrow {}^3\mathrm{H} + \mathrm{p},$$

9.2.6
$$\mathrm{d} + \mathrm{n} \rightarrow {}^3\mathrm{H} + \gamma,$$

9.2.7
$${}^3\mathrm{H} + \mathrm{p} \rightarrow {}^4\mathrm{He} + \gamma \text{ or } {}^3\mathrm{H} + \mathrm{d} \rightarrow {}^4\mathrm{He} + \mathrm{n}.$$

These reactions, and the nuclei they connect, are illustrated in figure 9.1.

Note that there are two ways to get from ^{3}He to ^{4}He; these will compete because, although the ^{3}He(d,p)^{4}He cross section is much larger than that for ^{3}He(n,γ)^{4}He at the temperatures at which BBN occurs, there are many less deuterons to initiate the former reaction than there are neutrons to initiate the latter. A similar situation exists for converting ^{3}H to ^{4}He. At the end of BBN the remaining ^{3}H will β-decay to ^{3}He ($T_{1/2} = 12.2$ years) and so will contribute to the ultimate BBN abundance of ^{3}He.

^{4}He

The abundance of neutrons can be tracked through the several minutes during which BBN occurred. It turns out that the neutron abundance during that

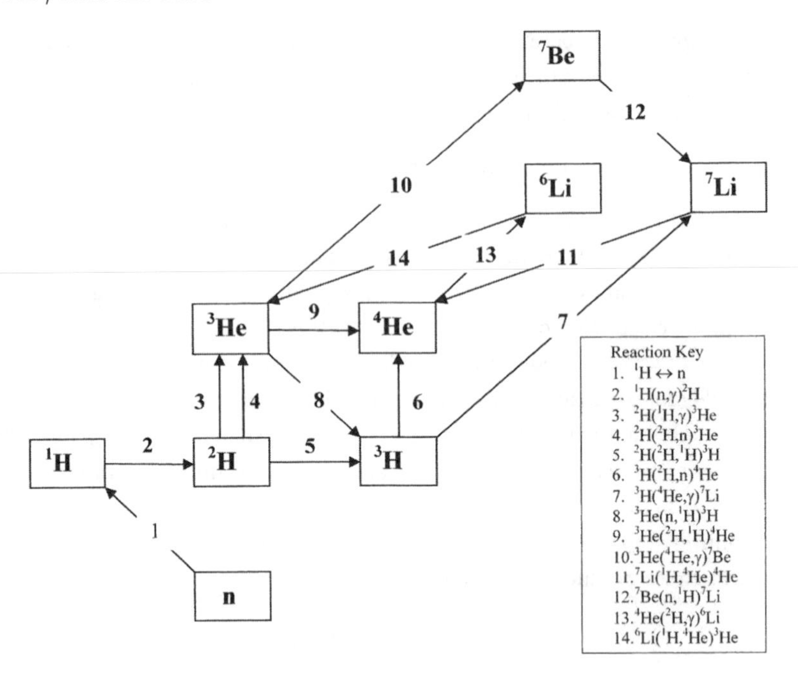

Fig. 9.1. The nuclides involved in big bang nucleosynthesis and the most important reactions that connect them.

time is about 12 %; virtually all of these neutrons will be combined with the more abundant protons to form ^{4}He. Thus the ratio of ^{4}He to ^{1}H from the BBN represents a very basic test of our understanding of the processes that occurred in the BBN. This would suggest that the amount of ^{4}He in matter that is thought to have been processed only in the BBN should be 24 % by mass. A major effort of astronomers has been directed toward determining this value. Since metals are produced in stars, astronomers can use the metallicity of a star to determine how close to the big bang the material of which it is made was produced. Thus the most metal-poor stars are the ones to which the astronomers turn for measuring BBN abundances; some of these have been found (see chap. 7) with metallicities of 1/10,000 of that of our Sun. In the case of ^{4}He, there are additional features that complicate the observations; these have also been dealt with in a variety of ways, some in selecting the objects to be observed and others by correcting the observed abundances. When a fairly large number of stars have been studied, their ^{4}He abundance can be extrapolated to zero metallicity to determine the primordial abundance. The value found by one group is $0.238 \pm 0.002 \pm 0.005$ (Fields and Olive 1998), where the first uncertainty is statistical and the second is systematic. This is at least in qualitative agreement with the BBN predicted value of 0.24. Another analysis, that of Olive, Steigman, and Walker (2004), also obtained a value of 0.238 ± 0.005. A study by another group (Isotov and Thuan 1998, 2004) suggested that the precision of the ^{4}He abundance might be much higher than that obtained by the other groups; they obtained a value of 0.2429 ± 0.0009. However, this uncertainty is not consistent with that obtained in the other studies.

Some interesting features do emerge when one looks at the details. The "standard model" of BBN, in which it is assumed that the baryon distribution was uniform and homogeneous during that period, is described by one parameter, η, the baryon-to-photon ratio, $\eta = n_B/n_\gamma$ and $\eta_{10} = 10^{10}\eta$. Because the ratio of ^{4}He to ^{1}H varies slowly with η, it is not so useful in determining η, although, if the abundance were really determined with the accuracy claimed by Isotov and Thuan (2004), it could determine η very accurately. However, Olive and Skillman (2004) reanalyzed the data used by Isotov and Thuan and concluded that major systematic uncertainties had not been included, especially with respect to the treatment of underlying stellar absorption. Olive and Skillman argue that an uncertainty of ± 0.013 would be more realistic when more of these systematic uncertainties are taken into account. Their reanalysis of the data produced a value for the primordial ^{4}He abundance (Y) of $0.232 < Y < 0.258$. This suggests that the indicated ^{4}He abundance indicated in figure 9.2 should be shifted upward somewhat and that the uncertainties should be increased, which would bring it into excellent agreement with the *WMAP* result (see section 9.4). Even this requires a measurement of the ^{4}He primordial abundance to 5 %, a formidable task!

^{2}*H*, ^{3}*He*

However, ^{2}H and ^{3}He are also produced in the BBN; these are considerably more sensitive to η than the ^{4}He abundance is. ^{3}He is both produced and destroyed in stellar burning, so its primordial value is difficult to determine from those observed (but see Balser et al. 1999). However, ^{2}H does have a well defined primordial value, at least within caveats concerning some of its recent determinations. This, together with its steep dependence on η, does let ^{2}H become a useful, perhaps the most important, "baryometer." It is generally thought that ^{2}H is only destroyed in stars, so its detection in environments that have not been processed by stars should produce a value that does reflect BBN. Such environments are provided by low-metallicity clouds of interstellar gas that exist between quasars and Earth; one can look for absorption lines from light produced from objects more distant from Earth than the clouds,

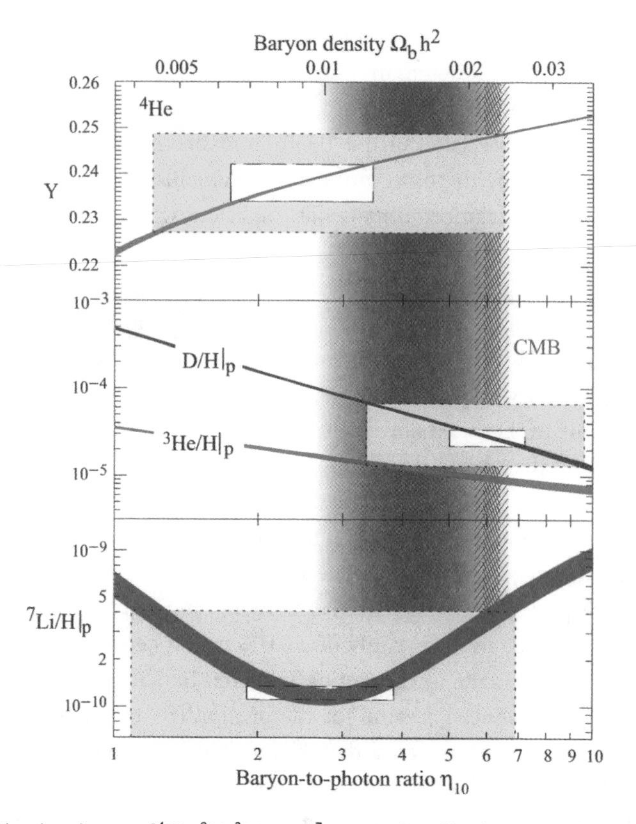

Fig. 9.2. The abundances of ^{4}He, ^{2}H, ^{3}He, and ^{7}Li as predicted by the standard model of BBN, along with indicated uncertainties in the theoretical predictions. Boxes indicate the observed light element abundances (smaller boxes, 2σ statistical errors; larger boxes, 2σ statistical and systematic errors added in quadrature). The ^{4}He data are from Fields and Olive (1998), the ^{2}H data are from Kirkman et al. (2000) and Linsky (2003), and the ^{7}Li data are from Ryan et al. (2000) and Pinsonneault et al. (2002). No observations are indicated for ^{3}He, as its primordial value is difficult to obtain. The narrow vertical band along the right side indicates the CMB measure of the cosmic baryon density. From Fields and Sarkar (2004). Copyright 2004, with permission from Elsevier.

that is, the clouds are "back lit." Unfortunately, this approach has produced an uncomfortably large range of values for the primordial ^{2}H abundance; these values do appear to vary over a wider than statistical range. Although the reasons for the large range may be physical, they may also be resolved as higher quality data using modern instrumentation are obtained. It should be noted that the ^{3}He abundance indicated in figure 9.2 is the sum of the abundances of the two species ^{3}H and ^{3}He.

^7Li

Another nuclide, ^{7}Li is also produced (see below) in the BBN, but using it as a test of BBN is somewhat more complicated than use of ^{4}He or ^{2}H, as its allowed value is not unique. It is fragile; it is easily destroyed in stars. Furthermore, it is also worth noting that determination of the primordial abundance of ^{7}Li has had some difficult questions of interpretation associated with it in times past (see, e.g., Deliyannis and Pinsonneault 1997), although these do not affect the above qualitative statement.

The reactions involved in the production and destruction of ^{7}Li are

9.2.8 $$^3\mathrm{H} + {}^4\mathrm{He} \rightarrow {}^7\mathrm{Li} + \gamma,$$

9.2.9 $$^3\mathrm{He} + {}^4\mathrm{He} \rightarrow {}^7\mathrm{Be} + \gamma,$$

9.2.10 $$^7\mathrm{Be} + \mathrm{e}^- \rightarrow {}^7\mathrm{Li} + \nu_e,$$

9.2.11 $$^7\mathrm{Be} + \mathrm{n} \rightarrow {}^7\mathrm{Li} + \mathrm{p},$$

9.2.12 $$^7\mathrm{Li} + \mathrm{p} \rightarrow 2\,{}^4\mathrm{He}.$$

The β-decay of ^{7}Be does not occur until much later than the period of BBN, as it occurs by electron capture, and the ^{7}Be will remain fully ionized in the rare environment following BBN until the time of recombination several hundred thousand years later. Nonetheless, it will ultimately contribute to the observed ^{7}Li abundance.

The ^{7}Li abundance comparison between theory and observation is somewhat more difficult. Two types of measurements have been made. One measures the Li abundance in metal-poor stars; a recent study of this type is given in Ryan et al. (2000). The standard observation is that the ^{7}Li abundance flattens out as metallicity decreases, producing the "Spite plateau" (Spite and Spite 1982); the abundance at this plateau has long been interpreted as the primordial abundance. The other type of study utilizes the Li abundances from globular clusters (see, e.g., Bonifacio et al. 2002 and sec. 9.5); these can also be chosen to have low enough metallicity that they should produce the same reading of the primordial Li abundance. Unfortunately, the two approaches appear to give results that differ by a factor of 2.

The complicated shape of the ^{7}Li abundance as a function of η is the result of competing processes. At low η, considerably more ^{3}H is made than ^{3}He, resulting in there being more ^{7}Li produced than ^{7}Be and more ^{7}Li than is produced at higher η. Conversely, at low η, little ^{7}Be is produced, but as η increases, more ^{3}He is produced than ^{3}H, resulting in more ^{7}Be than ^{7}Li.

Since ^{7}Be ultimately turns into ^{7}Li, the total mass 7 production is the sum of these two rather different production modes. But their sum produces the shape indicated in figure 9.2; there it can be seen that the predicted and observed ^{7}Li BBN abundances disagree by a factor of 3–4. This represents a significant unsolved problem for the standard model of BBN.

From the Spite plateau, as galactic age and metallicity increase, the ^{7}Li abundance builds up mostly from spallation reactions (reactions induced by relatively high-energy particles in which the heavier nucleus is broken up into two or more lighter nuclei) in the interstellar medium mostly between high-energy protons and ^{12}C. The nuclear physics aspects of this buildup have been studied for many years (see, e.g., Reeves, Fowler, and Hoyle 1970; Meneguzzi, Audouze, and Reeves 1971; and Walker, Mathews, and Viola 1989), and the cross sections have been well measured with nuclear physics accelerators to provide the inputs to the chemical evolution models. In addition, one reaction that could possibly produce ^{7}Li early in the history of the Galaxy would be

9.2.13 $$^4\text{He} + {}^4\text{He} \rightarrow {}^7\text{Li} + \text{p or } {}^7\text{Be} + \text{n},$$

following which (in the second reaction) the ^{7}Be would β-decay to ^{7}Li.

Although there are reactions that might have produced observable elements slightly heavier than ^{7}Li, the predicted abundances within the standard model of BBN are thought to be too small by several orders of magnitude to be observable at present (see, e.g., Kajino and Boyd 1990) unless the baryon density is allowed to assume values much larger than are now known to exist in the early universe. Although ^{9}Be has been observed in metal-poor stars (Gilmore et al. 1992), these observations have been interpreted as representing the buildup of ^{9}Be via spallation reactions, primarily from protons on ^{12}C. Of course, there are no stable mass 5 or 8 nuclei, so there are no such nuclei produced in BBN, and the reactions indicated in the above equations, although possibly appearing to the student to be somewhat convoluted, are necessary to produce the nuclides that do provide the tests of BBN.

^{6}Li

The most striking disagreement between the predictions of BBN and the observed results has developed recently: this is in the ^{6}Li abundance. If one assumes the baryon-to-photon ratio indicated by *WMAP*, the predicted ^{7}Li abundance is about a factor of 3 larger than that observed. But recent work by Asplund et al. (2006) has indicated an even worse problem for the ^{6}Li BBN abundance. The data appear to suggest a plateau at low metallicity similar to that observed for ^{7}Li. This is especially striking, as the standard BBN model predicts a ^{6}Li abundance about 3 orders of magnitude smaller than the level of the plateau. ^{6}Li is produced in BBN primarily by the reaction

9.2.14 $$^2\text{H} + {}^4\text{He} \rightarrow {}^6\text{Li} + \gamma,$$

the weakness of which explains the tiny BBN abundance of ^{6}Li. However, it is destroyed by the very strong reaction

9.2.15 $$^6\text{Li} + \text{p} \rightarrow {}^4\text{He} + {}^3\text{He}.$$

Further exacerbating the theoretical-observational discrepancy, ^{6}Li is strongly processed in stars, which suggests that its BBN abundance may be in even greater discord with the predictions of the standard BBN model, assuming that the ^{6}Li plateau holds up. The discrepancies presented by the two Li isotopes are currently unsolved questions.

Uncertainties and Caveats

The uncertainties on the theoretical predictions indicated in figure 9.2 are dominated by the uncertainties on the measured nuclear reactions rates. However, the uncertainty for ^{4}He depends very little on the reaction rates; it is mostly dependent on the neutron lifetime. And, of course, there is the assumption that the standard BBN model, with its single parameter, provides a correct description thereof.

A longstanding concern in BBN has been that the observed ^{4}He and ^{2}H abundances, when compared with the predicted abundances of BBN, do not give quite the same value of η. This can be seen in figure 9.2; the two allowed regions do not line up. However, the problem may have been solved by Olive and Skillman (2004), who found that the value often quoted for the BBN ^{4}He abundance is slightly too low and that the uncertainties quoted for that abundance are much too small in some analyses. However, if the discrepancy persists, it may also be telling us of new physics. This was discussed by Barger et al. (2003); one possibility (among many!) may be that there is a "neutrino asymmetry," that is, there may be more ν_es than $\bar{\nu}_e$s. This would have the effect of driving the reactions in equations 9.1.13 and 9.1.14 to a lower neutron abundance, which would clearly result in less ^{4}He.

Although the uncertainties in the observations, and therefore the limitations to refinement of any discrepancies between them and the theoretical predictions, are almost certainly systematic, it is still within the purview of nuclear astrophysicists to determine the nuclear reaction rates with the best precision possible. In this context, Nollett and Burles (2000) and subsequently

Burles, Nollett, and Turner (2001) considered the effects of the nuclear physics uncertainties on the predictions of the BBN abundances. Krauss and Romanelli (1990), in particular, and subsequently Nollett and Burles (2000) and others used a Monte Carlo approach to study the effect of each of the reactions that went into their reaction network, shown in figure 9.1, on each of the BBN abundances as a function of the baryon-to-photon ratio. They tried to identify where additional data were needed or what level of precision would be required to improve on the contribution of each to the total uncertainty of the BBN abundances. For the most part, they concluded that the largest uncertainties were still in the observed BBN abundances but that there were improvements in the nuclear physics that would improve the theoretically predicted abundances. While the neutron lifetime has long been the limitation on the theoretical predictions, that measurement (see, e.g., Spivak 1988; Mampe et al. 1989; Nezvizhevskii et al. 1992; Mampe et al. 1993; Byrne et al. 1996; Arzumanov et al. 2000; Nico et al. 2005) is now approaching the precision of the nuclear reactions. The weighted average of these values for the neutron lifetime is 885.3 ± 0.4 s, in agreement with the value given by the Particle Data Group (2004).

In another approach to precision cosmology, Fields and Prodanovic (2005) have assumed that the value of η determined by *Wilkinson Microwave Anisotropy Probe* (*WMAP*) (see sec. 2.2 and, for a more detailed discussion, sec. 9.4 below) is the correct one and have analyzed the discrepancies between the BBN abundances therefrom and those determined by astronomical observations. What they show is that the ^{2}H abundances do seem to give the same value for η as *WMAP* to within uncertainties but that is not the case for any of the other BBN abundances, in some cases by large factors. They note that this may be due to systematic uncertainties in some cases, but in others it may be suggesting the new physics noted by Barger et al. (2003). Most notably, the ^{4}He abundance, which is very accurately determined, at least statistically, appears to disagree with the *WMAP* result, as can be seen in figure 9.2. Until these possible discrepancies are resolved to the satisfaction of everyone working in the field, BBN will continue to be an active area of research. Of course, even their resolution at the current level of the uncertainties does not mean that future improvements in the measurements will not again produce discrepancies between the different types of measurements of the BBN abundances.

With all these caveats, it should nonetheless be noted that, with the exception of ^{6}Li, the predictions of BBN do a remarkable job of predicting the light element abundances over 10 orders of magnitude with the choice of a single parameter.

9.3 The Parameters of Cosmology

Two sets of observations, the Supernova Cosmology Project (Perlmutter et al. 1999) and the High-z Program (Riess et al. 1998), designed to measure the Hubble constant, were performed in the 1990s and were followed by another set of observations, using the *Hubble Space Telescope*, to observe Cepheid variables (see chap. 1) (Freedman et al. 2001). These measurements followed a decades-old conflict that had been waged between two groups to determine the Hubble constant, with one group, led by Gerard deVaucouleurs getting values around 100 km/s/Mpc and the other, led by Allan Sandage and Gustav Tamann, getting about half that, but with uncertainties much smaller than the difference. Freedman et al. (2001) achieved a value for H_o of $72 \pm 3 \pm 7$, which somewhat curiously is close to the average of the older values. In such measurements, the redshift of an object is compared with its distance from Earth. Of course, the difficult part lies in determining the distance to Earth, that is, defining standard candles that would allow that distance determination. Cepheid variables (see chap. 1) are good indicators but are not especially bright, so they limit the distance to which they can be used.

The approach of the Supernova Cosmology Project and the High-z Program was to use observations of type Ia supernovae at distances as far away as they could be observed to measure several of the cosmological parameters, including Ω_M and Λ, but also surely including H_o (see, e.g., Perlmutter et al. 1997). These could be identified by careful examination of their light curves, which suggested that type Ia supernovae have a nearly uniform energy output (and the differences are correctable), thereby allowing determination of the distance to them by measuring their apparent brightness. Unfortunately, type Ia supernovae do not occur frequently, so patience is necessary in observing them and in determining where they are in their light curve. The data obtained in the Supernova Cosmology Project, and from another previous set of observations, the Calan/Tololo data, which observed lower redshift data, are shown in figure 9.3.

The new data shown in figure 9.3 were taken with the 4-m Cerro Tololo telescope and, in a few cases, with the *Hubble Space Telescope* (to achieve data at higher redshift) and the *Keck I* and *II* 10-m telescopes. The supernovae studied were determined to be type Ia supernovae by fitting their light curves with type Ia supernovae templates. However, these had to be time dilated by $1 + z$ (see chap. 1) before fitting them to the light curves for the individual supernovae to accommodate the cosmological lengthening of the supernova timescale (Goldhaber et al. 2001; Liebundgut et al. 1996; Riess et al. 1997).

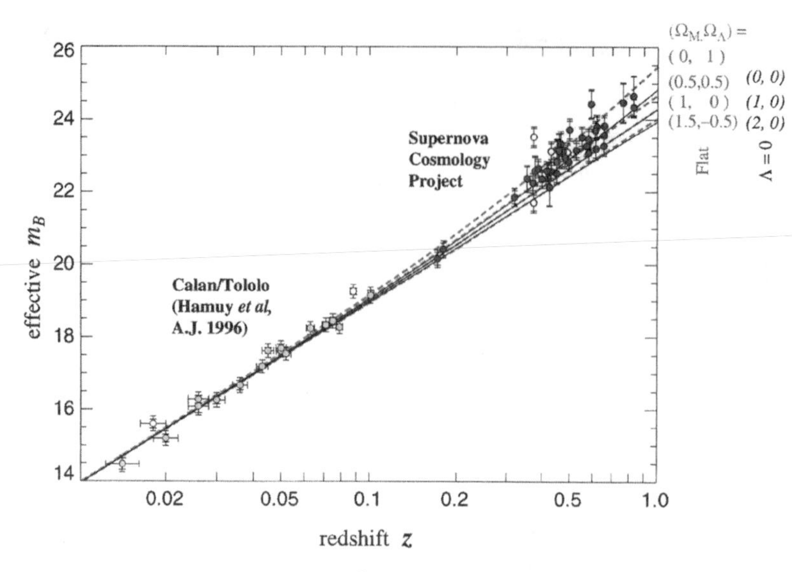

Fig. 9.3. Hubble diagram, showing $m_B^{effective}$, the effective peak magnitudes of the supernovae, versus redshift z, for 42 high-redshift type Ia supernova from the Supernova Cosmology Project and 18 lower redshift type Ia supernovae from the Calan/Tololo Supernova Survey (from Hamuy et al. 1996). Several outliers were not included in determining the fits to the data (although their inclusion did not affect the conclusions); these are indicated as open circles. The horizontal error bars represent the estimated peculiar velocity uncertainty of 300 km/s. The solid curves are the theoretical $m_B^{effective}(z)$ for a range of cosmological models with zero cosmological constant $(\Omega_M, \Omega_\Lambda) = (0, 0)$ on top, (1, 0) in the middle, and (2, 0) on the bottom. The dashed curves are for a range of flat cosmological models: $(\Omega_M, \Omega_\Lambda) = (0, 1)$ on top, (0.5, 0.5) second from top, (1, 0) third from top, and (1.5, −0.5) on the bottom. The high-redshift data are clearly seen to favor a nonzero Ω_Λ. From Perlmutter et al. (1999).

The data shown in figure 9.3 can be seen to exhibit the following features.

- The m_B values are larger than would be expected for a given z.
- The high-z supernovae are observed to be fainter than would be expected.
- Thus they are more distant than would be expected.
- Therefore the universe has since expanded more than would be expected.

The collection of studies using type Ia supernovae as standard candles produced the following conclusions:

- The data are strongly inconsistent with a cosmology in which the cosmological constant, Λ, is zero.
- The expansion of the universe is accelerating.
- The Hubble constant, determined from the fits to the data, produces an age of the universe of around 14 Gyr, with an uncertainty of about 1 Gyr.

- The cosmological constant must be a significant fraction of the energy density of the universe, in a "flat" universe, that is, in a universe in which $\Omega = \Omega_M + \Omega_\Lambda = 1$. The best-fit solution to the data has $\Omega_M = 0.28 \pm$ about 0.1. Thus $\Omega_\Lambda = 0.72$.

Notable in these two projects was the care that was taken to eliminate systematic errors. The light curves, that is, the time evolution of the luminosities, of the supernovae were carefully studied to be sure that the data were truly from type Ia supernovae. Consideration was given to the possibility that gravitational lensing could distort the brightness of the supernovae. The possibility that the photons might be influenced by environmental effects such as metallicity was considered, as was the possibility that the metallicity of the progenitor might affect the carbon-to-oxygen ratio of the exploding supernova and, hence, its energy output. The data were also carefully corrected for reddening due to extinction from dust from both our Galaxy and the host galaxy. With all these, and other, considerations, the corrections of the data and their attendant uncertainties were determined. In figure 9.3, both data sets were corrected for the type Ia supernova light curve width-luminosity relation. Two sets of error bars were considered, one giving the uncertainty when the measurement errors are added in quadrature and the other giving the total uncertainty when the intrinsic luminosity dispersion, 0.17 magnitudes of light-curve-width-corrected type Ia supernovae, is included in quadrature.

9.4 The Cosmic Microwave Background (CMB) Radiation

Following the epoch of BBN, as the universe continued to expand and cool, it eventually reached a point at which the electrons could combine with the nuclei formed. At this point, the universe, which had been completely opaque to the photons because of the large cross section for photons scattering from electrons, suddenly became transparent, and the resulting photons scattered little thereafter. Thus the spectrum of the photons that suddenly became free streaming will forever characterize the density distribution at the time of recombination, about 300,000 years after the big bang and at a redshift of slightly more than 1000.

Several measurements of the CMB radiation have been made in the decade prior to the writing of this book that have revolutionized cosmology. The first was from *COBE* (Smoot et al. 1991,1992), alluded to above, that measured the temperature of the blackbody CMB spectrum. These data have produced a stunning confirmation of the basic concept of the big bang. The photons detected by

COBE are the photons that were produced shortly after the big bang, but their energy/frequency spectrum reflects their distribution, now redshifted by many orders of magnitude by the expanding universe, at the time of recombination.

Another interesting feature of these photons is that they carry the imprint of their last scattering event forever, even to incredible detail, and those scattering events will, of course, reflect the density distribution at that point in time. Thus a number of experiments have been designed to search for fluctuations in the CMB radiation to a level of one part in 10^5. These include the Millimeter Anisotropy Experiment Imaging Array (MAXIMA; Lee et al. 2001), the Cosmic Background Imager (CBI; Pearson et al. 2003), the Arcminute Cosmology Bolometer Array Receiver (ACBAR; Kuo et al. 2002), the Differential Microwave Radiometer (*COBE*-DMR; Smoot et al. 1992), and the Balloon Observations of Millimetric Extragalactic Radiation and Geophysics (BOOMERANG; Netterfield et al. 2002). These projects have culminated in *WMAP* (Bennett et al. 2003; see sec. 2.2). The results of these experiments showed that the fluctuations on the CMB radiation, which represented the density fluctuations that existed at the time of recombination, could be detected to high precision. As noted by Pearson et al. (2003), the features that were detected by the CBI ranged in angular size from $\sim 6'$ to $\sim 15'$, corresponding to mass scales at the surface of last scattering of $\sim 5 \times 10^{14}$ to $8 \times 10^{15}\ M_{\odot}$, roughly the mass scale of clusters of galaxies. Thus, these data had the potential to test theories that purport to describe how the universe began and are related to theories that describe the subsequent galaxy formation, the seeds of which are thought to have been the density fluctuations that existed even in the very early universe.

All of the anisotropy data obtained from these experiments indicated the presence of "acoustic" peaks in the angular power spectrum of the CMB, which are the result of density fluctuations in the universe just prior to recombination and are now manifested as temperature fluctuations in the CMB. These structures were predicted many years prior to their detection by Sunyaev and Zeldovich (1970) and by Peebles and Yu (1970) and appear to be predicted in most cosmological models (Bond and Efstathiou 1987). However, cosmological models that are generated from topological defects, for example, would not produce such effects (see, e.g., Hu, Sugiyama, and Silk 1997; Liddle 1999).

In order to detect these fluctuations, one would clearly like to have a detector capable of measuring temperature fluctuations on an angular scale over which the fluctuations occur. A generic prediction of the models that do predict temperature fluctuations in the angular power spectrum is a dominant fundamental peak at an angular scale of about 1°, as can be seen in figure 9.4.

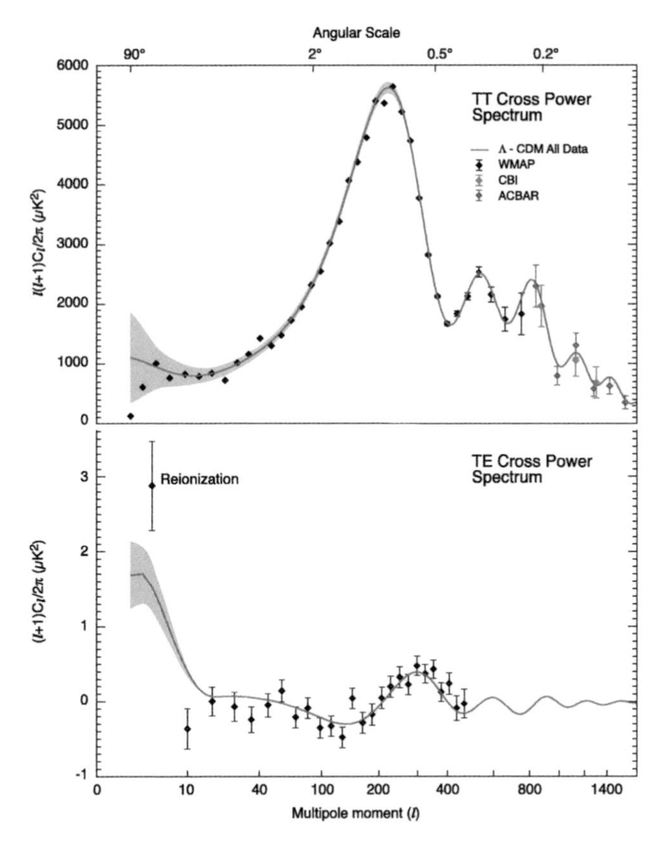

Fig. 9.4. The *WMAP* angular power spectrum. The *WMAP* temperature (TT) results are consistent with the ACBAR and CBI results, as shown at the top. The best-fit cold dark matter with cosmological constant model is shown. The temperature-polarization (TE) cross-power spectrum is shown at the bottom. "Cross Power," refers to the fact that the data shown were obtained from the "off-diagonal differencing assemblies," that is, the data were obtained from differences in temperatures, an approach that minimizes the spectral noise. The peak in the TE spectrum near $\ell = 300$ is seen to be out of phase with the TT power spectrum, as predicted for adiabatic initial conditions. The level of information contained in the details of these data is extraordinary. For example, the antipeak in the TE spectrum near $\ell = 150$ is evidence for "superhorizon modes" at decoupling, as predicted in inflationary models. From Bennett et al. (2003).

This decreases in angular scale as Ω, the universal energy density, decreases. The clear presence of this peak provides strong evidence for a low-curvature (i.e., essentially flat) universe (Bennett et al. 2003), a generic prediction of inflation models. If the fluctuations are fitted with a spherical harmonic expansion, this fundamental peak corresponds to $\ell \approx 200$. A second peak, which is the first harmonic of the fundamental peak, also exists in many models at $\ell \approx 550$. The height of this peak is related to the gravitational effects of the baryons at the time of recombination and so determines the baryonic density of the universe (see, e.g., Hu and White 2004). The height of the third peak, the second harmonic of the fundamental, is especially sensitive to the gravitational effects of the dark matter, so it determines the dark matter content of the universe. The data from the highest precision of these experiments, *WMAP*, are shown in figure 9.4. In those data, the ordinate is proportional to C_ℓ, which arises from the multipole expansion of the temperature anisotropy $T(\bar{n})$ as

$$T(\bar{n}) = \sum_{\ell,m} a_{\ell,m} Y_{\ell,m}(\bar{n}), \tag{9.4.1}$$

and

$$C_\ell = (2\ell + 1)^{-1} \sum_m |a_{\ell,m}|^2. \tag{9.4.2}$$

The *WMAP* data, shown in figure 9.4, are sufficiently precise that they allow a remarkable determination of several of the parameters that define cosmology. In the *WMAP* analysis (Bennett et al. 2003), the value for the total energy density Ω_{tot} was found to be equal to 1 to well within the uncertainties, that is, the solutions were found to be consistent with a flat universe. Because of this, as well as the fact that this value is consistent with a strong theoretical prejudice from inflation, the *WMAP* analysis shown above was constrained to this value, and searches were conducted on the other cosmological parameters. In these searches, the following parameters were determined:

- The Hubble constant was found to be $H_o = 71 + 4/{-3}$, in excellent agreement with the values obtained from the *Hubble Space Telescope* search value of $72 \pm 3 \pm 7$ km/s/Mpc (Freedman et al. 2001) but with greater accuracy.
- The age of the universe was determined to be 13.7 ± 0.2 Gyr. This clearly must be greater than ages determined from stars and so is consistent with universal ages determined from
  - Globular clusters 12 ± 1 Gyr (Reid 1997) or 10–14 Gyr (VandenBerg 2002; Brown et al. 2004)
  - The temperatures of the oldest white dwarfs 12.7 ± 0.7 Gyr (Hansen et al. 2002), and
  - Nucleosynthesis age dating, 15.6 ± 4.6 Gyr (Cowan et al. 1999).
- In addition, the value for Ω_B, the baryonic density, was found to be about 0.0444 ± 0.0018, in remarkable agreement with the value obtained from BBN of about 0.05, especially given that those two values were determined in completely different ways.

- The matter density was found to be $\Omega_m = 0.27 \pm 0.04$, which includes both baryonic matter and dark matter.
- The dark energy density, that is, that associated with the cosmological constant, was found to be $\Omega_\Lambda = 0.73 \pm 0.04$, in excellent agreement with the results from the type Ia supernova determination of 0.72.

The results of the observations of the type Ia supernovae along with those of the microwave background radiation have produced a stunningly accurate value of the Hubble constant. However, they also produced several other important signatures of cosmology: the age of the universe, the baryonic density (which was nicely consistent with that determined from BBN), a confirmation that there was dark—nonbaryonic—matter, and for the first time an indication that there was a dark energy, that is, something that acted in the opposite direction of gravity. Furthermore, this effect was playing off against other effects to produce a Hubble constant that actually was nearly constant over time, but only because of a delicate balance of factors that could have made it very nonconstant! Not surprisingly, this has produced a sudden flurry of theoretical activity to attempt to explain the dark energy, an effort that is continuing at the time this book is being written. There are also numerous discussions taking place to determine the best way to map out the properties of the dark energy, hopefully producing some clues as to its nature. Future experimental and observational projects will certainly yield important information on its effects. Such projects as the Atacama Cosmology Telescope (http://www.hep.upenn.edu/act/) and the South Pole Telescope (http://spt.uchicago.edu/) are designed to do precisely that. Hopefully, their results will enable the next-generation devices to take aim at the nature of the dark energy.

Regardless of the nature of the dark energy, it must have consequences on future cosmological measurements. For example, if it is possible for the dark energy to cluster, that would certainly affect the formation of structure at some level, possibly very strongly, in the universe (Bean and Dore 2003, 2004). However, whatever effects it has are not necessarily independent of other entities that can affect the distribution of the CMB. For example, the Sunyaev-Zeldovich effect is expected to redistribute the energies of the CMB. This is a shift from lower to higher energies of the CMB that results from scattering of the CMB photons from the energetic electrons in whatever matter the CMB encounters. Since the electrons are higher energy than the photons, they would see a net gain of energy in scattering from the matter. This has been observed by a number of investigators (see, e.g., Birkenshaw 1990; Herbig et al. 1995; Carlstrom, Joy, and Grego 1996; Myers et al. 1997).

The next generation of CMB measurements is of the polarization fluctuations of the microwave background radiation. Indeed, some of these already exist, notably from the Degree Angular Scale Interferometer (DASI; Kovac et al. 2002) and *WMAP*, but the next generation of such measurements promises to provide much greater detail about these fluctuations than those that exist at present. The polarization results from the last scatter of the photons from electrons just before their recombination with the nuclei from the big bang. Scattering does produce polarization, so even though the anticipated effects are small, the polarization does have the potential to produce an even more stringent test of the big bang model, even as far back in time as the inflationary epoch, than currently exists.

9.5 Globular Clusters

An independent confirmation of our description of cosmology is afforded by globular clusters: gravitationally bound concentrations of roughly 10,000 to 1 million stars, spread over a volume of several tens to about 200 light-years in diameter. One such object, M2 in Aquarius, is shown in figure 9.5. The Hertzsprung-Russell (HR) diagrams from these globular clusters can, as we will see below, provide a determination of the ages of these old systems of stars, which in turn provide a lower limit on the age of the universe.

The first globular cluster discovered, but then taken for a nebula, was M22 in Sagittarius. It was probably discovered by Abraham Ihle in 1665. This discovery was followed by that of the southern Omega Centauri (NGC 5139) by Edmond Halley on his 1677 journey to St. Helena. This "nebula" had been known but classified previously as a star. Charles Messier was the first to resolve one globular cluster, M4, but still referred to the other 28 of these objects in his catalog as "round nebulae." By summer 1782, before William Herschel started his comprehensive deep-sky survey with large telescopes, there were 33 globular clusters known. Herschel himself discovered 37 new globulars, bringing the number known to 70. He was the first to resolve virtually all of them into stars and coined the term "globular cluster" in the discussion adjacent to his second catalog of 1000 deep-sky objects (1789). These early known globulars all belong to our Milky Way Galaxy; the total number of known globular clusters in the Milky Way is now 151.

Since their discovery and resolution, globular clusters were always assumed to be swarms of stars held together by their mutual gravity. The proof for this correct guess came with spectroscopy, which showed that the stars of these clusters have radial velocities, as expected for such swarms, and perfectly

Fig. 9.5. Globular cluster M2 in Aquarius. Photo by D. Williams and N.A. Sharp with the 0.9 m telescope of Kitt Peak National Observatory. From http://www.noao.edu/image_gallery. Courtesy of NOAO/AURA/NSF and N.A. Sharp.

match in color-magnitude or HR diagram location, that is, they represent a population of stars of about the same age.

Of the 151 known Galactic clusters, 138 are concentrated in the hemisphere centered on Sagittarius, while only 13 globulars (8.6 %) are on the opposite side of us (among them M79). Of these 13, four (including M79) are suspected to be members of the remnant globular cluster system of the Canis Major Dwarf galaxy discovered in 2003, which is in the process of being integrated into the Milky Way's Galactic halo. Indeed, this concentration of the globulars in the Galactic center, producing this distribution asymmetry, is what led Shapley in his famous debate (see, e.g., http://antwrp.gsfc.nasa.gov/htmltest/gifcity/cs_gloss.html) with Curtis in 1920 to argue that the Sun lies far from the Galactic center.

Radial velocity measurements have revealed that most globulars are moving in highly eccentric elliptical orbits that take them far outside the Milky Way; they form a halo of roughly spherical shape that is highly concentrated to the Galactic center but that reaches out to a distance of several hundreds of thousands of light-years, much more than the dimension of the Galaxy's disk.

As they do not participate in the Galaxy's disk rotation, they can have high relative velocities of several hundred kilometers per second with respect to our solar system; this is what shows up in the radial velocity measurements.

Spectroscopic study of globular clusters shows that they are much lower in heavy-element abundance than stars such as the Sun that form in the disks of galaxies. Thus, globular clusters are believed to be very old and produced from the stellar debris from an earlier generation of stars (population II), which formed from the more primordial matter present in the young galaxy just after (or even before) its formation. The disk stars, by contrast, have evolved through many cycles of star birth and supernovae, which have enriched the heavy-element concentration in star-forming clouds and may also trigger their collapse.

The HR diagram shown in figure 9.6 illustrates the situation for a typical globular cluster (note that "B − V" is "blue minus visual magnitudes," or luminosities, which is another way astronomers have of specifying the temperature of a star; a bluer star is hotter, so this is consistent with the HR diagram shown in fig. 1.7); the data are for the globular cluster M5. Since more massive stars burn their hydrogen more rapidly than do less massive stars, they will end their life on the main sequence (MS in fig. 9.6) sooner than will the less massive stars. Since all the stars in a globular cluster were produced at the same time, this produces the main-sequence turnoff, which is indicated in figure 9.6 as TO, the stars of just the mass—and age—that are just now evolving beyond the main sequence. Thus the point at which the main-sequence turnoff occurs determines the age of the globular cluster, provided the correlation between location on the main sequence and age can be established, as is elaborated below.

When the central fuel is gone, hydrogen starts to burn in an envelope around a dense helium core (see chap. 7). The star's outer regions expand because of this new energy input. As the emitting surface area of the star's photosphere increases so does its apparent brightness; also, as it expands the photosphere cools (as it becomes cooler, the color of the star becomes redder). The star thus moves up and to the right on the HR diagram, climbing the red giant branch (RGB in fig. 9.6).

At the tip of the RGB, the helium-rich core ignites and helium fusion begins. This ignition of the core causes the star to move rapidly down the HR diagram to the horizontal branch (HB) region. The HB is the region inhabited by stars that are converting helium to carbon in their cores. A strong feature of the HB in this particular globular cluster, M5, is a gap in the HB. In M5, the HB is separated into two sections—a blue HB and a red HB—separated by a

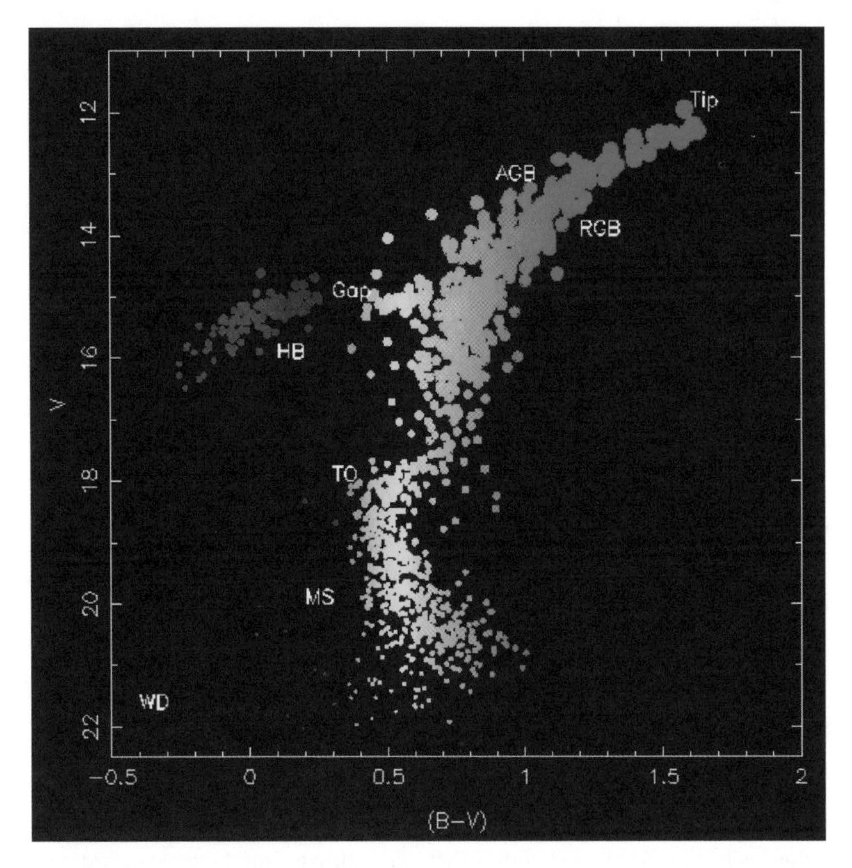

Fig. 9.6. Shown is an example of an HR diagram for the globular cluster M5. Various regions of the HR diagram are identified: main sequence (MS); turn off (TO); red giant branch (RGB); helium flash, occurring here at tip of RGB (tip); horizontal branch (HB); Schwarzschild gap in the HB (gap); asymptotic giant branch (AGB); and the final stellar remnants, white dwarfs (WD), which will lie off the bottom of the diagram. These regions show the main phases of stellar evolution. Figure from http://www.dur.ac.uk/ian.smail/gcCm/gcCm_intro.html. Reprinted with permission from I. Smail.

gap. The gap indicates a region of instability in the physics of the stellar envelope that results in stars in this region causing them to evolve quickly either onto the red or blue flavors of HB. HB morphology differs between globular clusters; some globular clusters show both blue and red HBs, and some show one and not the other.

As the central helium fuel runs out, shell burning starts again around the core. This time, however, two shells are formed (again, see chap. 7), the inner one burning helium and the outer burning hydrogen. The star now moves off the HB and up the asymptotic giant branch (AGB), possibly blowing off its outer layers. Ultimately, a star of medium mass will suffer the loss of its complete envelope and will eventually evolve from the AGB to the white dwarf region, leaving the central hot white dwarf in the middle of a planetary nebula.

Comparison of the measured point of departure from the main sequence of each globular cluster with that predicted from theoretical models, which are derived from the theory of stellar evolution, provides the possibility to estimate the age of that particular cluster. Surprisingly, almost all globular clusters seem to be about the same age; there seems to be a physical reason that they all formed in a short period of time in the history of the universe, and this period was apparently shortly after the galaxies formed. Early estimates yielded an age of 12–20 billion years; but more recent determinations give values of perhaps 10–14 Gyr (see, e.g., VandenBergh 2002; Brown et al. 2004). Their age provides a lower limit for the age of our universe. In early 1997, the discussion of the age of the globular clusters was revived by the general modifications of the distance scale of the universe, implied by results of the European Space Agency's astrometrical satellite *Hipparcos*. These results suggest that galaxies and many galactic objects, including the globular clusters, may be at a 10 % larger distance than previously thought; therefore, the intrinsic brightness of all their stars must be about 20 % higher. This implies that they may be roughly 15 % younger than previously thought, thus reducing the globular cluster age estimate. However, since the globular cluster age has traditionally been (embarrassingly) greater than the age of the universe determined by other techniques, this does bring that age estimate into better concordance with the age determined by other means.

For considerably more discussion of the issues associated with globular clusters, see Djorgovski and Meylan (1993). Much of the information for this discussion came from the Web site, http://www.dur.ac.uk/ian.smail/gcCm/gcCm_intro.html, which was created by Ian Smail.

This discussion of globular clusters is a fitting way to conclude our nuclear astrophysical considerations, since it combines all the features of stellar evolution that result from the nuclear astrophysical details we have discussed with the basics of cosmology. Indeed, a recent measurement of the $^{14}N(p,\gamma)^{15}O$ reaction (Formicola et al. 2004), discussed in chapter 5 in the context of solar hydrogen burning, also has a significant impact on cosmology via its impact on the ages of the globular clusters. This measurement, performed with the *LUNA* facility (see sec. 2.12.6), deals with the reaction that is the bottleneck reaction for the transition from the main sequence to the RGB. Since the mass of the stars at the turnoff point from the main sequence, and therefore the age of the globular clusters, depend on this reaction, it is crucial to the determination of the globular cluster ages. The work of Formicola et al. (2004) necessitated a revision of the reaction rate of that reaction enough that it imposed an upward revision on the ages of the globular clusters of 0.7 to 1.0 Gyr (Imbriani et al. 2004). This certainly focuses all of the issues discussed in this and in the preceding eight chapters on the nuclear reactions on which all these considerations are ultimately based; it would be difficult to emphasize in any greater way the importance of nuclear physics to all of astrophysics!

CHAPTER 9 PROBLEMS

1. Assuming that [n] = 12 % and [p] = 88 % from the big bang (those are the mass fractions just before BBN began), that all those neutrons ultimately were combined to form ^{4}He, and that everything except ^{1}H and ^{4}He had a negligible abundance, calculate the mass abundance of ^{4}He from BBN.
2. Assuming the distribution of photon density n is given by

$$\varphi(E) = (k_B T)^{-1} \exp(-E/k_B T),$$

and

$$dn/dE = n_{tot}\varphi(E),$$

determine the fraction of photons having energies greater than some cutoff energy $E_\odot$. Assume $E_o << k_B T$. Then, assuming that there are 10^{10} as many photons as neutrons at the temperature at which big bang nucleosynthesis begins, and that the number of photons having energy above 2.2 MeV must fall below the number of neutrons before deuterium can form, estimate the temperature at which deuterium begins to form.
3. A Physics 101 way to determine the critical density of the universe is as follows: Assuming that the velocity of recession follows Hubble's law, first determine the escape velocity v of a particle at the edge of a sphere of uniform density ρ and radius R. Then show that a particle of this velocity will be prevented from escaping to infinity if the density inside the sphere is at least as great as

$$\rho = 3H_o^2/8\pi G,$$

where H_o is the Hubble constant and G is the gravitational constant.
4. Extrapolate Hubble's law to a distance at which the recessional speed is equal to the speed of light. What is that distance?
5. One of the lines of sodium is emitted at a wavelength of 590.0 nm. If the light from a distant galaxy shows that line to be at 602.0 nm, what is the distance to the galaxy, assuming Hubble's law is valid?

6. There exists on the Internet a code that is publicly available by which one can perform BBN calculations: http://www-thphys.physics.ox.ac.uk/users/SubirSarkar/bbn.html. Try your own BBN calculation, with variations first by an order of magnitude each way in baryonic density, then by a factor of 2 each way, to obtain some estimate of the sensitivity of the BBN to η.
7. There are four ways to get from the mass 3 nuclides to ^{4}He in BBN: ^{3}H(p, γ), ^{3}He(n, γ), ^{3}H(d, n), and ^{3}He(d, p). One way to determine the importance of each to determining the ^{4}He abundance is to vary each to see how much of an effect that has on the results. Try varying each by 10 % (roughly the level to which they are currently known) to see how much the ^{4}He abundance is affected.
8. Use the *WMAP* data shown in figure 9.4 for the fundamental peak in the temperature fluctuations, along with equations 9.4.1 and 9.4.2, to estimate the magnitude of the temperature fluctuations in the CMB.

Appendix

ADDITION OF ANGULAR MOMENTA, CLEBSCH-GORDON COEFFICIENTS, AND ISOSPIN

A.1 Vector Addition of Angular Momenta

Angular momenta obey the following commutation property, which can be written as

A.1.1 $$[J_i, J_j] = J_iJ_j - J_jJ_i = iJ_k\varepsilon_{ijk}, i, j, k = 1, 2, 3; \text{ or } x, y, z.$$

In any quantum mechanics applications, these operators act on wave functions that we may label by the eigenvalues of the operators $\bar{J}^2 = \Sigma_i J_i^2$ and one of the components, usually the operator J_z. The axis along which J_z points is the quantization axis. This axis makes sense as a direction in the usual space if $\bar{J}$ is an actual angular-momentum vector operator—orbital, spin, or a combination, with a spatial orientation. However, we will also discuss a different interpretation below, isospin, which will formally be identical to this angular-momentum scenario but which will have its orientation in a different "space." Angular momentum wave functions will be written as $|j\, m_z\rangle$, or just $|j\, m\rangle$, as it is usually assumed that the axis of quantization is the z-axis. These wave functions are eigenvectors of $\bar{J}^2$ and J_z, that is,

A.1.2 $$J^2|j\,m\rangle = j(j+1)|j\,m\rangle, \text{ and } J_z|j\,m\rangle = m|j\,m\rangle.$$

Note that, technically, the first of the equations in equation A.1.2 has a factor of $\hbar^2$ on the right-hand side and the second a factor of $\hbar$ on the right-hand side. However, for convenience we have adopted a system of units in which $\hbar = 1$ for this discussion. We can also define "creation and annihilation" operators, or "raising and lowering operators," $J_\pm = J_x \pm iJ_y$ such that

A.1.3 $$J_\pm|j\,m\rangle = [j(j+1) - m(m \pm 1)]^{1/2}|j\,m \pm 1\rangle.$$

There are many situations in physics in which we need to add (vectorially) two or more angular momenta. For example, the spin-orbit force in both atoms and nuclei will couple a particle's spin with its orbital angular momentum, which requires eigenfunctions of the composite operators. So if we have two angular momenta, $\bar{J}_1$ and $\bar{J}_2$, we require a formalism to combine them just as one does with vectors to form total angular momentum $\bar{J}$, that is,

A.1.4 $$\bar{J} = \bar{J}_1 + \bar{J}_2.$$

Since equation A.1.4 is an equation that involves vectors, it must also hold for any one of its components, specifically,

A.1.5 $$J_z = J_{1z} + J_{2z}.$$

A.2 Formation of Wave Functions of the Total Angular Momentum: Coupling of Two Angular Momenta

Writing the eigenfunctions for the angular-momentum vectors $\bar{J}_1$ and $\bar{J}_2$ as $|j_1\, m_{1z}\rangle$ and $|j_2\, m_{2z}\rangle$, we may write the wave function that accommodates ("spans the space of") both quantum mechanical operators as the product wave function $|j_1\, m_{1z}\rangle\, |j_2\, m_{2z}\rangle$, so that (dropping the subscript z)

A.1.6
$$J_z|j_1m_1\rangle|j_2m_2\rangle = (J_{1z} + J_{2z})|j_1m_1\rangle|j_2m_2\rangle$$
$$= (m_1 + m_2)|j_1m_1\rangle|j_2m_2\rangle.$$

Thus the projection of the total angular momentum equals the sum of the projections of the individual angular momenta. For J_i, there are $2j_i + 1$ possible projections of m_i, and so there are $(2j_1 + 1)(2j_2 + 1)$ possible sums of $m = m_1 + m_2$, some of which can occur with a multiplicity greater than one. For example, if $j_1 = j_2 = 1/2$, $m = 0$ occurs twice: once as $0 = 1/2 - 1/2$ and once as $0 = -1/2 + 1/2$. The total number of multiplicities of the m values is $(2{\cdot}1/2 + 1)(2{\cdot}1/2 + 1) = 4$: $m = -1, 0, 0, +1$.

What we are seeking is linear combinations of the product wave functions $|j_1\, m_{1z}\rangle\, |j_2\, m_{2z}\rangle$ that are eigenstates of $\bar{J}^2$ and J_z. It can be shown that $\bar{J}_1^2$ and $\bar{J}_2^2$ commute with $\bar{J}^2$ and J_z, so that a basis of simultaneous eigenfunctions of the four operators $\bar{J}^2$, J_z, $\bar{J}_1^2$, and $\bar{J}_2^2$ can be found. It is also straightforward to show that J_{1z} and J_{2z} do not commute with $\bar{J}^2$. The new basis is typically written as $|j_1\, j_2\, j\, m\rangle$ or often just $|j\, m\rangle$, since j_1, j_2 remain fixed in all the wave functions, while the previous product basis is typically written either as $|j_1\, m_1\rangle\, |j_2\, m_2\rangle$ or $|j_1\, j_2\, m_1\, m_2\rangle$; both forms are complete sets of eigenfunctions.

Now we address the question, given a state $|j_1\, j_2\, m_1\, m_2\rangle$, how do we express it in terms of the $|j_1\, j_2\, j\, m\rangle$? Since the collection of wave functions $|j_1\, j_2\, m_1\, m_2\rangle$

is by definition complete, we can form an expansion

A.1.7
$$|j_1 j_2 j m\rangle = \sum_{m_1, m_2} |j_1 j_2 m_1 m_2\rangle c^{jm}_{j_1 j_2 m_1 m_2},$$

which must always be possible; the coefficients $c^{jm}_{j_1 j_2 m_1 m_2}$ are called the Clebsch-Gordan coefficients or, within a slight redefinition, 3-j coefficients.

Clearly, j and m must be able to assume all allowed values, and in doing so, another complete set of wave functions is obtained: the product-basis wave functions are simultaneous eigenvalues of the operators $\bar{J}^2_1, \bar{J}^2_2, J_{1z}$, and J_{2z}, while the new-basis wave-functions are simultaneous eigenvalues of the operators $\bar{J}^2, \bar{J}^2_1, \bar{J}^2_2$, and J_z. The vectors of this new complete basis can be made orthonormal via the Gram-Schmidt orthonormalization procedure, which will be assumed to have been employed.

A.3 Clebsch-Gordan Coefficients

The orthonormality of the $|j_i\ m_i\rangle$ can be used to determine the Clebsch-Gordan coefficients. To do so, multiply equation A.1.8 by $\langle j_1 j_2 m'_1 m'_2|$,

A.1.8
$$\langle j_1 j_2 m'_1 m'_2 || j_1 j_2 j m\rangle = c^{jm}_{j_1 j_2 m'_1 m'_2},$$

where j_1 and j_2 are fixed. For a variety of historical reasons, these coefficients are also written as

A.1.9
$$c^{jm}_{j_1 j_2\, m_1 m_2} = \langle j_1\ j_2\ m_1 m_2 | j m\rangle,$$

and also

A.1.10
$$c^{jm}_{j_1 j_2 m_1 m_2} = (-1)^{j_1 - j_2 + m}(2j+1)^{1/2} \begin{pmatrix} j_1 & j_2 & j \\ m_1 & m_2 & -m \end{pmatrix},$$

where the last, matrix-like symbol is called the 3-j symbol.

As both bases, $|j_1 j_2\ m_1\ m_2\rangle$ and $|j_1 j_2 j\, m\rangle$ are orthonormal by construction, the transformation given in equation A.1.7 must be unitary, and so its inverse is

A.1.11
$$|j_1 j_2 m_1 m_2\rangle = \sum_{j,m} \langle j m | j_1 j_2 m_1 m_2\rangle |j_1 j_2 j m\rangle,$$

where

A.1.12
$$\langle j m | j_1 j_2 m_1 m_2\rangle = \langle j_1 j_2 m_1 m_2 | j m\rangle^*.$$

The Clebsch-Gordan coefficients can and are always chosen to be real, so we may drop the asterisk here. Thus, using equations A.1.7 or A.1.11, one can freely switch back and forth between the two bases, provided the Clebsch-Gordan coefficients are known.

Clearly $\langle j_1\, j_2\, m_1\, m_2 | j\, m\rangle = 0$ unless $m = m_1 + m_2$; this simply reflects the fact that $J_z = J_{1z} + J_{2z}$. In addition, the vector equation A.1.4 may be represented by the expression

A.1.13 $$|j_1 - j_2| \le j \le |j_1 + j_2|.$$

This can be seen readily by observing that the maximum possible value for m is $m_{1\max} + m_{2\max} = j_1 + j_2$. Since this must be the (maximal) projection of the maximal possible J, it must be true that $j \le j_1 + j_2$. On the other hand, the minimal possible value for j is $|m_{1\max} - m_{2\max}| = |j_1 - j_2|$. Since this must be the (minimal positive) projection of the minimal possible value of j, it must be true that $|j_1 - j_2| \le j$.

Therefore, $\langle j\, m | j_1\, j_2\, m_{1z}\, m_{2z}\rangle = 0$ unless the conditions specified in equation A.1.13, the "triangle inequality," as well as $m = m_1 + m_2$ are satisfied.

One now has all the machinery necessary to calculate the Clebsch-Gordan coefficients in a deliberate way. If the task is to couple angular momenta $\bar{J}_1$ and $\bar{J}_2$, and to write the eigenfunctions of $\bar{J}^2$ and J_z, we can begin by writing the eigenfunction for the maximum value of j and its maximum projection; there is only one such term in its wave function. It is

A.1.14 $$|j_1 j_2 j j\rangle = |j_1 j_1\rangle |j_2 j_2\rangle,$$

since everything has to assume its maximum value, that is, the Clebsch-Gordan coefficient for this case is 1.0. One can now generate the rest of the members of the wave functions for the $|j_1\, j_2\, j\, m\rangle$ eigenfunctions for which $j = j_1 + j_2$ by application of the lowering, or annihilation, operators. For example,

A.1.15 $$\begin{aligned} |j_1 j_2 j j-1\rangle &= (J_{1-} + J_{2-})|j_1 j_1\rangle |j_2 j_2\rangle \\ &= \{J_{1-}|j_1 j_1\rangle\}|j_2 j_2\rangle + |j_1 j_1\rangle\{J_{2-}|j_2 j_2\rangle\} \\ &= [j_1(j_1+1) - j_1(j_1-1)]^{1/2}|j_1 j_1 - 1\rangle|j_2 j_2\rangle \\ &\quad + |j_1 j_1\rangle[j_2(j_2+1) - j_2(j_2-1)]^{1/2}|j_2 j_2 - 1\rangle. \end{aligned}$$

Since the j_1 operators operate only on the $|j_1\, m_1\rangle$ space, and similarly for the j_2 operators, they identify the Clebsch-Gordan coefficients for the two terms in this wave function. All subsequent wave functions will also have two terms except the last one, since the last term, that with $|j_1\, j_2\, j\, {-j}\rangle$, must have the configuration

A.1.16 $$|j_1 j_2 j - j\rangle = |j_1 - j_1\rangle |j_2 - j_2\rangle,$$

since the most negative projection of j must be that given by the sum of the most negative projections of j_1 and j_2: $-j_1 - j_2$.

The next task is to determine the wave functions for the eigenfunctions having j less than $j_{\max}$ by one unit, that is, $|j_1\, j_2\, j-1\, m\rangle$. Here one uses the orthogonality of the coefficients for the wave functions having the same j_1, j_2, and m but different j. Thus, having calculated the wave function as done above for $|j_1 j_2 j j-1\rangle$, one can calculate the wave function for $|j_1 j_2 j-1\, j-1\rangle$ by assuming that it must have the general form

A.1.17 $$|j_1 j_2 j-1 j-1\rangle = a|j_1 j_1 - 1\rangle|j_2 j_2\rangle + b|j_1 j_1\rangle|j_2 j_2 - 1\rangle,$$

then by taking the inner product,

A.1.18 $$\langle j_1 j_2 j j-1 | j_1 j_2 j-1 j-1\rangle = 0.$$

The inner product will give one equation, so with the required orthonormality of the expression for $|j_1\, j_2\, j-1\, j-1\rangle$, one thus has the two equations needed to calculate a and b. The remaining eigenfunctions $|j_1\, j_2\, j-1\, m\rangle$ can then be determined by applying the lowering operator to $|j_1\, j_1\, j-1\, j-1\rangle$.

This procedure can clearly be followed through for all the possible values of j, from its maximum value $j_1 + j_2$ to its minimum value $|j_1 - j_2|$.

There does exist a general formula for the Clebsch-Gordan coefficients, which is possibly more difficult to use than the procedure outlined above, but which is useful for programming:

A.1.19a $$c^{jm}_{j_1 j_2 m_1 m_2} = \delta_{m m_1 + m_2} \rho^{j}_{j_1 j_2} \sigma \tau,$$

A.1.19b $$\delta_{m, m_1 + m_2} = \begin{cases} 1, & m = m_1 + m_2, \\ 0, & m \neq m_1 + m_2; \end{cases}$$

A.1.19c

$$\rho^{j}_{j_1, j_2} = \{(j_1 + j_2 - j)!(j + j_1 - j_2)!(j_2 + j - j_1)!(2j+1)/[(j_1 + j_2 + j + 1)!]\}^{1/2}$$

A.1.19d $$\sigma = \{(j+m)!(j-m)!(j_1+m_1)!(j_1-m_1)!(j_2+m_2)!(j_2-m_2)!\}^{1/2},$$

A.1.19e $$\tau = \sum_r (-1)^r \{(j_1 - m_1 - r)!(j_2 + m_2 - r)!(j - j_2 + m_1 + r)! \\ (j - j_1 - m_2 + r)!(j_1 + j_2 - j - r)!r!\}^{-1},$$

where the sum over r extends over all values for which none of the factors in the parentheses vanishes. This makes the sum finite. Alternatively, use that $0! = 1$

and that $(-n)! = \Gamma(1 - n) = \infty$ for $n = 1, 2, \ldots$, so such terms contribute nothing to the sum (http://string.howard.edu/~tristan/QM2/QM2WE.pdf).

Important relationships between the Clebsch-Gordan coefficients include the following:

Completeness

A.1.20
$$\sum_{m_1 m_2} c^{jm}_{j_1 j_2 m_1 m_2} c^{j'm'}_{j_1 j_2 m_1 m_2} = \delta_{jj'} \delta_{mm'},$$

A.1.21
$$\sum_{jm} c^{jm}_{j_1 j_2\, m_1 m_2} c^{jm}_{j_1 j_2 m'_1 m'_2} = \delta_{m_1 m'_1} \delta_{m_2 m'_2}.$$

Symmetry

A.1.22
$$c^{jm}_{j_1 j_2 m_1 m_2} = c^{j-m}_{j_2 j_1 - m_2 - m_1} = (-1)^{j - j_1 - j_2} c^{jm}_{j_2\, j_1 m_2 m_1}$$

2.1 Isospin

Nuclear Reactions

Another application of this vector coupling formalism has been to nuclear and particle physics reactions. This is the concept of isospin. We will find that we can apply the formalism involving Clebsch-Gordan coefficients that was developed for angular momentum to isospin and apply it to reactions in nuclear physics and particle physics.

In the isospin formalism, replace the operator J with the operator T and quantum number t, and the "3" component thereof with T_z, with quantum number t_z. In this system, the proton is given an isospin $t = 1/2$ and a component along the z-axis (in isospin space!) of $t_3 = +1/2$. The neutron is part of the isospin doublet with the proton, so its isospin $t = 1/2$ also, but its 3 component is $t_3 = -1/2$. Note that this represents a convention; many authors assume that t_3 for the proton is $-1/2$ and that it is $+1/2$ for the neutron. Nuclear physicists tend to use this latter formalism, in which $(t, t_3) = (1/2, +1/2)$ for the neutron (so that nuclei with a neutron excess, which applies to most of the nuclides in the periodic table, will have a positive $T_3 = \Sigma_i t_{3i}$), whereas particle physicists tend to use the $(t, t_3) = (1/2, +1/2)$ formalism for the proton.

If we go with the particle physics convention for the moment, then the diproton, or ^{2}He, has isospin values of $(T, T_3) = (1, +1)$, while the dineutron has isospin values of $(T, T_3) = (1, -1)$. The third member of this isospin triplet is the deuteron. It surely does have $T_3 = 0$ under any convention, but it must also be the third member of the $T = 1$ isospin triplet if we are to treat the proton and neutron as identical particles but with different T_3-values. However, its $T = 1$ state is actually its first excited state; its ground state is $(T, T_3) = (0, 0)$. We will use these values shortly to confirm that the member of this triplet in the deuteron is indeed the excited state.

First, however, we discuss an interesting application of the isospin formalism to nuclear reactions, which was given by Barshay and Temmer (1964). They noted that in any reaction $A + B \rightarrow C + C'$, "where C and C′ are members of the same isospin multiplet, i.e., they are isobaric analogs, and where the isospin of either A or B is zero, then, if isospin is strictly conserved, and if C and C′ are exactly connected by a rotation in isospin space, the differential cross section of the reaction products will exhibit symmetry about 90 degrees in the center of mass system independent of reaction mechanism."

Barshay and Temmer prove this assertion by noting that, subject to the assumption about strict isospin conservation, C and C′ are identical particles, so since they will be either bosons or fermions, the product of the space and spin components of the wave function will have a definite symmetry. Hence, even angular momenta and odd angular momenta are associated with orthogonal spin wave functions, which, when their absolute square is taken, cannot introduce interference terms. Thus, the angular distribution will contain only even powers of $\cos\theta$, leading to the symmetry asserted in the Barshay-Temmer theorem above.

A number of tests of this theorem have been conducted in the years since it was proposed. One was of the $d + {}^4He \rightarrow {}^3H + {}^3He$ reaction (Gross et al. 1970). The data for this reaction are shown in figure A.1.

Gross et al. observe that the Barshay-Temmer theorem apparently fails in their test. They note that the uncertainties in the measurements are considerably less than the observed violation of the theorem. They discuss several possibilities for this violation, most notably, effects of the Coulomb interaction. That interaction does violate isospin conservation (the $1/r$ potential, which technically can be written as $\Sigma_{ij}\ (e^2/r_{ij})(1/2 + t_{3i})(1/2 + t_{3j})$, does not commute with T^2, and therefore T is not a good quantum number), so this is a plausible explanation. Note, however, that the violation is not large; qualitatively the Barshay-Temmer theorem does give the observed result.

Particle Physics Reactions

Other tests of isospin conservation have been conducted in high-energy physics. To note one of them, a comparison was made between the yields of the $p + p \rightarrow \pi^+ + d$ and $n + p \rightarrow \pi^0 + d$ reactions. As noted above, the

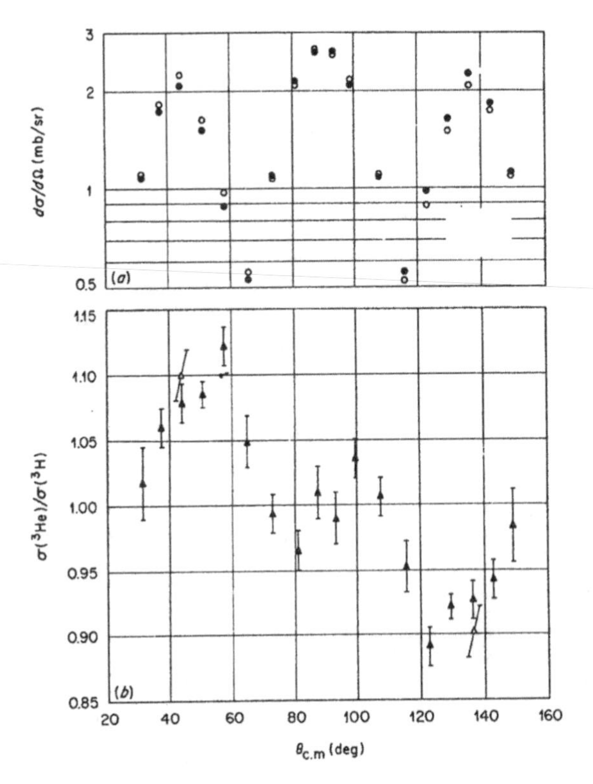

Fig. A.1. The upper half shows the measured differential cross sections for the process ^{4}He + ^{2}H → ^{3}H + ^{3}He using an 82-MeV ^{4}He beam. Open circles represent the ^{3}He yields, and closed circles represent the ^{3}H yields. Relative errors are smaller than the size of the data symbols, and the absolute scale is uncertain to ±10%. The lower half shows the angular dependence of the ratio of ^{3}He to ^{3}H yields. The Barshay-Temmer theorem requires this ratio to be 1.0 at all angles. Reprinted with permission from Gross et al. (1970). Copyright 1970 by the American Physical Society.

deuteron in its ground state is an isospin singlet, that is, $T = T_z = 0$. The pion is an isospin triplet, with the π^+ having isospin quantum numbers $T = T_z = 1$ and the π^0 having $T = 1$, $T_z = 0$.

The cross sections for these two reactions are seen to be in a nearly exact ratio of two at all energies. This is exactly what would be predicted from isospin conservation. To see this, consider the isospin components of the wave functions for the various entrance and exit channels. They are, since the pion is an isospin triplet, with three charge states of +, 0, and −,

A.2.1 $$\text{p} + \text{p}: \quad |1/2 + 1/2\rangle|1/2 + 1/2\rangle = |1/2\ 1/2\ 1\ 1\rangle$$

(with a Clebsch-Gordan coefficient of 1 .0);

A.2.2 $n + p$: a state with isospin that can be either 0 or 1, so it will be dealt with below;

A.2.3 $d + \pi^+$: $|00\rangle|11\rangle = |0\ 1\ 1\ 1\rangle$ (with a Clebsch-Gordan coefficient of 1 .0);

A.2.4 $d + \pi^0$: $|00\rangle|10\rangle = |0\ 1\ 1\ 0\rangle$ (with a Clebsch-Gordan coefficient of 1 .0).

The n + p system is not an eigenfunction of the T^2 operator. To determine the wave function of that system, consider the isospin triplet of systems ^{2}He, ^{2}H, and 2n. To get the $T = 1$ state of the deuteron, we can operate with the $T_- = T_{1-} + T_{2-}$ operator, the lowering (or annihilation) operator, on the ^{2}He state. This will produce the state

A.2.5
$$(1/2)^{1/2}[|1/2 + 1/2\rangle_1|1/2 - 1/2\rangle_2 + |1/2 - 1/2\rangle_1|1/2 + 1/2\rangle_2] = |1/2\ 1/2\ 1\ 0\rangle.$$

As noted above, this is the isospin part of the wave function for the first excited state of the deuteron.

To obtain the wave function for the ground state of the deuteron, use the procedure outlined above for determining the next lower j, or in this case, t, state; this gives

A.2.6
$$(1/2)^{1/2}[|1/2 + 1/2\rangle_1|1/2 - 1/2\rangle_2 - |1/2 - 1/2\rangle_1|1/2 + 1/2\rangle_2] = |1/2\ 1/2\ 0\ 0\rangle.$$

It was asserted above that the np system is not an eigenfunction of isospin. Indeed, from equations A.2.5 and A.2.6 it can easily be seen that it is a linear superposition of the wave functions of the ground state ($t = 0$) and first excited state ($t = 1$), equally weighted, of the deuteron:

A.2.7
$$n + p: \quad (1/2)^{1/2}[|1/2\ 1/2\ 1\ 0\rangle - |1/2\ 1/2\ 0\ 0\rangle].$$

We are now in a position to calculate what isospin conservation predicts for the relative probabilities of these two reactions; this will simply be the ratio of the squares of the isospin conserving parts products between the entrance and exit channels for the two reactions. Thus, for $\sigma(pp \to d\pi^+)$, both entrance and exit channels have isospin of 1, so that isospin conservation dictates that the isospin amplitudes, hence probabilities, will be the same, 1.0, in entrance and exit channels. However, For $\sigma(np \to d\pi^0)$, the exit channel is entirely isospin 1, but the entrance channel has probabilities of half isospin 1 and isospin 0 (from

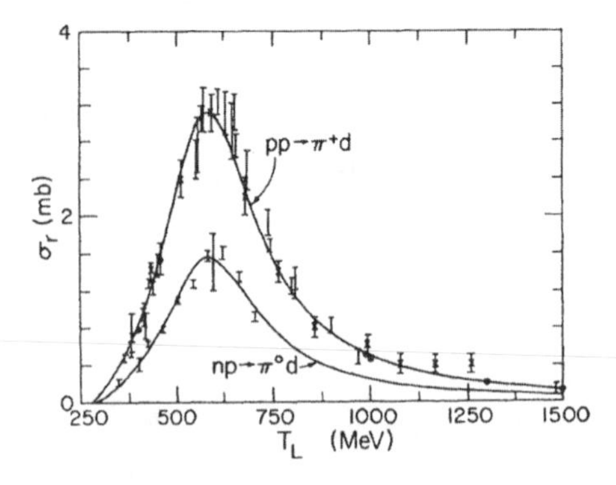

Fig. A.2. Cross sections for reactions $p + p \to d + \pi^+$ and $n + p \to d + \pi^0$. Data are from a compilation in, and are reprinted with permission from, VerWest and Arndt [1982]. Copyright 1982 by the American Physical Society.

the squares of the coefficients). But only the isospin 1 component, that is, half the wave function (after taking the absolute squares of the amplitudes), can couple to the exit channel if isospin is conserved, so the cross section should be half that of the first reaction if isospin is conserved.

Thus the ratio of the cross section $\sigma(pp \to d\pi^+)$ to $\sigma(np \to d\pi^0)$ is predicted from isospin conservation to be 1/2; this is seen in figure A.2 to be the case, certainly to within the uncertainties.

For additional discussion of isospin see Park (1992).

3.1 Some Nuclear Physics

β-Decay

As another example, consider the β-decay of an $N = Z$ nucleus ^{56}Ni. This nucleus is unstable, as its Coulomb forces have begun to dominate over its intrinsic tendency to equalize the number of protons and neutrons (see sec. 3.1). However, it is a doubly closed shell nucleus, as both the protons and neutrons have just filled the $f_{7/2}$ shell. The rules for Fermi or Gamow-Teller (GT) transitions in nuclei are repeated here for convenience:

Fermi:

$$\Delta J = 0, \text{ no change in parity;}$$
$$\Delta T = 0, \Delta T_3 = \pm 1.$$

Gamow-Teller:

$$\Delta J = 0, \pm 1 (0 \to 0 \text{ not allowed}), \text{ no change in parity;}$$
$$\Delta T = 0, \pm 1 (0 \to 0 \text{ not allowed}) \; \Delta T_3 = \pm 1.$$

Recall from chapter 3 that these are the rules for allowed transitions; the rules become more complicated if changes in angular momentum are required.

The ground state of ^{56}Ni has $N = Z$, so must be the $T = 0$ state. The only possible β-decay of this state is to ^{56}Co, since the mother nucleus needs to reduce its charge to reach stability. However, the ^{56}Co ground state is $T = 1$, so the decay cannot proceed by a Fermi transition. Thus it must be a GT transition. There is another complication for this β-decay; one of the highest lying protons in ^{56}Ni cannot decay to a neutron in the same ($f_{7/2}$) state because that state is completely filled (to a good approximation, anyway). Thus this $f_{7/2} \rightarrow f_{7/2}$ decay would violate the Pauli principle.

However, ^{56}Co will have states that are $T = 1$ states that are analog states of those that exist at higher excitation energy in ^{56}Ni. One can determine the quantum numbers of these states by examining the GT matrix element, given in equation 3.5.21; it shows that the isospin raising or lowering operator will change the T_3 component of the isospin, but not the T component, while the σ component will perform a "spin flip." Thus, the $f_{7/2}$ orbital becomes, in a GT transition, an $f_{5/2}$ orbital, which, in ^{56}Ni, is completely (to a good approximation) unfilled. In nuclei such as ^{56}Ni ($N = Z$, doubly closed shell), the β-decay has been dubbed the "superallowed" GT β-decay (see, e.g., Hamamoto and Sagawa 1993).

Nuclear Structure

Consider another nuclear physics problem. A nucleus like $^{49}{}_{21}$Sc (a radioactive isotope of scandium) has 21 protons and 28 neutrons. It may be described as an inert (spinless) droplet formed of the 20 protons and 28 neutrons (these being the numbers of particles in closed nuclear shells) and a single proton orbiting around it. Consider the case when the proton (with spin-1/2) orbits with total angular momentum $j = 3/2$, that is, the proton is in a $p_{3/2}$ orbital, the total angular momentum being the sum of orbital and spin angular momentum, $\bar{J} = \bar{\ell} + \bar{s}$. This would not be expected to be the ground state of ^{49}Sc; the shell model orbitals (see chap. 3) would suggest that the ground state would have the unpaired proton in an $f_{7/2}$ orbital. Incident on this proton is an X-ray, and the proton absorbs it. Considering only allowed transitions, the (dipole transition) matrix element for the interaction process will be proportional to $\langle j_f m_f | \bar{r} | 3/2 m_i \rangle$.

1. What are the possible total angular momenta of the proton after absorption? The answer must be that $j_f = 5/2, 3/2, 1/2$, since the vector operator

$\bar{r}$ can add to the existing angular momentum (vectorially) with a value of one unit, so that $\langle \bar{j}_f = \bar{1}_f + \bar{3}/2$, so that $|3/2 - 1|\,|j_f| \leq 3/2 + 1$.

2. What are the possible projections of the total angular momentum? $m_f = m \pm 1$ or $m_f = m$, depending on the angle between the quantization axis and $\bar{r}$.

This Clebsch-Gordan coefficient section follows closely, albeit with significant modifications, the write-up given on the Web site, http://string.howard.edu/~tristan/QM2/QM2WE.pdf.

Bibliography

Abbasi, R., et al. 2004, Phys. Rev. Lett., 92, 151101.
Abdurashitov, J. N., et al. 1999, Phys. Rev. Lett., 83, 4686.
Aggouras, G., et al. 2005, Astropart. Phys., 23, 377.
Ahmad, I., et al. 1998, Phys. Rev. Lett., 80, 2550.
Ahmad, Q. R., et al. 2001, Phys. Rev. Lett., 87, 071301.
———. 2002, Phys. Rev. Lett., 89, 011301.
Ahmed, S. N., et al. 2004, Phys. Rev. Lett., 92, 181301.
Akerib, D., et al. 2004, Phys. Rev. Lett., 93, 211301.
Akimune, H., et al. 1995, Phys. Rev. C, 52, 604.
Alford, M., Brady, M., Paris, M., and Reddy, S. 2005, Astrophys. J., 629, 969.
Aliotta, M., et al. 2001, Nucl. Phys. A, 690, 790.
Alpher, R. A., Bethe, H. A., and Gamow, G. 1948, Phys. Rev., 73, 803.
Alpher, R. A., and Herman, R. C. 1949, Phys. Rev., 75, 1089.
Altmann, M., et al. 2000, Phys. Lett. B, 490, 16.
Amari, S., Zinner, E., and Lewis, R. S. 1999, Astrophys. J. Lett., 517, 59.
Amari, S., et al. 2001, Astrophys. J., 551, 1065.
Anders, E., and Grevesse, N. 1989, Geochim. Cosmochim. Acta, 53, 197.
Arnett, D. 1996. *Supernovae and Nucleosynthesis: An Investigation of the History of Matter from the Big Bang to the Present*. Princeton: Princeton University Press.
Araki, T., et al. 2005, Phys. Rev. Lett., 94, 081801.
Argast, D., Samland, M., Thielemann, F.-K. and Qian, Y.-Z. 2004, Astron. & Astrophys., 416, 997.
Arnould, M., and Goriely, S. 2003, Phys. Rep., 384, 1.
Arpesella, C., et al. 1995, Nucl. Instr. Meth. A, 360, 607.
———. 1996, Phys. Lett. B, 389, 452.
Arzumanov, S., et al. 2000, Phys. Lett B, 483, 15.
Asplund, A., Lambert, D. L., Nissen, P. E., Primas, F., and Smith, V. V. 2006, Astrophys. J., 644, 229.
Austern, N. 1970, *Direct Nuclear Reaction Theories*. New York: John Wiley.
Ayasli, S., and Joss, P. C. 1982, Astrophys. J., 256, 637.
Azhari, A., et al. 1999, Phys. Rev. Lett., 82, 3960.
Azuma, R., et al. 2003, Nucl. Phys. A, 718, 119c.
Azuma, R. E., et al. 1994, Phys. Rev., 50, 1194.
Baby, L. T., et al. 2003, Phys. Rev., Lett., 90, 022501.

———. 2003, Phys. Rev. C, 67, 065805.

Bahcall, J. N. 1962, Phys. Rev., 128, 1297.

———. 1989, *Neutrino Astrophysics*. Cambridge, UK: Cambridge University Press.

———. 1994, Phys. Rev. D, 49, 3923.

———. 1997, Phys. Rev. C, 56, 3391.

Bahcall, J. N., Brown, L. S., Gruzinov, A., and Sawyer, R. F. 2002, Astron. & Astrophys., 383, 291.

Bahcall, J. N., Gonzales-Garcia, M. C., and Pena-Garay, C. 2003, JHEP02, 1.

Bahcall, J. N., and Pena-Garay, C. 2004, New J. Phys., 6, 63.

Bahcall, J. N., and Pinsonneault, M. 1998, Phys. Lett. B, 433, 1.

Bahcall, J. N., Pinsonneault, M., and Basu, S. 2001, Astrophys. J., 555, 990.

Bahcall, J. N., Serenelli, A. M., and Basu, S. 2005, Astrophys. J. Lett., 621, 85.

Bahcall, J. N., and Ulrich, R. 1988, Rev. Mod. Phys., 60, 297.

Balser, D. S., et al. 1999, Astrophys. J., 510, 759.

Bardayan, D., et al. 1999, Phys. Rev., Lett., 83, 45.

———. 2000, Phys. Rev. C, 62, 055804.

Barger, V., Kneller, J. P., Lee, H.-S., Marfatia, D., and Steigman, G. 2003a, Phys. Lett. B, 566, 8.

Barger, V., Kneller, J. P., D. Marfatia, Langacker, P., and Steigman, G. 2003b, Phys. Lett. B, 569, 123.

Barklem, P. S., et al. 2005, astro-ph/0505050.

Baumgarte, T. W., Janka, H.-Th., Keil, W., Shapiro, S., and Teukolsky, S. 1996, Astrophys. J., 468, 823.

Baur, G., and Rebel, H. 1996, Ann Rev. Nucl. Part. Sci., 46, 321.

Beacom, J., Boyd, R. N., and Mezzacappa, A. 2000, Phys. Rev., Lett., 85, 3568.

———. 2001, Phys. Rev. D, 63, 073011.

Bean, R., and Dore, O. 2003, Phys. Rev. D, 68, 023514.

———. 2004, Phys. Rev. D, 69, 063503.

Beer, H., Kaeppeler, F., Reffo, G., and Venturini, G. 1983, Astrophys. & Space Sci., 97, 95.

Belczynski, K., et al. 2006, Astrophys. J., 648, 1110.

Bennett, C. L., et al. 2003, Astrophys. J. Suppl., 148, 1.

Berger, M. S., and Jaffe, R. L. 1987, Phys. Rev. C, 35 213.

———. 1991, Phys. Rev. C, 44, 566E.

Bernas, M., et al. 1997, Phys. Lett. B, 415, 111.

Bernatowicz, T. J., Akande, O. W., Croat, T. K., and Cowsik, R. 2005, Lunar & Planet. Sci., 36.

Bethe, H. A. 1939, Phys. Rev., 55, 434.

———. 1990, Rev. Mod. Phys., 62, 801.

Bethe, H. A., Brown, G. E., Applegate, J., and Lattimer, J. M. 1979, Nucl. Phys. A, 324, 487.

Bildsten, L. 1995, Astrophys. J., 438, 852.

———. 1998, The Many Faces of Neutron Stars, NATO ASIC Proc. 515, 419.

———. 2000, in Cosmic Explosions, ed. S.S. Holt and W.W. Zhang. New York: AIP, 359.

Bildsten, L., and Brown, E. F. 1997, Astrophys. J., 477, 897.

Bildsten, L., Chang, P., and Paerels, F. 2003, Astrophys. J. Lett., 591, 29.

Bionta, R. M., et al. 1987, Phys. Rev., Lett., 58, 1494.

Birkenshaw, M. 1990, *The Cosmic Microwave Background : 25 Years Later*, ed. M. Mandeles and N. Vittorio. Dordrecht: Kluwer, 77.

Bjornstad, T., et al. 1986, Nucl. Phys. A, 453, 463.

Blackmon, J., et al. 2001, Nucl. Phys. A, 688, 142.

———. 2003, Nucl. Phys. A, 718, 127c.

Blank, B., et al. 1995, Phys. Rev., Lett., 74, 4611.

Blatt, J. M., and Weisskopf, V. F. 1962, *Theoretical Nuclear Physics.* New York: Wiley.

Blondin, J. M., Mezzacappa, A., and DeMarino, C. 2003, Astrophys. J., 584, 971.

Bollen, G., and Schwarz, S. 2003, J. Phys. B: At. Mol. Opt. Phys. 36, 941.

Bond, J. R., and Efstathiou, G. 1987, Month. Not. Royal Astron. Soc. 226, 655.

Bonifacio, P., et al. 2002, Astron. & Astrophys., 390, 91.

Bosch, F. 2003, J. Phys. B: At. Mol. Opt. Phys. 36, 585.

Bosch, F., et al. 1996, Phys. Rev., Lett., 77, 5190.

Boyd, R. N. 1994, Int. J. Mod. Phys. E3, 249.

———. 2000, Proc. PROCON99, ed. J. Batchelder, AIP Conf. Proc. 518. Melville, NY: AIP, 239.

Boyd, R. N., and Saito, T. 1993, Phys. Lett. B, 298, 6.

Boyd, R. N. 1999, *Heavy Elements and Related New Phenomena,* ed. R. Gupta and W. Greiner. Singapore: World Scientific, 893.

Brown, E. F. 2000, Astrophys. J., 531, 988.

———. 2004, Astrophys. J. Lett., 614, 57.

Brown, E. F., Bildsten, L., and Rutledge, R. E. 1998, Astrophys. J. Lett., 504, 95.

Brown, L. S., and Gabrielse, G. 1986, Rev. Mod. Phys., 58, 233.

Brown, L. S., and Sawyer, R. F. 1997, Rev. Mod. Phys., 69, 411.

Brown, T. M., et al. 2004, Astrophys. J. Lett., 613, 125.

Buchmann, L., Azuma, R. E., Barnes, C. A., Humblet, J., and Langanke, K. 1996, Phys. Rev. C, 54, 393.

Buchmann, L., Hilgemeier, M., Krauss, A., Redder, A., Rolfs, C., and Trautvetter, H. P. 1984, Nucl. Phys. A, 415, 93.

Budker, G. I. 1967, At. Energy 22, 346.

Burbidge, E. M., Burbidge, G. R., Fowler, W. A., and Hoyle, F. 1957, Rev. Mod. Phys., 29, 547.

Burles, S., Nollett, K. M., and Turner, M. S. 2001, Astrophys. J. Lett., 552, 1.

Burrows, A. 1990, Ann. Rev. Nucl. Part. Sci. 40, 181.

Burrows, A., Livne, E., Dessart, L., Ott C. D., and Murphy, J. 2006, Astrophys. J., 640, 878.

Busso, M., Gallino, R., and Wasserburg, G. J. 1999, Ann. Rev. Astron. & Astrophys. 37, 239.

Byrne, J., et al. 1996, Europhys. Lett., 33, 187.

Carlstrom, J. E., Joy, M., and Grego, L. 1996, Astrophys. J. Lett., 456, 75.

Carroll, S. 2005, Sky & Telescope, March, 32.

Caughlan, G. R., and Fowler, W. A. 1988, At. Data Nucl. Data Tables 40, 283.

Caughlan, G. R., Fowler, W. A., Harris, M. J., and Zimmerman, B. A. 1985, At. Data Nucl. Data Tables 32, 197.

Cayrel, R., et al. 2001, Nature, 409, 691.

Champagne, A. E., and Wiescher, M. 1992, Ann. Rev. Nucl. Part. Sci. 42, 39.

Chang, P., Bildsten, L., and Wasserman, I. 2005, Astrophys. J., 629, 998.

Chartier, M., et al. 1998, Nucl. Phys. A, 637, 3.

Chen, B., Dobaczewski, J., Kratz, K.-L., Langanke, K., Pfeiffer, B., Thielemann, F.-K., and Vogel, P. 1995, Phys. Lett. B, 355, 37.

Chen, H. 1985, Phys. Rev., Lett., 55, 1534.

Chiappini, C., Romano, D., and Matteucci, F. 2003, Month. Not. Royal Astron. Soc. 339, 63.

Chiste, V., et al. 2001, Phys. Lett. B, 514, 233; http://www.dfn.if.usp.br/pagina-.dfn/relatorios/ Relatofnc99/current/Experiment/node6.html.
Chloupek, F., et al. 1999, Nucl. Phys. A, 652, 391.
Clark, J. A., et al. 2005, Eur. Phys. J. A 25, 629.
———. 2004, Phys. Rev., Lett., 92, 192501.
Clayton, D. D. 1983, Principles of Stellar Evolution and Nucleosynthesis. Chicago: University of Chicago Press.
———. 1984, Astrophys. J., 280, 144.
Clayton, D. D., and Leising, M. 1987, Phys. Rep., 144, 1.
Clayton, D. D., and Nittler, L. R. 2004, Ann. Rev. Astron. & Astrophys., 42, 39.
Clement, R. R. C., et al. 2004, Phys. Rev. Lett., 92, 172502.
Cole, A. L., et al. 2006, Astrophys. J., 652, 1763.
Colgate, S. A., and White, R. H. 1966, Astrophys. J., 143, 626.
Colpi, M., Geppert, U., Page, D., and Possenti, A. 2001, Astrophys. J. Lett., 548, 175.
Corbelli, E., and Salpeter, E. E. 1988, Astrophys. J., 326, 551.
Cornelisse, R., Kuulkers, E., in't Zand, J. J. M., Verbunt, F., and Heise, J. 2002, Astron. & Astrophys., 382, 174.
Coszach, R., et al. 1995, Phys. Lett. B, 353, 184.
Cottam, J., Paerels, F., and Mendez, M. 2002, Nature, 420, 51.
Cottle, P. D., and Kemper, K. W. 1998, Phys. Rev. C, 58, 3761.
Cowan, J. J., McWilliam, A., Sneden, C., and Burris, D. L. 1997, Astrophys. J., 480, 246.
Cowan, J. J., and Sneden, C. 2005, Neutron-Capture Element Abundances ini Halo Stars, ASP Conf. Ser.Cosmic Abundances as Records of Stellar Evolution and Nucleosynthesis, ed. T. Barnes and F. Bash. San Francisco; ASP, 313.
Cowan, J. J. and Thielemann, F.-K. 2004, Phys. Today, October.
Cowan, J. J., Thielemann, F.-K., and Truran, J. W. 1991, Ann. Rev. Astron. & Astrophys., 29, 447.
Cowan, J. J. et al. 1999, Astrophys. J., 521, 194.
———. 2005, Astrophys. J., 627, 238.
Cumming, A. 2003, Astrophys. J. Lett., 583, 87.
Cumming, A., and Bildsten, L. 2000, Astrophys. J., 544, 453.
———. 2001, Astrophys. J. Lett., 559, 127.
Cumming, A., and Macbeth, J. 2004, Astrophys. J. Lett., 603, 37.
D'Auria, J., et al. 2004, Phys. Rev. C, 69, 065803.
Davids, B., et al. 2001a, Phys. Rev., Lett., 86, 2750.
———. 2001b, Phys. Rev. C, 63, 065806.
Davis, Jr., R., Harmer, D. S., and Hoffman, K. C. 1968, Phys. Rev., Lett., 20, 1205.
DeCrock, P., et al. 1991, Phys. Rev., Lett., 67, 808.
Deliyannis, C. P., and Pinsonneault, M. H. 1997, Astrophys. J., 488, 836.
Descouvement, P., and Baye, D. 1994, Nucl. Phys. A, 567, 341.
DeShalit, A., and Feshbach, H. 1974, *Theoretical Nuclear Physics.* Vol. 1. Nuclear Structure. New York: John Wiley.
de-Shalit, A., and Talmi, I. 1963, *Nuclear Shell Theory.* Mineola, N.Y.: Dover Publications.
Diehl, R., et al. 1995, Astron. & Astrophys., 298, 445.
———. 1997, Nucl. Phys. A, 621, 79C.
———. 2005, http://www.arcetri.astro.it/iaus227/poster/diehl_r.pdf.
Dighe, A. S., and Smirnov, A. Y. 2000, Phys. Rev. D, 62, 033007.

Dillmann, I., et al. 2002, Eur. Phys. J. A, 13, 281.
———. 2003, Phys. Rev., Lett., 91, 162503.
Djorgovski, S., and Meylan, G. 1993, in Structure and Dynamics of Globular Clusters, ed. G. Meylan and S. Djorgovski, ASP Conf. Series 50. San Francisco: ASP, p. 325.
Dobaczewski, J., W. Nazarewicz, and T. R. Werner, 1996, Phys. Rev. C, 53, 2809.
Drake, J. J., et al. 2002, Astrophys. J., 572, 996.
Duncan, R., and C. Thompson, C. 1992, Astrophys. J. Lett., 392, 9.
Dupraz, C., et al. 1997, Astron. & Astrophys., 324, 683.
Ebisuzaki, T., and Nakamura, N. 1988, Astrophys. J., 328, 251.
Edmonds, A. R. 1963, *Angular Momentum in Quantum Mechanics.* Princeton: Princeton University Press.
Edvardsson, B., et al. 1993, Astron. & Astrophys., 275, 101.
Eichler, D. 2002, Monthly Notices Royal Astron. Soc. 335, 883.
Elliott, S. 2000, Phys. Rev. C, 62, 065802.
Ellis, J., Fields, B., and Schramm, D. 1996, Astrophys. J., 470, 1227.
Engel, J., McLaughlin, G. C., and Volpe, C. 2003, Phys. Rev. D, 67, 013005.
Esbensen, H., Bertsch, G. F., and Snover, K. A. 2005, Phys. Rev., Lett., 94, 042502.
Faber, J. A., Baumgarte, T. W., Shapiro, S. L., and Taniguchi, K. 2006, Astrophys. J. Lett., 641, 93.
Fabian, A., et al. 1994, Pub. Astron. Soc. Japan 46, L59.
Fesen, R. A., Pavlov, G. G., and Sanwal, D. 2006, Astrophys. J., 636, 848.
Fields, B. D., and Olive, K. A. 1998, Astrophys. J., 506, 177.
Fields, B. D., and Sarkar, S. 2004, Phys. Lett B, 592, 1.
Fields, B. D., Hochmuth, K. A., and Ellis, J. 2005, Astrophys. J., 621, 902.
Fields, B. D., and Prodanovic, T. 2005, Astropohys. J. 623, 877.
Filippone, B., Elwyn, A. J., Davids, C. N., and Koetke, D. D. 1983, Phys. Rev. C, 28, 2222.
Formicola, A., et al. 2000, Eur. Phys. J., A8, 443.
———. 2004, Phys. Lett. B, 591, 61.
Fowler, W. A., Caughlan, G. R., and Zimmerman, B. A. 1967, Ann. Rev. Astr. Astrophys., 5, 525.
———. 1975, Ann. Rev. Astr. Astrophys. 13, 69.
Fowler, W. A., and Hoyle, F. 1964, Astrophys. J., Suppl. 9, 201.
Fox, D. B., et al. 2005, Science 437, 845.
Freedman, W., et al. 2001, Astrophys. J., 553, 47.
Frohlich, C., et al. 2006, Phys. Rev., Lett., 96, 142502.
Fryer, C. L., Woosley, S. E., and Heger, A. 2001, Astrophys. J., 550, 372.
Fryxell, B. A., and Woosley, S. E. 1982, Astrophys. J., 261, 332.
Fukuda, S., et al. (Super-Kamiokande Collaboration) 2001, Phys. Rev., Lett., 86, 5651.
Fukuda, Y., et al. 1998, Phys. Rev., Lett., 81, 1562.
Fuller, G. M., and Meyer, B. S. 1995, Astrophys. J., 453, 792.
Fuller, G. M., and Qian, Y.-Z. 2005, Phys. Rev. D73, 023004
Fulop, Z., et al. 1996, Z. Phys. A, 355, 203.
Fynbo, H. O. U., et al. 2005, Nature, 433, 136.
Gaensler, B. M., et al. 2005, astro-ph/0502393.
Gallino, R., et al. 1998, Astrophys. J., 497, 388.
Gamow, G. 1952, *The Creation of the Universe.* New York: Viking Press. Reprint, Mineola, N.Y.: Dover Publications, 2004.

Gehrels, N., et al. 2004, Astrophys. J., 611, 1005.
———. 2005, Science, 437, 851.
Geissel, H., et al. 1992, Nucl. Instrum. Methods B, 70, 247.
Gilmore, G., Gustafsson, B., Edvardsson, B., and Nissen, P. E. 1992, Nature, 357, 379.
Goldhaber, G., et al. 2001, Astrophys. J., 558, 359.
Goode, P., and Boyd, R. N. 1976, Phys. Rev. C, 14, 379.
Goriely, S. 2001, in Tours Symposium on Nuclear Physics III, ed. M. Arnould et al. AIP Conf. Proc. 561. New York: AIP, 53.
Goriely, S., Arnould, M., Borzov, I., and Rayet, M. 2001, Astron & Astrophys. Lett., 375, 35.
Gorska, M., et al. 1997, Phys. Rev., Lett., 79, 2415.
Gorres, J., et al. 1998, Phys. Rev., Lett., 80, 2554.
Graulich, J. S., et al. 1997, Nucl. Phys. A, 626, 751.
Greife, U., et al. 1994, Nucl. Instr. Methods A, 350, 327.
Greiner, J., et al. 1995, Astron. Astrophys. 302, 121.
Gribov, V. N., and Pontecorvo, B. 1969, Phys. Lett. B, 28, 493.
Gross, E. E., Newman, E., Roberts, W. J., Rutkowski, R. W., and Zucker, A. 1970, Phys. Rev. Lett., 24, 473.
Gyurky, G., et al. 2003, Nucl. Phys. A, 718, 599C.
Hagemann, M., et al. 2004, Phys. Lett. B, 579, 251.
———. 2005, Phys. Rev. C, 71, 014606.
Hammache, F., et al. 2001, Phys. Rev., Lett., 86, 3985.
Hamuy, M., Phillips, M. M., Suntzeff, N. B., Schommer, R. N., Mazo, J., and Aviles, R. 1996, Astron. J., 112, 2398.
Hannawald, M., et al. 2000, Phys. Rev. C, 62, 054301.
Hansen, C. J., and Van Horn, H. M. 1975, Astrophys. J., 195, 735.
Harissopulos, S., et al. 2005, J. Phys. G: Nucl. Part. Phys., 31, S1417.
Harrison, E. R. 1964, Proc. Phys. Soc., 84, 213.
Harss, B., et al. 1999, Phys. Rev., Lett., 82, 3964.
———. 2002, Phys. Rev. C, 65, 035803.
Hauser, W., and Feshbach, H. 1952, Phys. Rev., 87, 366.
Haustein, P. 1988, At. Data Nucl. Data Tables, 39, 185.
Haxton, W. C., Langanke, K., Qian, Y.-Z., and Vogel, P. 1997, Phys. Rev. Lett., 78, 2694.
Heger, A., Kolbe, I., Haxton, W. C., Langanke, K., Martinez-Pinedo, G., and Woosley, S. E. 2005, Phys. Lett. B, 606, 258.
Hellstrom, M., et al. 1996, Zeits. Phys. A, 356, 229.
Hencheck, M., et al. 1994, Phys. Rev. C, 50, 2219.
Herbig, T., et al. 1995, Astrophys. J. Lett., 449, 5.
Hill, V., et al. 2002, Astron & Astrophys., 387, 560.
Hillebrandt, W., Nomoto, K., and Wolff, R. G. 1984, Astron. & Astrophys., 133, 195.
Hirata, K., et al. 1987, Phys. Rev., Lett., 58, 1490.
Hoffman, J. A., Lewin, W. H. G., and Doty, J. 1977, Astrophys. J. Lett., 217, 23.
Hoffman, R. D., Woosley, S. E., Fuller, G. M., and Meyer, B. S. 1996, Astrophys. J., 460, 478.
Hoffman, R. D., Woosley, S. E., and Qian, Y.-Z. 1997, Astrophys. J., 482, 951.
Holtzman, J. 2005, http://astronomy.nmsu.edu/holtz/a616/images/.
Hoppe, P., and Ott, U. 1997, in *Astrophysical Implications of the Laboratory Study of Presolar Materials*, ed. T. J. Bernatowicz and E. K. Zinner. AIP Conf. Proc. 402. New York: AIP.
Hoppe, P., Amari, S., Zinner, E., and Lewis, R. S. 1995, Geochim. Cosmochim. Acta, 59, 4029.

Horoi, M., Jora, R., Zelevinsky, V., Murphy, A. S. J., and Boyd, R. N. 2002, Phys. Rev. C, 66, 015801.

Howard, W. M., Meyer, B. S., and Woosley, S. E. 1991, Astrophys. J. Lett., 373, 5.

Hoyle, F., Dunbar, D. N. F., Wenzel, W. A., and Whaling, W. 1953, Phys. Rev., 92, 1095.

Hoyle, F. 1946, Monthly Notices Royal Astron. Soc., 106, 343.

———. 1954, Astrophys. J. Suppl., 1, 121.

Hoyle, F., and Fowler, W. A. 1960, Astrophys. J., 132, 565.

Hurley, K., et al. 2005, Nature, 434, 1098.

Hu, W., Sugiyama, N., and Silk, J. 1997, Nature, 386, 37.

Hu, W., and White, M. 2004, Sci. Am. Spec. Rep., Feb., 44.

Iben, I., Jr. 1975, Astrophys. J., 196, 525.

———. 1976, Astrophys. J., 208, 165.

———. 1977, Astrophys. J., 217, 788.

Iben, I., Jr., and Renzini, A. 1982, Astrophys. J. Lett., 259, 79.

Ireland, T. 2004, http://www.sciencemag.org/cgi/content/full/286/5448/2289.

Isotov, Y. I., and Thuan, T. X. 1998, Astrophys. J., 500, 188.

———. 2004, Astrophys. J., 602, 200.

Iwasa, N., et al. 1999, Phys. Rev., Lett., 83, 2910.

Jacoby, G. H., Hunter, D. A., and Christian, C. A. 1984, Astrophys. J. Suppl., 56, 257.

Jading, Y., et al. 1997, Nucl. Instrum. Methods B, 126, 76.

Jin, L., Meyer, B. S., The, L.-S., and Clayton, D. D. 1997, Nucl. Phys. A, 621, 319c.

Jordan, G. C., IV, and Meyer, B. S. 2004, Astrophys. J. Lett., 617, 131.

Joss, P. C., and Li, F. L. 1980, Astrophys. J., 238, 287.

Junghans, A. R., et al. 2003, Phys. Rev. C, 68, 065803.

Junker, M., et al. 1998, Phys. Rev. C, 57, 2700.

Kajino, T., and Boyd, R. N. 1990, Astrophys. J., 359, 267.

Kaeppeler, F. 2004, ECT Workshop, May 24–28, Trento, Italy. http://pntpm3.ulb.ac.be/Trento/talks/pdf/fkaeppeler.pdf.

Kaeppeler, F., Gallino, R., Busso, M., Picchio G., and Raiteri, C. M. 1990, Astrophys. J., 354, 630.

Kaeppeler, F., et al. 2004, Phys. Rev. C, 69, 055802.

Kautzsch, T., Walters, W. B., and Kratz, K.-L. 1998, AIP Conf. Ser., 447, 1183.

Kautzsch, T., et al. 1996, Phys. Rev. C, 54, R2811.

———. 2000, Eur. Phys. J. A, 9, 201.

Kawabata, K., et al. 2004, Inst. Astron. Union Circ. 8410.

Kayser, B. hep-ph/0506165.

Kiener, J., et al. 1966, Nucl. Phys. A, 552, 66.

Kikuchi, T., et al. 1977, Phys. Lett. B, 391, 261.

———. 1998, Eur. Phys. J. A, 3, 213.

Kirkman, D., et al. 2000, Astrophys. J., 529, 655.

Klepper, O. 1997, Nucl. Phys. A, 626, 199c.

Kluge, H. J., ed. 1986, ISOLDE Users Guide, CERN 86-05.

Knie, K., et al. 1999, Phys. Rev., Lett., 83, 18.

———. 2004, Phys. Rev., Lett., 93, 171103.

Koehler, P., et al. 1996, Phys. Rev. C, 54, 1463.

Kolbe, E., and Langanke, K.-H. 2001, Phys. Rev. C, 63, 025802.

Kolhinen, V. S., et al. 2004, Nucl. Instr. Methods A, 528, 776.

Koonin, S. E., Dean, D. J., and Langanke, K. H. 1997, Phys. Rep. 278, 1.
Kouveliotou, C. 1997, Science, 277, 1257.
Kouveliotou, C., et al. 1998, Nature, 393, 235.
Kovac, J. M., et al. 2002, Nature, 420, 772.
Kratz, K.-L., Bitouzet, J.-P., Thielemann, F.-K., Möller, P., and Pfeiffer, B. 1993, Astrophys. J., 403, 216.
Kratz, K.-L., Gabelmann, H., Hillebrandt, W., Pfeiffer, B., Schlosser, K., Thielemann, F.-K., & the ISOLDE Collaboration. 1986, Z. Phys. A325, 489.
Kratz, K.-L., Pfeiffer, B., Thielemann, F.-K., and Walters, W. B. 2000, Hyperfine Interactions 129, 185.
Kratz, K.-L., Thielemann, F.-K., Hillebrandt, W., Möller, P., Harms, V., and Truran, J. 1988, J. Phys. G, 14, 331.
Krauss, L., and Romanelli, P. 1990, Astrophys. J., 358, 47.
Kubono, S., et al. 1988, Z. Phys. A, 331, 359.
———. 1992, Nucl. Phys. A, 537, 153.
Kuo et al., C L. astro-ph/0212289.
Kuulkers, E., den Hartog, P. R., in 't Zand, J. J. M., Verbunt, F. W. M., Harris, W. E., and Cocchi, M. 2003, Astron. & Astrophys., 399, 663.
Kuulkers, E., 't Zand, J. Homan, S., van Stratten, D., Altamirano, P., and van der Klis, M. 2004, astro-ph/0402076.
Lambert, D. 1992, Astron. & Astrophys. Rev., 3, 201.
Lane, A. M., and Thomas, R. G. 1958, Rev. Mod. Phys., 30, 257.
Langanke, K., and Martinez-Pinedo, G. 2003, Rev. Mod. Phys.,75, 819.
Langanke, K., Vogel, P., and Kolbe, E. 1996, Phys. Rev., Lett., 76, 2629.
Langanke, K., Wiescher, M. W., Fowler, W. H., and Goerres, J. 1986, Astrophys. J., 301, 629.
Lalazissis, G. A., Vretenar, D., Poschl, W., and Ring, P. 1998, Phys. Lett. B, 418, 7.
Lattanzio, J. C. 1989, Astrophys. J. Lett., 344, 25.
Lattanzio, J. C., and Lugaro, M. A. 2005, Nucl. Phys. A, 758, 477c.
Lattimer, J. M., and Schramm, D. N. 1976, Astrophys. J., 210, 549.
Leahy, J. P. 2003, University of Manchester http://www.jb.man.ac.jk/~jpl/cosmo/friedman .html.
Lee, T., Wasserburg, G. J., and Papanastassiou, D. A. 1976, Geophys. Res. Lett., 3, 41.
———. 1977, Astrophys. J., Lett., 211, 107.
Lee, A. T., et al. 2001, Astrophys. J. Lett., 561, 1.
Lesko, K., Norman, E., Larimer, R.-M., Bacelar, J., and Beck, E. 1989, Phys. Rev. C, 39, 619.
Liddle, A. R. 1999, astro-ph/9901124.
Liebundgut, B., et al. 1996, Astrophys. J. Lett., 466, 21.
Lin, L.-M., Cheng, K. S., Chu, M.-C., and Suen, W.-M. 2006, Astrophys. J., 639, 381.
Linsky, J. 2003, Space Sci. Rev. 106, 49.
Lu, Z.-T., Holt, R. J., Mueller, P., O'Connor, T. P., Schiffer, J. P., Wang, L.-B. 2005, Nucl.Phys. A, 754, 361.
Lugaro, M., Herwig, F., Lattanzio, J. C., Gallino, R., and Straniero, O. 2003, Astrophys. J., 586,1305.
Lunney, D., Pearson, J. M., and Thibault, C. 2003, Rev. Mod. Phys., 75, 1021.
Madsen, J. 1988, Phys. Rev., Lett., 61, 2909.
Maeder, A. 1992, Astron. & Astrophys., 264, 105.
Maeder, A., and Maynet, G. 1989, Astron. & Astrophys., 210, 155.

Mahoney, W. A., Ling, J. C., Wheaton, W. A., and Jacobson, A. S. 1984, Astrophys. J., 286, 578.

Mampe, W., Ageron, P., Pendlebury, J. M., and Steyerl, A. 1989, Phys. Rev., Lett., 63, 593.

Mampe, W., et al. 1993, JETP Lett., 57, 82.

Mathews, G. J., Bazan, G., and Cowan, J. J. 1992, Astrophys. J., 391, 719.

Matsuoka, M., Piro, L., Yamauchi, M., & Makashima, M. 1990, Astrophys. J., 361, 440.

Mazzali, P. A., et al. 2005, Science, 308, 1284.

McLaughlin, G. C., Fetter, J. M., Balantekin, A. B., and Fuller, G. M. 1999, Phys. Rev. C, 59, 2873.

McLaughlin, G. C., and Surman, R. 2005, Nucl. Phys., 758, 189c.

Meneguzzi, M., Audouze, J., and Reeves, H. 1971, Astron. & Astrophys., 15, 337.

Messenger, S., Keller, L. P., and Lauretta, D. S. 2005, Science, 309, 737.

Meyer, B. S. 1989, Astrophys. J., 343, 254.

Meyer, B. S., and Brown, J. S. 1997, Astrophys. J., Suppl., 112, 199.

Meyer, B. S., Krishnan, T. D., and Clayton, D. D. 1996, Astrophys. J., 462, 825.

———. 1998, Astrophys. J., 498, 808.

Meyer, B., Mathews, G. J., Howard, W. M., Woosley, S. E., and Hoffman, R. D. 1992, Astrophys. J., 399, 656.

Meyer, B. S., McLaughlin, G. C., and Fuller, G. M. 1998, Phys. Rev. C, 58, 3696.

Myers, S. T., et al. 1997, Astrophys. J., 485, 1.

Mezzacappa, A., et al. 1998, Astrophys. J., 493, 848.

Michel, F. C. 1991, *Theory of Neutron Star Magnetospheres*. Chicago: University of Chicago Press.

Michotte, C., et al. 1996, Phys. Lett. B, 381, 402.

Mikheyev, S. P., and Smirnov, A. Y. 1985, Sov. J. Nucl. Phys., 42, 913.

Miller, G. E., and Scalo, J. M. 1979, Astrophys. J., Suppl. 41, 513.

Möller, P., Nix, J. R., and Kratz, K.-L. 1997, At. Data Nucl. Data Tables 66, 131.

Motizuki, Y., Takahashi, K., Janka, H.-T., Hillebrandt, W., and Diehl, R. 1999, Astron. & Astrophys., 346, 831.

Motobayashi, T., et al. 1991, Phys. Lett. B, 264, 259.

Mukhamedzhanov, A. M., and Timofeyuk, N. K. 1990, JETP Lett., 51, 282.

Myra, E. S., and Burrows, A. 1990, Astrophys. J., 364, 222.

Nandra, P., and Pounds, K. 1992, Nature, 359, 215.

Nasser, H., et al. 2006, Phys. Rev. Lett., 96, 041102.

Netterfield, C. B., et al. 2002, Astrophys. J., 571, 604.

Nezvizhevskii, V. V., et al. 1992, JEPT 75, 405.

Nico, J. S., et al. 2005, Phys. Rev. C, 71, 055502.

Nittler, L. R. 2003, Earth & Planet. Sci. Lett., 209, 259.

Nollett, K. M., and Burles, S. 2000, Phys. Rev. D, 61, 123505.

Nomoto, K., and Hashimoto, M. 1988, Phys. Rep., 163, 13.

Nomoto, K., Thielemann, F.-K., and Yokoi, K. 1984, Astrophys. J., 286, 644.

Nomoto, K., Tominaga, N., Umeda, H., Maeda, K., Ohkubo, T., and Deng, J. 2005, Nucl. Phys. A, 758, 263c.

Norman, E. B., et al. 1998, Phys. Rev. C, 57, 2010.

Notani, M., et al. 2004, Nucl. Phys. A, 746, 113c.

———. 2002, http://www.cns.s.u-tokyo.ac.jp/ann03/online/pdfs/a20_notani.pdf.

Oberlack, U., et al. 1996, Astron. & Astrophys. Suppl., 120, 311.

Ohtsubo, T., et al. 2005, Phys. Rev. Lett., 95, 052501.

Olive, K. A., and Skillman, E. D. 2004, Astrophys. J., 617, 29.

Olive, K. A., Steigman, G., and Walker, T. P. 2004, Phys. Rep. 333, 389.

Omtvedt, J. P., et al. 1995, Phys. Rev., Lett., 75, 3090.

Opik, E. J. 1951, Proc. Royal Irish Acad. A, 54, 49.

Ozkan, N., et al. 2002, Nucl. Phys. A, 710, 469.

Paczynski, B. 1998, Astrophys. J. Lett., 494, 45.

Paczynski, B., and Proszynski, M. 1986, Astrophys. J., 302, 519.

Pagel, B. E. J. 1997, Nucleosynthesis and Chemical Evolution of Galaxies. Cambridge: Cambridge University Press.

Particle Data Group. 2004, http://pdg.lbl.gov/2004/listings/s017.pdf.

Park, D. 1992, *Introduction to Quantum Theory*, 3d ed. New York: McGraw-Hill.

Peacock, J. A. 1999, *Cosmological Physics*. Cambridge: Cambridge University Press.

Pearson, J. M., Nayak, R. C., and Goriely, S. 1996, Phys. Lett. B, 387, 455.

Pearson, T. J., et al. 2003, Astrophys. J., 591, 556.

Peebles, J., and Yu, J. T. 1970, Astrophys. J., 162, 815.

Perlmutter, S., et al. 1997, Astrophys. J., 483, 565.

———. 1999, Astrophys. J., 517, 565.

Pfeiffer, B., Kratz, K.-L., Thielemann, F.-K., and Walters, W. B. 2001, Nucl. Phys. A, 693, 282.

Pinsonneault, M. H., Steigman, G., Walker, T. P., and Narayanan, V. K. 2002, Astrophys. J., 574, 398.

Piro, A., and Bildsten, L. 2004, Astrophys. J., Lett., 616, 155.

———. 2005a, Astrophys. J., 619, 1054.

———. 2005b, Astrophys. J., 629, 438.

———. 2006, Astropyhs. J., 638, 968.

Plaga, R., et al. 1987, Nucl. Phys. A, 465, 291.

Pounds, K., et al. 1986, Month. Not. Royal Astron. Soc., 221, 7.

Preston, M. A. 1962, *Physics of the Nucleus*. Reading, Palo Alto, London: Addison Wesley.

Prantzos, N., Hashimoto, M., and Nomoto, K. 1990, Astron. & Astrophys., 234, 211.

Prantzos, N., Hashimoto, M., Rayet, M., and Arnould, M. 1990, Astron. & Astrophys., 238, 455.

Pruet, J., et al. 2006, Astrophys. J., 644, 1028.

Qian, Y.-Z., Fuller, G. M., Mathews, G. J., Mayle, R. W., Wilson, J. R., and Woosley, S. E. 1993, Phys. Rev., Lett., 71, 1965.

Qian, Y., Haxton, W., Langanke, K., and Vogel, P. 1997, Phys. Rev. C, 55, 1532.

Raffelt, G. 2001, Astrophys. J., 561, 890.

Rauscher, T., Thielemann, F.-K., and Kratz, K.-L. 1997, Phys. Rev. C, 56, 1613.

Raisbeck, G. M., Yiou, F., Bourles, D., Lorius, C., Jouzel, J., and Barkov, N. I. 1987, Nature, 326, 273.

Ravenhall, D. G., Pethick, C. J., and Wilson, J. R. 1983, Phys. Rev., Lett., 50, 2066.

Ravn, H. L., Sundell, S., Westgaard, L., and Roeckl, E. 1978, Nucl. Instrum. Methods 123, 217.

Rayet, M., Prantzos, N., and Arnould, M. 1990, Astron. & Astrophys., 227, 271.

Rayet, M., Arnould, M., M. Hashimoto, Prantzos, N., and Nomoto, K. 1995, Astron. & Astrophys., 298, 517.

Reeves, H., Fowler, W. A., and Hoyle, F. 1970, Nature, 226, 727.

Rehm, K. E., et al. 1996, Phys. Rev. C, 53. 1950.
Renaud, M., et al. 2006, Astrophys. J. Lett., 647, 41.
Ressler, J. J., et al. 2000, Phys. Rev., Lett., 84, 2104.
Riess, A. G., et al. 1997, Astron. J., 114, 722.
———. 1998, Astron. J., 116, 1009.
Rodriguez, D., et al. 2004, Phys. Rev., Lett., 93, 161104.
Rolfs, C., and Rodney, W. S. 1988, *Cauldrons in the Cosmos*. Chicago: University of Chicago Press.
Rosswog, S., et al. 1999, Astron. & Astrophys., 341, 499.
Rosswog, S., Davies, M. B., Thielemann, F.-K., and Piran, T. 2000, Astron & Astrophys., 360, 171.
Rubbia, C., et al. 1998, CERN/LHC/98-02 (EET) Geneva, May 30.
Ruderman, M. 1974, Science, 184, 1079.
Ryan, S. G., Beers, T. C., Olive, K. A., Fields, B. D., and Norris, J. E. 2000, Astrophys. J., Lett., 530, 57.
Saito, T., Hatano, Y., Fukuda, Y., and Oda, H. 1990, Phys. Rev., Lett., 65, 2094.
Salpeter, E. E. 1952, Astrophys. J., 115, 326.
———. 1953, Ann. Rev. Nucl. Sci., 2, 41.
———. 1954, Australian J. Phys., 7, 373.
Sarazin, F., et al. 2000, Phys. Rev., Lett., 84, 5062.
Sasaqui, T., Otsuki, K., Kajino, T., and Mathews, G. J. 2006, Astrophys. J., 645, 1345.
Satchler, G. R. 1983, *Direct Nuclear Reactions*. Oxford: Oxford University Press.
Sauter, T., and Kaeppeler, F. 1998, Phys. Rev. C, 55, 3127.
Savard, G., et al. 1997, Nucl. Phys. A, 626, 353.
Schaller, G.,Schaerer, D., Meynet, G., and Maeder, A. 1992, Astron. & Astrophys. Suppl. Ser., 96 269.
Schatz, H., Bildsten, L., and Cumming, A. 2003, Astrophys. J. Lett., 583, 90.
Schatz, H., Bildsten, L., Cumming, A., and Wiescher, M. 1999, Astrophys. J., 524, 1014.
Schatz, H., and Rehm, K. E. 2006, Nucl. Phys. A 777, 601.
Schatz, H., et al. 1998, Phys. Rep., 294, 167.
———. 2001, Phys. Rev., Lett., 86, 3471.
Schiavilla, R., et al. 1998, Phys. Rev. C, 58, 1263.
Schiffer, J. P. 1971, Ann. Phys., 66, 798.
Schirato, R., and Fuller, G. M., astro-ph/0205390.
Schmidt, B. P., et al. 1998, Astrophys. J., 507, 46.
Schramm, D. N., and Wasserburg, G. J. 1970, Astrophy. J., 162, 57.
Schumann, F., et al. 2003, Phys. Rev., Lett., 90, 232501.
Schwarz, S., et al. 2003, Nucl. Instr. Methods B, 204, 776.
Schwartz, S. J., et al. 2005, Astrophys. J. Lett., 627, 129.
Schwarzschild, B. 2005, Phys. Today, May 21.
Schwarzschild, M., and Harm, R. 1965, Astrophys. J., 145, 496.
Segre, E. 1977, *Nuclei and Particles*. Reading, Mass: Benjamin/Cummings.
Shergur, J., et al. 2002, Phys. Rev. C, 65, 034313.
Slane, P. O., Helfand, D. J., and Murray, S. S. 2002, Astrophys. J. Lett., 571, 45.
Smith, D., Morgan, E. H., and Bradt, H. V. 1997, Astrophys. J. Lett., 479, 137.
Smoot, G., et al. 1991, Astrophys. J., Lett., 371, 1.
———. 1992, Astrophys. J., Lett., 396, 1.

Sneden, C., et al. 2002, Astrophys. J. Lett., 566, 25.
———. 2003, Astrophys. J., 591, 936.
Snow, T. P. 1985, *The Dynamic Universe*, 2d ed. St. Paul, Minn.: West Publishing.
Somorjai, E., et al. 1998, Astron. & Astrophys., 333, 1112.
Spite, M., and Spite, F. 1982, Nature, 297, 483.
Spitkovsky, A., Levin, Y., and Ushomirsky, G. 2002, Astrophys. J., 566, 1018.
Spivak, P. E. 1988, Zh. Eksp. Fiz. 94, 1.
Stolzenberg, H., et al. 1990, Phys. Rev., Lett., 65, 3104.
Straniero, O., Chieffi, A., Limongi, M., Busso, M., Gallino, R., and Arlandini, C. 1997, Astrophys. J., 478, 332.
Straniero, O., Gallino, R., and Cristallo, S. 2005, Ann. Rev. Astron. & Astrophys., 37, in press.
Strieder, F., et al. 2001, Nucl. Phys. A, 696, 219.
Strohmayer, T. E., and Brown, E. F. 2002, Astrophys. J., 566, 1045.
Strohmayer, T. E., Zhang, W., and Swank, J. H. 1997, Astrophys. J. Lett., 487, 77.
Strohmayer, T. E., et al. 1996, Astrophys. J. Lett., 469, 9.
———. 1998, Astrophys. J. Lett., 498, 135.
Strohmayer, T., and Bildsten, L. 2005, in *Compact Stellar X-ray Sources*, eds. W. H. G. Lewin and M. van der Klis. Cambridge: Cambridge University Press.
Stuart, A., et al. 1999, *Kendall's Advanced Theory of Statistics*, vol. 2A. New York: Oxford University Press.
Sunyaev, R. A., and Zeldovich, Ya. B. 1970, Astrophys. Space Sci., 7, 20.
Surman, R., Engel, J., Bennett, J. R., and Meyer, B. S. 1997, Phys. Rev., Lett., 79, 1809.
Swank, J. H., et al. 1977, Astrophys. J. Lett., 212, 73.
Symbalisty, E. M. D., and Schramm, D. N. 1981, Rep. Prog. Phys., 44, 293.
Tagliente, G., and n_TOF Collaboration. 2004, Brazilian J. Phys., 34, 1033.
Takahashi, K., and Boyd, R. N. 1988, Astrophys. J., 327, 1009.
Takahashi, K., Boyd, R. N., Mathews, G. J., and Yokoi, K. 1987, Phys. Rev. C, 36, 1522.
Takahashi, K., Witti, J., and Janka, H.-Th. 1994, Astron. & Astrophys., 286, 857.
Takahashi, K., and Yokoi, K. 1983, Nucl. Phys. A, 404, 578.
Taam, R. E., Woosley, S. E., and Lamb, D. Q. 1996, Astrophys. J., 459, 271.
Thielemann, F.-K., Nomoto, K., and Yokoi, K. 1986, Astron. & Astrophys., 158, 17.
Thielemann, F.-K., Kratz, K.-L., Pfeiffer, B., Rauscher, T., van Wormer, L., and Wiescher, M. C. 1994, Nucl. Phys. A, 570, 329c.
Thomas, R. G. 1952, Phys. Rev., 88, 1109.
Thompson, L. G., et al. 1997, Science, 276, 1821.
Timmes, F. X., Woosley, S. E., Hartmann, D. H., and Hoffman, R. D. 1996, Astrophys. J., 464, 332.
Tinsley, B. M. 1975a, Astrophys. J., 198, 145.
———. 1975b, Astrophys. J., 216, 548.
Tischhauser, P., et al. 2002, Phys. Rev., Lett., 88, 072501.
Totani, T., Sato, K., Dalhed, H. E., and Wilson, J. R. 1998, Astrophys. J., 496, 216.
Toukan, K. A., Debus, K.,Kappeler, F., and Reffo, G. 1995, Phys. Rev. C, 51, 1540.
Trache, L., Carstoiu, F., Gagliardi, C. A., and Tribble, R. E. 2001, Phys. Rev., Lett., 87, 271102.
Tsirigotis, A., et al. 2004, Eur. Phys. J. C33, s956.
Tuli, J. K. 2000, Nucl. Wallet Cards, U.S. Nuclear Data Network.
Vandegriff, J., Raimann, G., Boyd, R. N., Caffee, M., and Ruiz, B. 1996, Phys. Lett. B, 365, 418.

van den Hoek, L. B., and de Jong, T. 2004, Astron. & Astrophys., 318, 231.
VandenBergh, D. A., Richard, O., Michaud, G., and Richer, J. 1992, Astrophys. J., 571, 487.
van Paradijs, J. 1978, Nature, 274, 650.
Van Wormer, L., Gorres, J., Iliadis, C., Wiescher, M., and Thielemann, F.-K. 1994, Astrophys. J., 432, 326.
VerWest, B. J., and Arndt, R. A. 1982, Phys. Rev. C, 25, 1979.
Villasenor, J. S., et al. 2005, Science, 437, 855.
Volpe, C., Auerbach, N., Colo, G., and Van Giai, N. 2002, Phys. Rev. C, 65, 044603.
Walker, T. P., Mathews, G., and Viola, V. 1989, Astrophys. J., 299, 745.
Walker, T. P., Steigman, G., Schramm, D. N., Olive, K. A., and Kang, H.-S. 1991, Astrophys. J., 376, 51.
Wallace, R. K., and Woosley, S. E. 1981, Astrophys. J., Suppl. Ser., 45, 389.
Wallerstein, G., et al. 1997, Rev. Mod. Phys., 69, 995.
Walter, F. M., and Lattimer, J. 2002, Astrophys. J. Lett., 576, 145.
Walters, W., et al. 2004, private communication, including the following references: Kautzsch et al.1996; Jading et al.1997; Kautzsch, Walters, and Kratz 1998; Hannawald et al. 2000; Kautzsch et al.2000; Kratz et al. 2000; Pfeiffer et al. 2001; Dillmann et al. 2002; Shergur et al. 2002; Wohr et al. 2002.
Wasserburg, G. J., and Qian, Y.-Z. 2000, Astrophys. J. Lett., 529, 21.
Wasserburg, G. J., Busso, M., and Gallino, R. 1996, Astrophys. J. Lett., 466, 109.
Wasserburg, G. J., Busso, M., Gallino, R., and Nollett, K. M. 2006, Nucl. Phys. A 777, 5.
Watanabe, G., Maruyama, T., Sato, K., Yasouka, K., and Ebisuzaki, T. 2005, Phys. Rev., Lett., 94, 031101.
Watts, A. L., and Strohmayer, T. E. 2006, Astrophys. J. Lett., 637, 117.
Weaver, T. A., and Woosley, S. E. 1980, *Supernova Spectra*. AIP Conf. Proc. No. 63, ed. R. E. Meyerott and G. H. Gillespie. New York: AIP, 15.
———. 1993, Phys. Rep., 227, 65.
Weaver, T. A., Zimmerman, G. B., and Woosley, S. E. 1978, Astrophys. J., 225, 1021.
Weigert, A. 1966, Z. Astrophys., 64, 395.
Weinberg, N. N., Bildsten, L., and Schatz, H. 2006, Astrophys. J., 639, 1018.
Werner, T. R., et al. 1996, Nucl. Phys. A, 597, 327.
Wheeler, J. C., Cowan, J. J., and Hillebrandt, W. 1998, Astrophys. J. Lett., 493, 101.
Wiescher, M., Gorres, J., and Thielemann, F.-K. 1988, Astrophys. J., 326, 384.
Wilson, J. R. 1974, Phys. Rev., Lett., 32, 849.
Winger, J. A., et al. 1993, Phys. Lett. B, 299, 24.
Winkler, C., et al. 2003, Astron. & Astrophys. Lett., 411, 1.
Witten, E. 1984, Phys. Rev. D, 30, 272.
Witti, J., Janka, H.-Th., and Takahashi, K. 1994, Astron. & Astrophys., 286, 841.
Wohr, A., et al. 2002, Proc. 11th Workshop on Nucl. Astrophys.
———. 2004, Nucl. Phys. A, 742, 349.
Wolfenstein, L. 1978, Phys. Rev. D, 17, 2369.
Woosley, S. E. 1986, *Nucleosynthesis and Chemical Evolution*, 16th Advanced Course of the Swiss Academy of Astronomy and Astrophysics, ed. B. Hauck, A. Maeder, and G. Meynet. Swiss Society of Astrophysics and Astronomy.
———. 1993, Astrophys. J., 405, 273.
———. 1998, RIKEN Winter School Lecture Notes, ed. Y. Motizuki and K. Sumiyoshi.
Woosley, S. E., Arnett, D., and Clayton, D. D. 1972, Astrophys. J., 175, 201.

Woosley, S. E., and Baron, E. 1992, Astrophys. J., 391, 228.

Woosley, S. Fowler, W., Holmes, J., and Zimmerman, B. 1978, At. Data Nucl. Data Tables 22, 371.

Woosley, S. E., Hartmann, D. H., Hoffman, R. D., and Haxton, W. C. 1990, Astrophys. J., 356, 272.

Woosley, S. E., Heger, A., and Weaver, T. A. 2002, Rev. Mod. Phys., 74, 1015.

Woosley, S. E., and Hoffman, R. D. 1992, Astrophys. J., 395, 202.

Woosley, S. E., and Howard, W. M. 1978, Astrophys. J., Suppl. Ser., 36, 285.

———. 1990, Astrophys. J. Lett., 354, 21.

Woosley, S. E., Langer, N., and Weaver, T. A. 1993, Astrophys. J., 411, 823.

———. 1995, Astrophys. J., 448, 315.

Woosley, S. E., and Weaver, T. 1995, Astrophys. J. Suppl., 101, 181.

Woosley, S. E., Wilson, J. R., Mathews, G. J., Hoffman, R. D., and Meyer, B. S. 1994, Astrophys. J., 433, 229.

Woosley, S. E., et al. 2004, Astrophys. J., Suppl., 151, 75.

Yanagisawa et al. 2005, Nucl. Instrum. Methods, A539, 74.

Yang, J., Turner, M. S., Steigman, G., Schramm, D. N., and Olive, K. A. 1984, Astrophys. J., 281, 493.

Zach, J. J., Murphy, A. S., Boyd, R. N., and Marriott, D. 2002, Nucl. Instr. Methods in Phys. Res. A, 484, 194.

Zegers, R. G. T., et al. 2005, Nucl. Phys. A, 758, 67c.

Zhang, C. T., et al. 1996, Phys. Rev., Lett., 77, 3743.

Zingale, M., et al. 2001, Astrophys. J. Suppl., 133, 195.

Zinner, E. 1998, Ann. Rev. Earth Planet. Sci., 26, 147.

Index